Corrosion Analysis

Michael Dornbusch
Niederrhein University of Applied Sciences
Krefeld, Germany

CRC Press is an imprint of the
Taylor & Francis Group, an **informa** business
A SCIENCE PUBLISHERS BOOK

CRC Press
Taylor & Francis Group
6000 Broken Sound Parkway NW, Suite 300
Boca Raton, FL 33487-2742

First issued in paperback 2020

ISBN-13: 978-1-138-63204-2 (hbk)
ISBN-13: 978-0-367-78081-4 (pbk)

Library of Congress Cataloging-in-Publication Data

Names: Dornbusch, Michael, author.
Title: Corrosion analysis / Michael Dornbusch.
Description: Boca Raton, FL : CRC Press, Taylor & Francis Group, [2018] | "A science publishers book." | Includes bibliographical references and index.
Identifiers: LCCN 2018029679 | ISBN 9781138632042 (hardback)
Subjects: LCSH: Corrosion and anti-corrosives. | Protective coatings.
Classification: LCC TA462 .D67 2018 | DDC 620.1/1223--dc23
LC record available at https://lccn.loc.gov/2018029679

Visit the Taylor & Francis Web site at
http://www.taylorandfrancis.com

and the CRC Press Web site at
http://www.crcpress.com

Preface

Corrosion is defined (DIN EN ISO 8044, DIN 50900) as the reaction of a metallic material with the environment that produces a measurable change of the material with an impairment of the material properties. Based on this definition corrosion is traditionally subdivided into two main topics [1]:

- High temperature corrosion, i.e., the attack of metal surfaces by hot and aggressive gases
- Atmospheric corrosion, i.e., the degradation of metals in contact with aqueous electrolytes

The focus of this book is on the analysis of atmospheric corrosion and the corrosion protection by means of coatings on the metal substrate. The analysis of atmospheric corrosion processes of materials is divided into two issues:

1. Investigation of mechanism of corrosion of the metal substrate
2. Investigation of degradation and transport processes of the protecting coating on the metal substrate

There, of course, is an interaction especially at the interface between substrate and coating during corrosion processes therefore both aspects cannot be separated.

My aim is to provide the readers, for the first time, with a book in which both the analytical methods for the substrate and the coating surface are presented, discussed and illustrated by several examples of application in corrosion and coating science.

Because of the large number of electrochemical and spectroscopic methods used in corrosion and coating science, a lot of basic information is necessary that cannot be provided by this book. Chapter one presents a very brief introduction in the basics of electrochemistry, corrosion processes and coating engineering to clarify the terms and definitions and give a short introduction in the principles of corrosion processes. Because of that further reading is always cited in the chapters and the definitive book for each topic is mentioned at the beginning of every chapter.

Chapter two deals with the major electrochemical methods used on the metal and coating surface. The focus on the metal substrate is understanding the mechanism of corrosion and the formation of passive layers whereas the transport processes of water and ions are the focus of the coating analysis. The interface between the substrate and the coating could be investigated with the Scanning Kelvin Probe and is also illustrated in this chapter.

Chapter three deals with the major spectroscopic methods used on the metal and coating surface. The focus is the investigation of the chemical composition of the metal surfaces and the chemical reactions or changes during corrosive treatment. The investigation of the chemical composition of the coatings structure, organic or inorganic, with the focus of corrosion protective properties and the changes during transport processes is also mentioned in this chapter.

Chapter four presents scientific and industrial test methods to investigate the corrosion rate of metal substrates and coated surfaces and characterize the advantages and limits of each method.

In the Appendix several thermodynamic and spectroscopic data are summarized for daily use in corrosion research.

Every method has limitations or uncertainties and therefore especially in corrosion research a second method is necessary to gain a clear picture of the corrosion process or the protection mechanism. Therefore a lot of spectra and graphs are illustrated in examples in every chapter to get an impression of the results of a method and the limitations and risk of artefacts.

The second aim of the book is to present a "second method" or at least ideas for it in order to validate an experiment.

Electrochemical methods do not detect chemical information and spectroscopic methods do not provide electrochemical data, and industrial methods are more of a quality test than they provide information on corrosion or protection mechanisms. Therefore a smart combination of methods could give authoritative results to understand corrosion processes or to develop new or optimized corrosion protective systems.

Düsseldorf, February 2018

---

[1] P. Marcus, F. Mansfeld, Analytical Methods in Corrosion Science and Engineering, CRC Press, 2006

Contents

1
Basics

1.1 Basics in Atmospheric Corrosion

The corrosion of metal substrates is often dominated by electrochemical reactions. The electrochemical principles of corrosion could be subdivided into the thermodynamic and the kinetic aspect. A very short introduction in both aspects is given below. Only the most important basics and fundamental equations are presented, necessary for the discussion of the analytical methods. For a detailed introduction see refs. [1, 2].

1.1.1 *Thermodynamic Aspects*

If a metal surface, e.g., iron, is immersed in an electrolyte or even water a double layer (Fig. 1.1) on the surface is generated caused by different potentials between the surface and the electrolyte. The following equilibrium occurs with a potential E^0 between the metal surface and the solution (standard potentials and the references of common substrates are summarized in Table 5.2 in the appendix (app.)):

$$Fe \Leftrightarrow \mathrm{Fe}^{2+} + 2e^{-} \wedge E^0 = -0.44\ V \qquad \text{Reac. 1.1}$$

The created double layer (Helmholtz layer) behaves at first sight like a capacitor (see also Chap. 2.1.3) and the thickness of the layer is defined by the size of the hydrated ions.

The radius r of the solvated ions is dominated by the hydration shell and could be calculated with the Debye-Hückel-Onsager theory as follows [2]:

$$\frac{1}{r} = \sqrt{\frac{8\pi e_0^2 N_A}{1000\ \varepsilon k_B T} I} \ \wedge\ I = \frac{1}{2}\sum_i z_i^2 c_i \qquad \text{Eq. 1.1}$$

e_0 : Elementary charge, $1.602*10^{-19}$ C
N_A : Avogadro constant, $6.022*10^{23}$ mol^{-1}
ε : Permittivity of the solvent, Water: 78.3(298.15 K)
k_B : Boltzmann constant, $1.381*10^{-23}$ J K^{-1}
T : Temperature in [K]
z_i : Charge of the ion
c_i : Concentration of the ion [mol/l]
I : Ion strength [mol/l]

Therefore the thickness of the double layer is controlled by the ion strength in the electrolyte. On closer inspection of the double layer the diffusion of the ions to and from the surface has to be received which produce a diffusion layer (Gouy Chapman layer) and the voltage drop between the surface and the electrolyte occurs in both layers and the diffusion layer could also be described as a capacitor (Fig. 1.1).

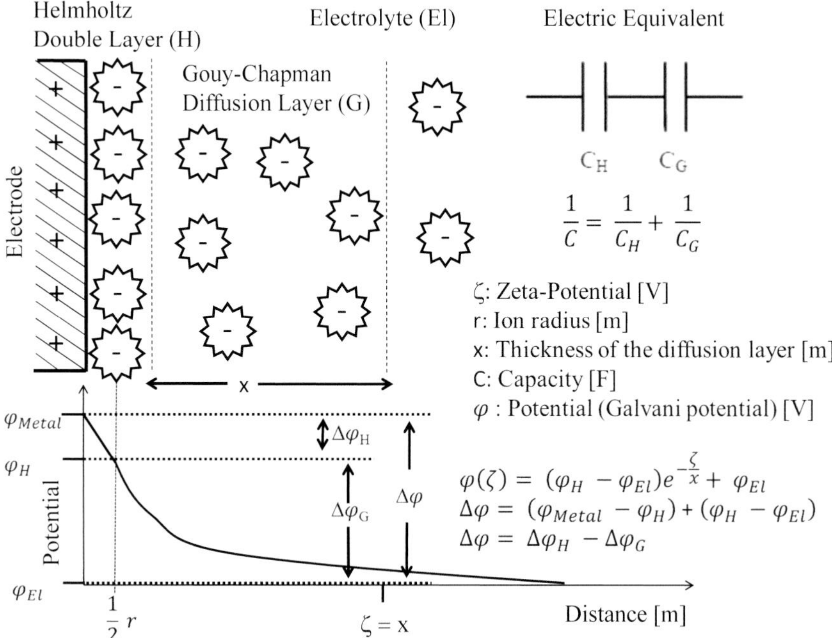

Fig. 1.1 Double layer on a metal electrode in an electrolyte with the relevant potentials, distances and electric equivalents.

The resulting concentration dependent potential could be calculated by the Nernst equation [1]:

$$E = E^0 + \frac{RT}{nF}\ln K \wedge K = \frac{[Ox]}{[\mathrm{Re}d]} \qquad \text{Eq. 1.2}$$

R : Gas constant, 8.314 J $K^{-1}mol^{-1}$

n : Number of electrons

F : Faraday constant, 9.648 10^4 C $mol^{-1} \wedge F = e_0 N_A$

K : Equilibrium constant

For standard conditions (T = 298.15 K) the Eq. 1.2 could be simplified to:

$$E = E^0 + \frac{0.0591V}{n}\log K \qquad \text{Eq. 1.3}$$

For Reac. 1.1 follows:

$$E = -0.44V + \frac{0.0591V}{2}\log\frac{[Fe^{2+}]}{[Fe]} = -0.44V + 0.0296\,\mathrm{V}\log[\mathrm{Fe}^{2+}] \qquad \text{Eq. 1.4}$$

Solid compounds could be ignored in the equation. A list of relevant equilibrium reactions and the corresponding potential are summarized in Table 5.1 in the app.

In this situation no corrosion, i.e., the continuous dissolution of iron, could occur. A second reaction which takes up the electrons is necessary with a higher potential because the resulting driving force ΔE, has to be positive because the resulting Gibbs free energy (ΔG) has to be negative, of course [1, 2]:

$$\Delta E = E_{Cathode} - E_{Anode} \qquad \text{Eq. 1.5}$$

$$\Delta G = -n\,F\,\Delta E \qquad \text{Eq. 1.6}$$

The dissolution of the metal, the oxidation, formed the anode electrode and the other reaction, the reduction, the cathode. The cathode reaction in atmospheric corrosion processes is dominated by the reduction of oxygen in neutral or alkaline conditions and by the hydrogen reduction under acidic conditions:

$$O_2 + 2H_2O + 4e^- \Leftrightarrow 4\mathrm{OH}^- \qquad \text{Reac. 1.2}$$

$$O_2 + 4H^+ + 4e^- \Leftrightarrow 2H_2O \quad \text{Reac. 1.3}$$

$$2H^+ + 2e^- \Leftrightarrow H_2 \quad \text{Reac. 1.4}$$

Both electrodes need a conductive contact for the electron transfer and an electrolyte contact for the ion transfer. If all these requirements are present the sine qua non for corrosion is fulfilled. If the corrosion process starts under neutral conditions, the following reactions on iron take place (see also [1]):

At the anode surface:

$$Fe \Leftrightarrow Fe^{2+} + 2e^- \quad \text{Reac. 1.1}$$

In the electrolyte near the anode surface:

$$Fe^{2+} + H_2O \Leftrightarrow FeOH^+ + H^+ \quad \text{Reac. 1.5}$$

In the electrolyte near the anode surface:

$$Fe^{2+} + \frac{1}{4}O_2 + \frac{1}{2}H_2O \Leftrightarrow Fe^{3+} + OH^- \quad \text{Reac. 1.6}$$

$$Fe^{3+} + 3H_2O \Leftrightarrow Fe(OH)_3 + 3H^+ \quad \text{Reac. 1.7}$$

Between the anode and the cathode surface:

$$Fe^{3+} + 3OH^- \Leftrightarrow FeOOH + H_2O \quad \text{Reac. 1.8}$$

At the cathode surface:

$$O_2 + 2H_2O + 4e^- \Leftrightarrow 4OH^- \quad \text{Reac. 1.2}$$

Because of Reacs. 1.5 and 1.7 the anode produces an acidic pH and because of Reac. 1.2 the cathode an alkaline pH. Therefore in between the precipitation of corrosion products is possible (Reac. 1.8) and a distance between anode and cathode, i.e., a separation between the acidic and the alkaline environment accelerates the corrosion under neutral or alkaline conditions. Depending on the electrolyte different corrosion products precipitate on the surface followed by further reactions (for details see Fig. 4.1).

The precipitation of the corrosion product Goethite (α-FeOOH) according to Reac. 1.8 is described with the solubility constant [3]:

$$K_L = \left[Fe^{3+}\right]\left[OH^-\right]^3 = 10^{-40.4} \quad \text{Eq. 1.7}$$

A list of relevant precipitation reactions and the corresponding constants are summarized in Table 5.1 in the app. In general the solubility constant could be achieved with the Gibbs free enthalpy with the following equation [3]:

$$\Delta G = -RT \ln K \quad \text{Eq. 1.8}$$

The precipitation of the corrosion products generates a passive layer, because the electron and ion transfer during the oxide or hydroxide layer is reduced, therefore the passivation reduces or stops the corrosion process.

Pourbaix Diagram

The complete thermodynamic picture could be illustrated with the so called Pourbaix diagram [4, 5] that summarizes the pH and potential dependence of the dissolution, i.e., the active state, the precipitation, i.e., the passivation and the metal, i.e., the immunity state.

The Pourbaix diagram for zinc is provided below. The assumptions from Pourbaix for the diagram are the following:

- The concentration of all ions and compounds is 10^{-6} mol/l,
- The partial pressure of gaseous is 1 atm and
- The activity of solid compounds and water is constant and could be ignored in the law of mass action.

First of all the potential range of water has to be calculated because it is important to correlate the data for a standard environment:

$$O_2 + 4H^+ + 4e^- \Leftrightarrow 2H_2O \quad \text{Eq. 1.9}$$

Reac. 1.3 and E = 1.228V – 0.0591V pH + 0.0147 log p_{O2}

$$2H^+ + 2e^- \Leftrightarrow H_2 \quad \text{Eq. 1.10}$$

Reac. 1.4 and E = –0.0591V pH – 0.0296 log p_{H2}

The relevant data for the equations describing the thermodynamic behaviour of zinc and other metals are summarized in Tables 5.1 and 5.2 in the app. For zinc we focus here on the ε-$Zn(OH)_2$ and active ZnO as corrosion products or passive layer compounds and the dissolution in acidic and alkaline conditions. Therefore the following reactions have to be considered:

$$Zn^{2+} + 2e^- \Leftrightarrow Zn \quad \text{Eq. 1.11}$$

Reac. 1.9 and E = –0.763V + 0.0295V log[(Zn^{2+})]

With the assumptions from Pourbaix [4, 5] it follows:

$$E = -0.94V \qquad \text{Eq. 1.12}$$

$$\varepsilon - Zn(OH)_2 + 2H^+ + 2e^- \Leftrightarrow Zn + 2H_2O \qquad \text{Reac. 1.10}$$

With the data of the standard enthalpy of formation the reaction enthalpy could be calculated (data see Table 5.2 in the app.):

$$\Delta G = -2*237.1kJ/mol - (-553.5kJ/mol) = 79.3\ kJ/mol \qquad \text{Eq. 1.13}$$

With the use of Eq. 1.6 the potential is available:

$$E = -\frac{\Delta G}{nF} = -0.411V \qquad \text{Eq. 1.14}$$

and therefore with Eq. 1.3 the pH dependence of the potential could be calculated:

$$E = -0.411V - 0.0591V\ pH \qquad \text{Eq. 1.15}$$

$$Zn(OH)_4^{2-} + 2\ e^- \rightleftharpoons Zn + 4\ OH^- \qquad \text{Reac. 1.11}$$

With the same procedure for Reac. 1.10 and the use of pH + pOH = 14 the pH dependence of the potential could be calculated to:

$$E = 0.267V - 0.1182V\ pH \qquad \text{Eq. 1.16}$$

$$\varepsilon\text{-}Zn(OH)_2 \rightleftharpoons Zn^{2+} + 2OH^- \qquad \text{Reac. 1.12}$$

With the solubility constant (see Table 5.1) for Reac. 1.12 and with the assumptions mentioned above a certain pH value could be calculated to:

$$K_L = \left[Zn^{2+}\right]\left[OH^-\right]^2 = 10^{-16.47} \qquad \text{Eq. 1.17}$$

$$pH = 8.77 \qquad \text{Eq. 1.18}$$

Finally the solubility constant of the dissolution of ε-$Zn(OH)_2$ to the zincate anion could be investigated with the calculation of ΔG (in [kJ/mol]) and the use of Eq. 1.8:

$$\Delta G = -862.74 - (-553.5 + 2\cdot(-157.24)) = 5.24 \qquad \text{Eq. 1.19}$$

$$K = \frac{\left[Zn(OH)_4^{2-}\right]}{\left[OH^-\right]^2} = e^{-\frac{\Delta G}{RT}} = 0.121 \qquad \text{Eq. 1.20}$$

$$pH = 11.46 \qquad \text{Eq. 1.21}$$

The same procedure could be done for the reactions of active-ZnO to generate the Pourbaix diagram in Fig. 1.2 as the Eqs. 1.9, 1.10, 1.12, 1.15, 1.16, 1.18, and 1.21 for ε-$Zn(OH)_2$ and the corresponding for active-ZnO are drawn in the relevant range in a diagram:

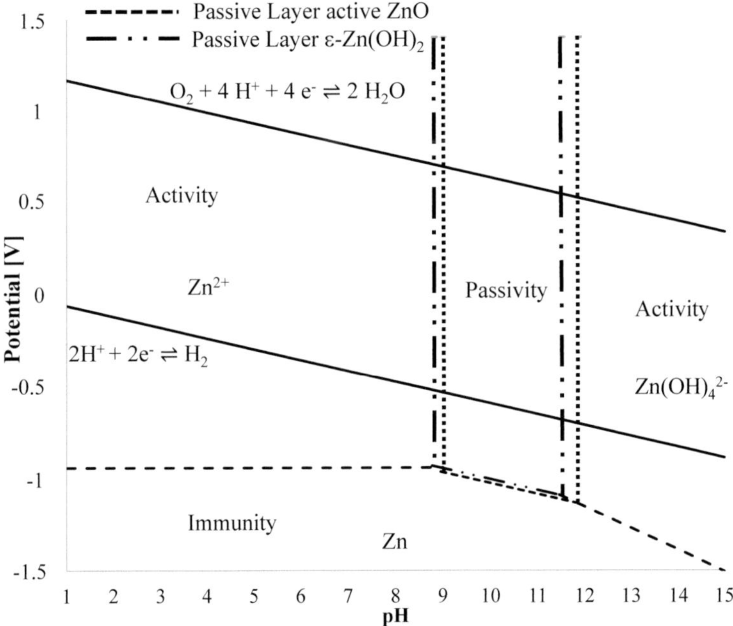

Fig. 1.2 Pourbaix diagram of zinc with a passive range based on ε-$Zn(OH)_2$ and active-ZnO.

In the range of water, zinc has three areas, two active in acidic and alkaline pH and one passive area. The passive range changes, if the composition is changed, therefore the chemical composition of the passive layer has to be investigated. For example in a CO_2 containing atmosphere the passive range is broadening [4] and carbonate compounds dominate the surface. The analysis of the surface could be done by means of infrared (Chap. 3.1.4), Raman (Chap. 3.1.5) or XPS (Chap. 3.1.1) spectroscopy to investigate the chemical composition and to define the passive layer of the metal in a certain atmosphere. The chemical composition of the corrosion products on zinc surfaces depending on the corrosive environment and changing with time (for details see Fig. 3.50).

The limitation of the Pourbaix diagram is the thermodynamic view. The diagram shows the complete thermodynamic situation of a metal but regardless of the kinetic aspect of the electrochemical and chemical reactions. Therefore the Pourbaix Diagram is a useful tool to summarize the thermodynamic data for a basic understanding of the metal in a certain atmosphere but the complete picture of the corrosion processes is only given with both the kinetic and thermodynamic aspect.

1.1.2 Kinetic Aspects

The rate v of an electrochemical reaction:

$$M^{n+} + n e^- \Leftrightarrow M \qquad \text{Reac. 1.13}$$

could be described as follows:

$$v = k[M] = \frac{d[M]}{dt} = -\frac{1}{n}\frac{d[e^-]}{dt} \wedge [e^-] = \frac{Q}{F} \wedge \qquad \text{Eq. 1.22}$$

Q: charge quantity ∧ k: rate constant

$$v = -\frac{1}{nF}\frac{dQ}{dt} = -\frac{1}{nF}i \qquad \text{Eq. 1.23}$$

Therefore the rate of an electrochemical reaction could be easily investigated by measuring the current I or better the current density i [A cm^{-2}] normalized to the surface area. In Fig. 1.3 the rate-determining steps in corrosion reactions, i.e., the charge transfer reaction, the diffusion to the surface and the passivation are illustrated, which will be discussed in detail in the following paragraph. The highest amount of corrosion products in a stable corrosion process is between the anode and cathode areas (see example (Ex.) 3.15) and the surface has to be able to transfer electrons in the metal to connect (galvanic coupling) anode and cathode and finally the electrolyte is necessary for the ion exchange between the electrodes.

Charge Transfer Controlled

If the rate-determining step in an electrochemical reaction is the transfer through the double layer (Fig. 1.1), Butler and Volmer [2] developed a

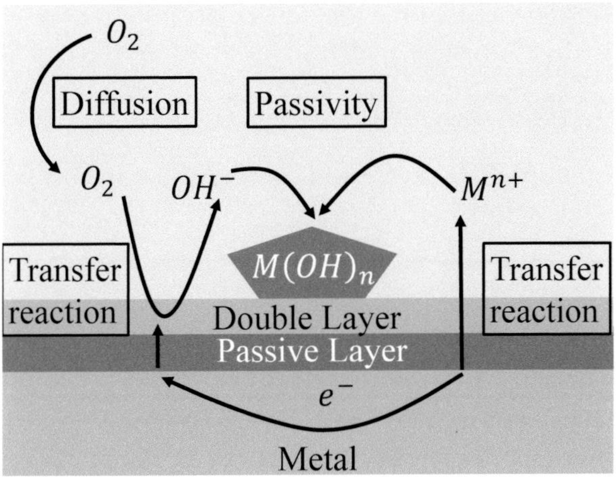

Fig. 1.3 Rate-determining steps in corrosion reactions on a metal surface.

kinetic model based on the transition state theory to describe this kind of electrochemical reactions presented in the following paragraph.

To develop a kinetic model for electrochemical reactions based on the Arrhenius [3] approach, i.e., the theory of the transition state, some information about the structure of the transition state is necessary. Because of the fact, that the transition state and the corresponding activation enthalpy $\Delta G^{\#}$ are more or less unknown, Butler and Volmer [2] described the change of the activation enthalpy by a parameter α as a similarity to the state of the ions in solution or to the state in the electrode. Furthermore the enthalpy with the use of Eq. 1.6 could be described as the sum of the over-voltage η and the equilibrium potential η_{eq} as follows:

Anode (Oxidation): $$\Delta G^{\#} = \Delta G_a^{\#} - \alpha n F(\eta + \eta_{Eq}) \quad \text{Eq. 1.24}$$

Cathode (Reduction): $$\Delta G^{\#} = \Delta G_c^{\#} + (1-\alpha) n F(\eta + \eta_{Eq}) \quad \text{Eq. 1.25}$$

The over-voltage η is the potential below or above the open circuit potential occurs in the equilibrium state. For the cathodic and the anodic reaction the current density could be described as (compare Eqs. 1.22 and 1.23):

$$i_a = F k_{Ox} [\mathrm{Re}d] \quad \text{Eq. 1.26}$$

$$i_c = -F k_{\mathrm{Re}d} [Ox] \quad \text{Eq. 1.27}$$

Because of the fact that the amount of electrons on the cathodic and the anodic reaction has to be similar in the absolute value, it follows:

$$i=|i_a|-|i_c|=F\,k_{Ox}\,[Red]-k_{Red}\,[Ox] \qquad \text{Eq. 1.28}$$

The rate constant could be expressed by the Arrhenius equation [6]:

$$k=B\,e^{-\frac{\Delta G^{\#}}{RT}} \qquad \text{Eq. 1.29}$$

and with the use of Eqs. 1.24 and 1.25 the term for the current density of an electrochemical reaction is achieved:

$$i=F\,B_a\,[Red]\,e^{-\frac{\left(\Delta G_a^{\#}-\alpha\,n\,F(\eta+\eta_{Eq})\right)}{RT}}-F\,B_c\,[Ox]\,e^{-\frac{\left(\Delta G_c^{\#}+(1-\alpha)\,n\,F(\eta+\eta_{Eq})\right)}{RT}} \qquad \text{Eq. 1.30}$$

$$i=F\,B_a\,[Red]\,e^{-\frac{\Delta G_a^{\#}}{RT}}\,e^{-\frac{(-\alpha\,n\,F\,\eta_{Eq})}{RT}}\,e^{\frac{\alpha\,n\,F\,\eta}{RT}}-F\,B_c\,[Ox]\,e^{-\frac{\Delta G_c^{\#}}{RT}}\,e^{-\frac{(\overline{1}-\alpha)\,n\,F\,\eta_{Eq}}{RT}}\,e^{-\frac{(1-\alpha)\,n\,F\,\eta}{RT}} \qquad \text{Eq. 1.31}$$

The summary of all terms describing the equilibrium state to an exchange current density for the anode i_a and the cathode i_c and the final summary of both to the total exchange current density $i_0 = i_a = i_c$ is possible, because in the equilibrium the current densities has to be equal. Therefore the Butler-Volmer equation could be written as (see also [2]):

$$i=i_a\,e^{\frac{\alpha\,n\,F\,\eta}{RT}}-i_c\,e^{-\frac{(1-\alpha)\,n\,F\,\eta}{RT}} \qquad \text{Eq. 1.32}$$

$$i=i_0\left(e^{\frac{\alpha\,n\,F\,\eta}{RT}}-e^{-\frac{(1-\alpha)\,n\,F\,\eta}{RT}}\right) \qquad \text{Eq. 1.33}$$

In a current density potential diagram (Evans diagram) the kinetic of a transfer controlled electrode reaction appears as a double exponential graph in the diagram (Fig. 1.4). Corrosion reactions consist of two electrode reactions—one is the dissolution of the metal and the other is the reduction of oxygen or hydrogen—which causes a mixture of two electrode reactions shown in Fig. 1.4. The resulting graph contains the anodic part of the metal and the cathodic part of the reduction process causes the corrosion potential E_{cor} and the corrosion exchange current density i_{cor}.

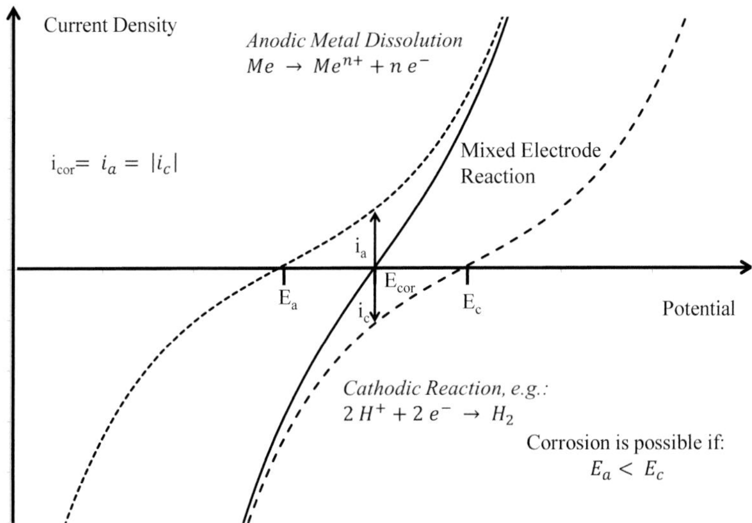

Fig. 1.4 Mixture of current-density-potential diagrams of the anodic metal dissolution and the cathodic reaction, if both reactions are transfer controlled.

The relevant parameters from the diagram are the exchange current density i_0 or the corrosion current density i_{cor}, and the corrosion potential E_{cor}. To investigate the data the Butler-Volmer equation could be simplified as follows.

If the over-voltage η is $>> \frac{RT}{nF} = \frac{25.7}{n}$ mV at 25°C [2], only one reaction has to be considered in the equation, therefore it follows:

Anode: $i = i_0\, e^{\frac{\alpha n F \eta}{RT}}$ Eq. 1.34

Cathode: $i = -i_0\, e^{-\frac{(1-\alpha) n F \eta}{RT}}$ Eq. 1.35

Both terms could be logarithmized to achieve the so called Tafel plots [7]:

Anode: $\log|i| = \log i_0 + \frac{\alpha n F}{2.3 R T}|\eta| \wedge \frac{1}{\beta_a} = \frac{\alpha n F}{2.3 R T}$ Eq. 1.36

Cathode: $\log i = \log i_0 + \frac{(1-\alpha) n F}{2.3 R T}\eta \wedge \frac{1}{\beta_c} = \frac{(1-\alpha) n F}{2.3 R T}$ Eq. 1.37

In addition to i_{cor} and E_{cor} from the intersection the slope β from the Tafel plots (Fig. 1.5) allows calculating the transfer coefficient α [7]:

With the current-density-potential diagram the corrosion rate (i.e., current density), the corrosion potential and some information about the transition state of the rate-determining step are available with the use of the Tafel plots. Some examples of the use in corrosion analysis are given in Chap. 2.1.1.

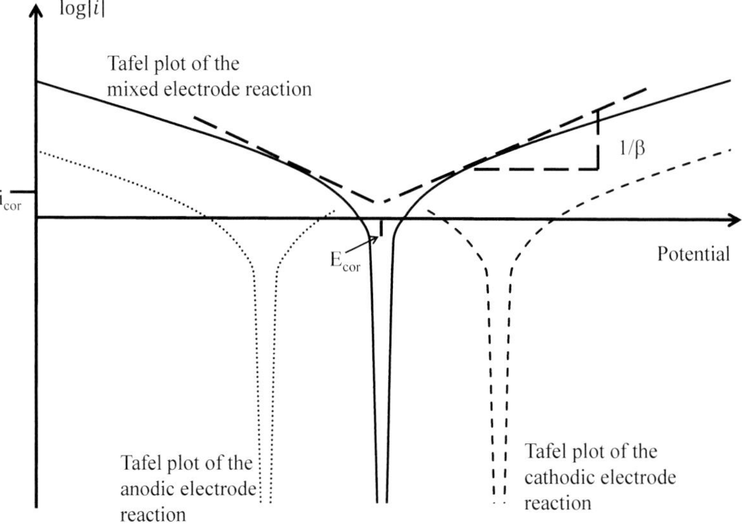

Fig. 1.5 Tafel plots of the current-density-potential graphs in Fig. 4.

Diffusion Controlled Reactions

The Butler-Volmer approach is useful, if the rate-determining step is the transfer process through the double layer. In electrochemical reactions with gaseous reactants or ions in low concentration (< 10^{-4} mol/l) the diffusion process to the surface of the electrode become the rate-determining step. Diffusion processes are described by the 1. Fickian law [8]:

$$J_z = \frac{1}{A}\left(\frac{dN}{dt}\right) = -D\left(\frac{dN}{dz}\right) \quad \text{or} \quad \left(\frac{dN}{dt}\right) = -DA\left(\frac{dN}{dz}\right) \qquad \text{Eq. 1.38}$$

J_z : Flux [$mol*m^{-2}*s^{-1}$]

D : Diffusion coefficient [m^2*s^{-1}]

N : Particle/molecule density [m^{-3}]

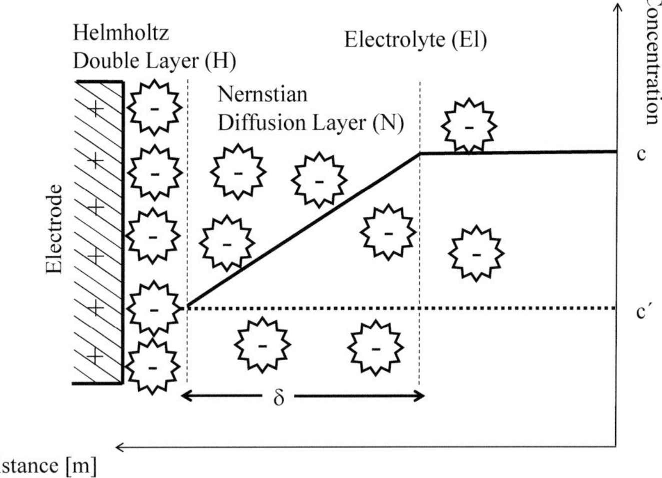

Fig. 1.6 Concentration gradient of a reactant in the Nernstian diffusion layer.

z : Distance in direction z [m]
A : Area [m²]
c : Concentration [mol*l⁻¹]

The concentration gradient in the Nernstian diffusion layer (Fig. 1.6) generates a flux J_δ through the distance δ as follows (see also Eqs. 1.1, 1.2 and 1.22):

$$J_\delta = -D\left(\frac{dN}{d\delta}\right) = -D\frac{(N-N')}{\delta} = -D\frac{(c-c')N_A}{\delta} \wedge i = e_0\, n\, J \qquad \text{Eq. 1.39}$$

The limiting current density i_{lim} (the maximum current density caused by the concentration of the reactant) could be achieved with the highest concentration:

$$i = -e_0\, n N_A\, D\frac{(c-c')}{\delta} = -n\, FD\frac{(c-c')}{\delta} \wedge i_{lim} = -n\, FD\frac{c}{\delta} \qquad \text{Eq. 1.40}$$

$$\Rightarrow c = -\frac{i_{lim}\,\delta}{nFD} \qquad \text{Eq. 1.41}$$

$$\Rightarrow c - c' = -\frac{i\delta}{n\, FD} \qquad \text{Eq. 1.42}$$

Solve the Eq. 1.42 to c´ with the use of Eq. 1.41, both concentrations could be expressed by a current density term as follows:

$$\Rightarrow c' = \frac{i\delta}{nFD} - \frac{i_{\lim}\delta}{nFD} \qquad \text{Eq. 1.43}$$

With the Nernst equation (Eq. 1.2) a correlation between the over-voltage and the current density could be achieved:

$$\eta = \frac{RT}{nF}\ln\frac{c'}{c} = \frac{RT}{nF}\ln\frac{\frac{i\delta}{nFD} - \frac{i_{\lim}\delta}{nFD}}{-\frac{i_{\lim}\delta}{nFD}} = \frac{RT}{nF}\ln\frac{i\delta - i_{\lim}\delta}{-i_{\lim}} = \frac{RT}{nF}\ln\frac{i_{\lim}-i}{i_{\lim}} = \frac{RT}{nF}\ln\left(1-\frac{i}{i_{\lim}}\right)$$

Eq. 1.44

In Fig. 1.7 the mixture of a transfer controlled metal dissolution and the diffusion controlled oxygen reduction is illustrated. The limiting current density cannot be exceeded and therefore the diffusion process limited the whole electrochemical process. The fact that the oxygen reduction as cathodic reaction is dominant in atmospheric corrosion and in addition to this the oxygen reduction is diffusion controlled, the transport phenomenon is important especially in corrosion protective coatings (see Chap. 2.2.2). Another important consequence is the surface size effect. The area for diffusion is a relevant part in Eq. 1.38 and therefore the reduction of the cathodic surface area is a main approach in corrosion

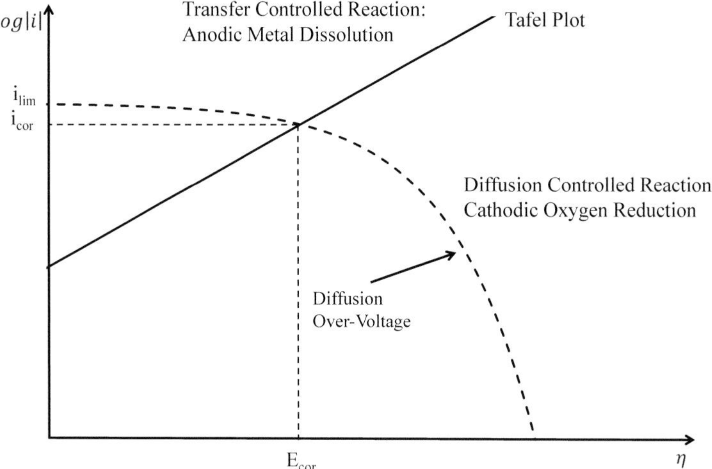

Fig. 1.7 Current-density-potential diagrams of a diffusion controlled cathode and transfer controlled anode reaction.

protection, too (see Chap. 2.1.1 and 2.2.2). Furthermore in a setup of an electrochemical experiment the ratio between the anode and cathode area receives attention (see Chap. 4.2.2).

Passivation

The formation of passive layers, i.e., the precipitation of hydroxide or oxide layers like Reacs. 1.7, 1.8 or 1.12 was already mentioned above. With sweeping the over-voltage on a metal electrode (Fig. 1.8) the metal dissolutes in the active range until a certain potential (Flade potential) is achieved.

At this potential the solubility product of the passive layer compound is reached or the existing but porous passive layer changed to a non-porous layer. In the passive range the corrosion process stops or is reduced to a very low rate. At a higher over-voltage the transpassive range is achieved and there are three different ways a metal could react. If the passive layer is stable and conductive, oxygen evaporation occurs on the surface. If the oxide layer is unstable, the metal dissolutes with the formation of oxo-anions like chromate and, if the passive layer is nonconductive and stable, there is no difference between the passive and the transpassive range. The metals with a stable and insulating passive layer are called valve metals (Al, Zr, Ti, Nb, Ta). To prevent corrosion processes, stable and nonconducting passive layers or modified passive layers—conversion layers—are a useful tool (see Chap. 1.3.1). The passive surface could be analyzed with electrochemical methods by means of cyclic voltammetry (Chap. 2.1.2) and electrochemical impedance spectroscopy (Chap. 2.1.3). The conductivity of a surface is defined by the distance of the valence band and the conductive band, i.e., the band gap (see Chap. 3.1.3). The

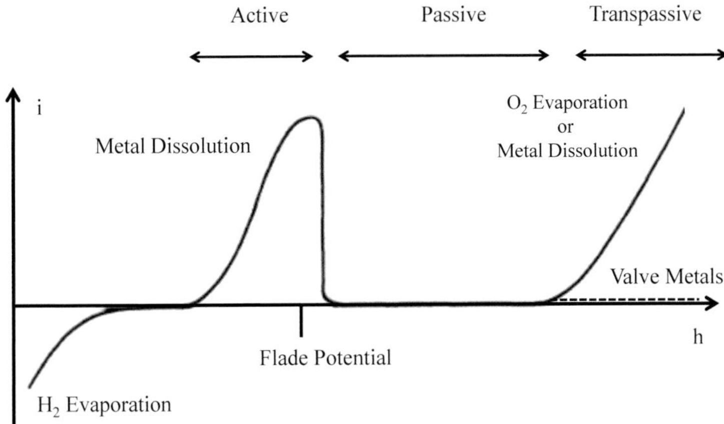

Fig. 1.8 Current-density-potential diagram of metal electrodes.

band gap of a passive layer could be investigated by means of UV-VIS spectroscopy (Chap. 3.1.3). Finally even in the passive range corrosion is still possible. The so called pitting corrosion is illustrated in the next chapter (see also Chap. 4.2.4).

1.2 Basics in Corrosion Processes

1.2.1 Corrosion Processes on Uncoated Metals

Figure 1.9 summarizes the most important corrosion types on an uncoated metal surface explained in some detail in the following paragraph. For a detailed description of the corrosion types see refs. [1, 9–11].

Uniform Corrosion

The corrosion attack occurs evenly over the whole surface and the metal produces no stable or insulating passive layer [10]. Therefore a homogenous thickness reduction takes place and the loss of material could be easily calculated by the corrosion rate (compare Figs. 1.5 and 1.7). Some data of corrosion rates of common metals are summarized in Table 5.4 in the app. Some corrosion current densities of metal substrates are summarized in Table 4.1.

Contact Corrosion (Galvanic Corrosion)

If a metal is in a conductive contact with a more noble metal, the corrosion rate increases. On the more noble metal the cathodic reactions takes place that accelerates the corrosion rate by two points:

The anodic reaction and the cathodic reaction are separated and the pH values of both do not inhibit each other to produce passive layers and secondly, if the cathodic part is the oxygen reduction, the reaction is diffusion controlled and because of the fact that often the surface area increases because of the additional metal surface the corrosion rate also increases (compare Eq. 1.38). Furthermore the necessary over-potential to pass the double layer often decreases and accelerates the reaction, too.

The final corrosion rate, i.e., the corrosion current density, could be estimated with the correlation according to Kaesche [1] based on the difference ΔE of the equilibrium potentials and the polarization resistance R_x of the surfaces as follows [1, 11, and 12]:

$$i_{corr} = \frac{\Delta E}{\frac{l}{\sigma} + R_a + R_c} \quad \text{Eq. 1.45}$$

l: Distance of the electrodes [m] σ: permittivity

The difference of the equilibrium potentials cannot be calculated with the thermodynamic data (Table 5.2) but with data from the so called practical galvanic series (see Table 5.3 in the app.) in certain environments. The polarization resistance of the surface could be calculated by means of the Stern-Geary [11, 12] approximation. For small over-voltages $\eta < 0.01V$ the exponent term in Eq. 1.33 is small $\frac{\alpha \eta F}{RT} << 1$ and the equation could be simplified according to the Taylor series:

$$e^x = 1 + x + \frac{x^2}{2!} + ... \qquad \text{Eq. 1.46}$$

We obtain (and use only the first term of the series) from Eq. 1.33 with the use of the slopes β (Eqs. 1.36 and 1.37):

$$i = i_{corr}\left(1 + \frac{\eta}{\beta_a} - \left(1 - \frac{\eta}{\beta_c}\right)\right) = i_{corr}\,\eta\left(\frac{\beta_c + \beta_a}{\beta_a \beta_c}\right) \qquad \text{Eq. 1.47}$$

We rearrange with the Ohmic law $R = \frac{\eta}{i}$ [13] and achieve the Stern-Geary equation:

$$i_{corr} = \frac{1}{R_{Pol}} \frac{\beta_a \beta_c}{\beta_a + \beta_c} \qquad \text{Eq. 1.48}$$

Therefore the polarization resistance could be investigated by means of current-density-potential diagrams. Some data from literature are summarized in Table 4.1. An alternative way to measure the polarization resistance is the use of the Electrochemical Impedance Spectroscopy (EIS) illustrated in Chap. 2.1.3.

Pitting Corrosion

Pitting corrosion is based on the dissolution of the passive layer in the presence of aggressive anions in an electrolyte on the passive surface (Fig. 1.8). All halides especially chloride and fluoride but also pseudo halides, sulphate, and perchlorate may cause pitting under certain conditions [14].

As shown in Fig. 1.9 the aggressive anion dissolves the passive layer (oxide, hydroxide, carbonate, etc.) and generates a local active area on the surface starting the anodic process. Both, insulating and conductive passive layers are able to work as a cathodic surface (caused by noble metal precipitates, defects or intermetallic phases) and the cathodic area is large in comparison to the local anodic process. The pit separates the

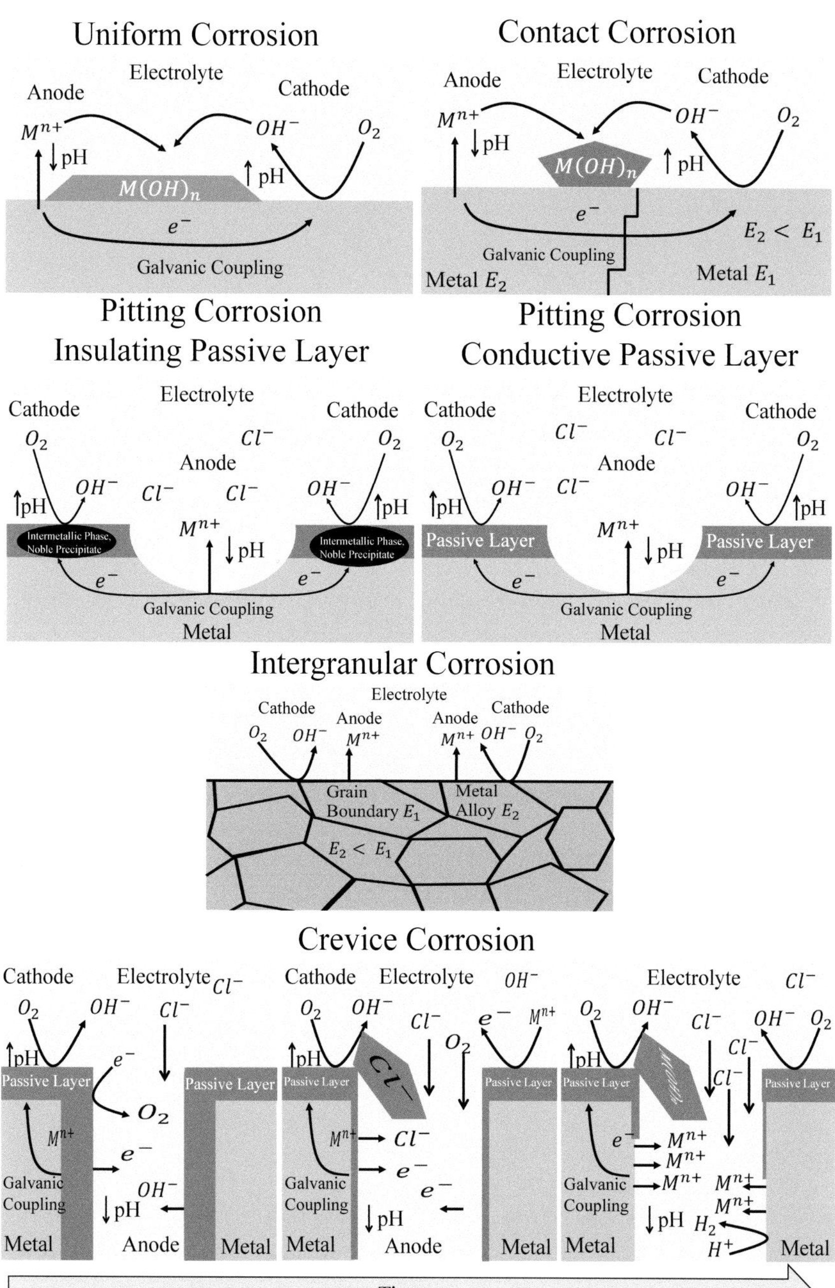

Fig. 1.9 Overview of corrosion types on uncoated metal surfaces.

anodic and cathodic process and in combination with the huge cathodic area the corrosion rate could be very high. A detailed description of pitting corrosion is given in ref. [14]. Because of the fact that corrosion processes are local phenomenons microscopic and space-resolved, spectroscopic methods are useful to analyze it and presented in Chaps. 2 and 3.

Intergranular Corrosion

Intergranular corrosion is based on a potential difference between the grain surface and the grain boundary in a polycrystalline metal alloy. The reason for the potential difference could be versatile. In principle corrosion starts at the grain boundary because it is always the most reactive area of a polycrystalline surface [15–17] but intergranular corrosion is based on an enrichment of impurities or compounds from the alloy at the grain boundary which decreases the passive properties on the grain or change the potential at the grain boundary. Therefore intergranular corrosion is a kind of contact corrosion on a microstructural scale. One important effect is a heating process, e.g., welding, and the subsequent cooling time on stainless steel surfaces [14]. If the final temperature or the cooling rate is not accurately controlled, a chromium enriched carbide phase is formed at the grain boundary and the chromium concentration in the grain decreases. Therefore galvanic corrosion is possible and promoted by the less stable passive layer on the grain because of the low chromium content (for the passive layer on chromium see Ex. 2.6). Intergranular corrosion could be based on other phenomena especially on stainless steels and aluminium alloys presented in detail in [14] (see also Chap. 4.2.4).

Crevice Corrosion

At a crevice formed by design or caused by a cracking process with a gap between 0.1 and 100 µm [14] corrosion could start, if oxygen and an electrolyte with an aggressive anion (chloride) are present. The tendency to separate cathode and anode as already mentioned is favoured because in the crevice the anode process is promoted by the decreasing oxygen content while on the surface the cathodic process is only limited by the oxygen diffusion to the surface (Fig. 1.9). In the crevice the pH value decreases (compare Reacs. 1.5 and 1.7) and the formation of a passive layer is not possible because of the low pH and the low oxygen concentration (compare Reac. 1.6). The chloride concentration increases in the crevice and the pH decreases so that a new cathodic process based on hydrogen evolution could accelerate the dissolution of the metal [10]. The situation in detail is more complex and could change depending on the metal alloy, the electrolyte and the geometry of the crevice. For further reading see refs. [10, 14, 18, and 19].

1.2.2 *Corrosion Processes on Coated Metals*

The corrosion phenomenon on a coated metal could be subdivided in principle into two processes the blistering and the delamination. Both effects described in principle in the following paragraph.

Blister

Blister could be generated under an intact organic coating as illustrated in Fig. 1.10 (compare [20, 21]). When the intact coating comes in contact with an electrolyte and oxygen, both water and oxygen and later ions from the electrolyte will diffuse into the coating (see Chap. 2.2.2). The water uptake or swelling of the polymer could generate detached areas because of tensile stress [22] at the interface or there are already generated during the application. In this area water is collected at the interface and with the present oxygen corrosion processes could be initiated. As already mentioned a separation of anode and cathode occurs forced by the precipitation of corrosion products in the blister. The cathode process forced the delamination of the anodic blister or, if possible, generates a second cathodic blister (not shown). Satellite blisters around the first blisters may be formed because of the tensile stress in the coating caused by the first blister [23].

Because of this corrosion mechanism the transport processes through the organic coating are relevant to understanding and to preventing corrosion processes at the interface. The transport of water could be investigated by means of Electrochemical Impedance Spectroscopy (EIS) (Chap. 2.2.2) and the ion transport with UV-VIS spectroscopy (Chap. 3.2.2). The electrochemical processes under the coating could be measured by means of the Scanning Kelvin Probe (SKP) [25] presented in Chap. 2.2.3.

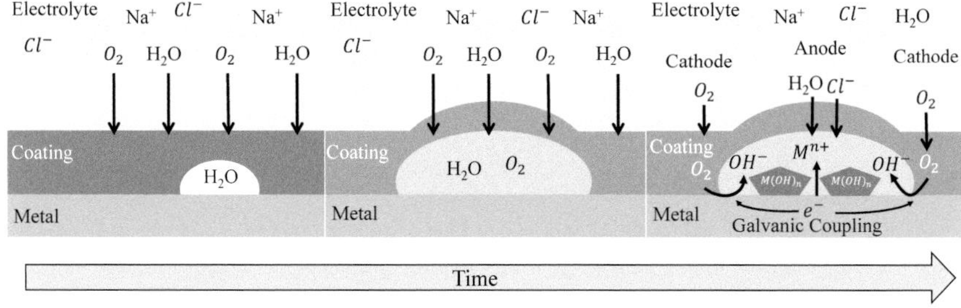

Fig. 1.10 Formation of blister corrosion according to [20, 24].

Delamination

Organic coatings have defects—sooner or later. The defects are caused by pores or pinholes in the coating during the film formation, cracks because of degradation processes by UV light or because of mechanical stress—scratches, stone chip on automotive car bodies or the like. If oxygen and an electrolyte are present, corrosion starts at the defect and the anode and cathode reaction are in the same area as the uniform corrosion in Fig. 1.9. As always in corrosion processes a separation from anode and cathode will occur and at a defect on an organic coated substrate in principle two mechanisms for the separation are possible (Fig. 1.11). If the cathode reaction moves under the coating, a cathodic delamination occurs and the pH value under the coating increases and the coating will be removed from the surface because of the electrochemical reactions (Reac. 1.2), the pH change or because of peroxide ions [26–28] produced during the reduction of the oxygen. To keep up the process the galvanic coupling between anode and cathode (Fig. 1.11) and the transport of oxygen, water and electrolyte through the coating and at the interface are necessary [29]. The transport processes through the coating could be investigated by means of EIS and at the interface by means of Scanning Kelvin Probe (SKP) illustrated in Chap. 2.2.2 and 2.2.3, respectively. The alternative mechanism is the anodic delamination. The corrosion products produced by the anodic reaction will deform the coating surface and act as a membrane to separate the pH values of the anode in front of the process and the following cathodic reaction area (Fig. 1.11). The anodic delamination especially on coated aluminium surfaces starts with blisters [24] and produce filaments on the coating surface called filiform worms and the corrosion type is the filiform corrosion [24, 30]. Actually there are more mechanisms especially of amphoteric metals such as zinc and aluminium explained in detail in Chap. 2.2.3. For further reading see refs. [31–34].

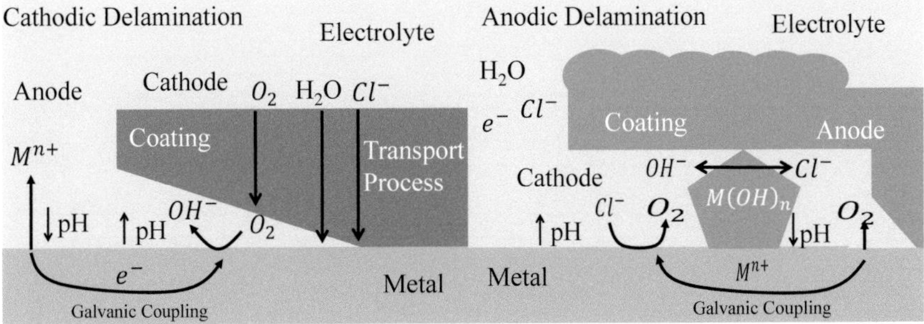

Fig. 1.11 Delamination processes under a coating.

1.3 Basics in Inorganic Coatings

1.3.1 Conversion Coatings

One approach to reduce the corrosion rate on a metal surface is the reduction of the electrochemical active surface for diffusion controlled reactions, e.g., atmospheric corrosion, and the insulating of the surface to inhibit the galvanic coupling between the anode and cathode area. Both could be achieved with so called conversion layers presented in the following paragraph. Furthermore the conversion layer increases the adhesion of an organic coating and reduces the delamination under the coating at defects.

Conversion layers with a thickness between a couple of nm and a couple of μm applied on the metal substrate to increase the corrosion protective properties of an organic coating or act as a temporary corrosion protection (transport corrosion protection). There are several terms for conversion coatings depending on the technical process:

Pre-treatment: A conversion coating applied before the application of an organic coating

Post-treatment: A conversion coating applied after a hot dip galvanizing process (see Chap. 1.3.2) on steel

Passivation: A conversion coating applied on a conversion coating (often a phosphatation)

Conversion layers consist of inorganic compounds but especially in the last two decades more and more organic modified [35–37] conversion layers have been developed. The function of a conversion layer could be summarized as follows:

- Temporary corrosion protection
- Enhancement of the adhesion
- Inhibition of the galvanic coupling
- Reduction of the surface for electrochemical reactions

The application process is based in all cases on the same electrochemical and chemical mechanism exemplary illustrated on the phosphatation process on steel (for details see refs. [23, 38–40]):

The metal substrate is dipped in an acidic (pH = 2–3) solution with solved metal cations or dispersed organic compounds with acidic groups. Acid corrosion starts immediately (Reacs. 1.1 and 1.4) and if nickel cations are present in the solution, the nickel is reduced and precipitated on the steel surface. Because of the fact, that the hydrogen over-voltage (the activation enthalpy, in principle) on the nickel surface is lower than on steel (Fig. 1.12), the acid corrosion rate is accelerated by the nickel precipitation

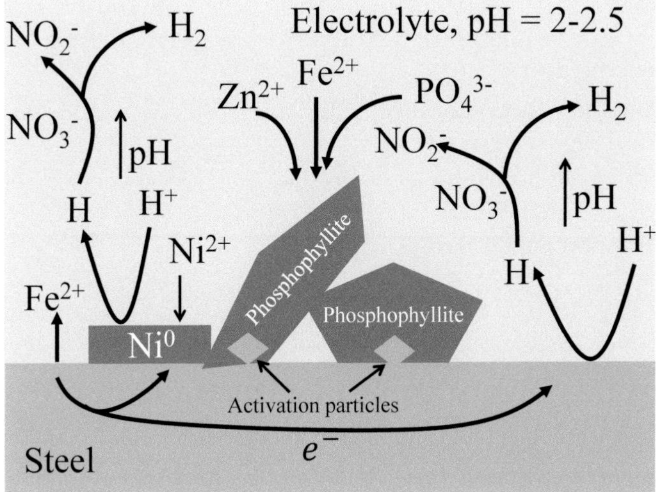

Fig. 1.12 Mechanism of the phosphate layer precipitation.

[41]. The acid corrosion process increases the pH value near the surface (1–3 μm [38–40]) and the solubility constant of the conversion layer compound is exceeded and the precipitation of it generates the conversion layer (Fig. 1.12). Further accelerating of the process is promoted by oxidizing accelerators like nitrate/nitrite, peroxide and other compounds. To optimize the crystal growth of zinc phosphate the metal surface is dipped in an activation bath before the phosphatation process. The activation consists of titanium or zinc phosphate with a particle size in the nm-range and the particles absorb on the metal surface working as seed crystals for the following phosphatation process.

Phosphatation

The first patent of a phosphatation process was claimed in 1869 [42]. Since that time a lot of innovations have been done to optimize the process. The pure phosphoric acid was replaced by zinc and manganese containing phosphoric acid baths with a lot of additives in the bath and finally the activation process, i.e., the deposition of nano sized titanium or zinc phosphate particles as seed crystals, has been established.

The phosphatation is based on the acid corrosion according to Reac. 1.4 in combination with the substrate dissolution as shown in Reac. 1.1 on steel and in Reac. 1.9 on zinc and just as on aluminium:

$$Al \Leftrightarrow Al^{3+} + 3e^- \wedge E^0 = -1.663\text{V} - 0.0197\,\text{V}\log[(\text{Al}^{3+})] \qquad \text{Reac. 1.14}$$

The increasing pH value near the surface exceeds the solubility constant of zinc or iron phosphate and the precipitation starts to produce the conversion layer:

Phosphophyllite reaction on steel [43]:

$$Fe^{2+} + 2Zn^{2+} + 3PO_4^{3-} + 4H_2O \Leftrightarrow FeZn_2\left(PO_4\right)_3 \cdot 4H_2O \qquad \text{Reac. 1.15}$$

Hopeite reaction on zinc coated steel [43]:

$$3Zn^{2+} + 2PO_4^{3-} + 4H_2O \Leftrightarrow Zn_3\left(PO_4\right)_2 \cdot 4H_2O \qquad \text{Reac. 1.16}$$

On aluminium Hopeite will also be deposited, but with a less coverage of the surface [23] as on zinc. There are many other processes generating other layers described in detail in ref. [41, 44].

The final conversion layer is based on phosphate crystals with a thickness between 1–3 µm (see Fig. 3.17). The coverage of the conversion layer is because of the mechanism (Fig. 1.12) lower than 100% but nearby in an optimized application process. The coverage could be investigated by means of cyclic voltammetry (see Ex. 2.5).

Zirconium Based Layers

The first use of zirconium based conversion layers was on tin (can coating) [45] followed by aluminium surfaces [46–48] but the transfer to steel and zinc coated steel surfaces had already been done [49, 50] and the implementation in industrial process is still in progress [51]. In contrast to the phosphatation process the zirconium conversion layer are amorphous and very thin in a range of 10–50 nm. The precipitation is based on the decomposition of H_2ZrF_6 near the surface caused by the increasing pH > 3 [52]:

$$H_2ZrF_6 + 4OH^- \Leftrightarrow Zr(OH)_4 \; or \; ZrO(OH)_2 + 2HF + 4F^- \; or + H_2O \qquad \text{Reac. 1.17}$$

In Chap. 3.1.1 (Ex. 3.2) the composition of the conversion layer will be discussed in detail. The advantages of this type of layer are the small amount of raw materials, the stability in acid and alkaline conditions and the prevention of mud during the application process.

Titanium Based Layers

H_2TiF_6 solutions are used in combination with phosphoric acid and manganese ions in a no rinse application of conversion layers for coil coating (see Chap. 1.4.4) applications [53]:

$$4Zn^{2+} + H_2TiF_6 + xMn^{2+} + yPO_4^{3-} \Leftrightarrow Zn_3(PO_4)_2 + Zn(OH)_2 + TiMn_x(PO_4)_y$$

Reac. 1.18

The complex conversion layer generates a good adhesion of the organic coating (primer) and produces a high corrosion protective coating system. The application similar to Reac. 1.17 often shows lower corrosion protective properties as the zirconium based process [54] and is therefore are rarely used for titanium based conversion layers.

Cerium Based Layers

Cerium based (nitrate or chloride) conversion layers are used on aluminium and magnesium surfaces generated with the following reaction [55, 56]:

$$Ce^{3+} + 3H_2O \Leftrightarrow Ce_2O_3 + 6H^+$$ Reac. 1.19

If H_2O_2 is added to the solution the amount of CeO_2 increases [55]. Cerium and other rare earth metals have been used to replace the chromate based conversion layers because of the high toxicity of chromate compounds. The thickness of cerium based conversion layers is in the nm range.

Molybdenum Based Layers

The molybdenum based conversion layers were another approach to replace chromate in conversion coatings [57]. The film formation process is very complex and depends on the concentration, substrate, immersion time and additives in the solution. The precipitation on a zinc surface could be based on the following reactions [57]:

$$Zn^{2+} + HMoO_4^- \Leftrightarrow ZnMoO_4 + H^+$$ Reac. 1.20

$$14Zn^{2+} + 3Mo_7O_{24}^{6-} + 32H^+ + 42e^- \Leftrightarrow 7Zn_2Mo_3O_8 + 16H_2O$$ Reac. 1.21

The molybdenum compounds have been found on the surface but also MoO_3 and $Mo_7O_{24}^{6-}$ signals are mentioned in literature [58]. All detected compounds are soluble in chloride containing electrolytes and therefore the corrosion protective properties in comparison to chromate are low. If the molybdate is combined with organic compounds [36–37], the performance of the resulting composite layer could be improved. The thickness of molybdenum based conversion layers depends on the immersion time between the nm- and the μm-range [57].

There are a lot of more conversion layers based on the deposition of other metals (e.g., yttrium) and many combinations with organic polymers or adhesion promotors and finally other deposition methods like the plasma polymerization process are mentioned in literature. For all illustrated conversion layers spectroscopic data are summarized in Tables 5.5, 5.7 and 5.9 in the app.

1.3.2 *Metal Coatings*

A common way to improve the corrosion protective properties of steel surfaces is to coat the surface with another metal layer. The dominated metal is zinc applicable in different ways:

Electrolytic Applied Zinc Layer

Zinc could be reduced in aqueous solutions because of the high hydrogen over-voltage and could be applied from acid and alkaline solutions [59]:

$$Zn^{2+} + 2e^{-} \Leftrightarrow Zn \qquad \text{Reac. 1.22}$$

$$\left[Zn(OH)_4\right]^{2-} + 2e^{-} \Leftrightarrow \left[Zn(OH)_2\right]^{2-} + 2OH^{-} \Leftrightarrow Zn + 2OH^{-} \qquad \text{Reac. 1.23}$$

The crystal size of the zinc layer could be controlled with additives in the electrolyte and therefore very smooth and glossy surfaces are available. The thickness of the metal layer is often between 5–20 µm and the process is used in coil coating application (automotive industry), parts for marine environment [60] and especially for small parts like fasteners and bolts.

Hot Dip Galvanized Zinc Layers

For the most part steel is covered with zinc in a hot dip galvanization process. The application could be subdivided into two processes:

Continuous Hot Dip Galvanizing

The steel strip moves through hot zinc melt often containing a low amount of aluminium [61]. The thickness of the zinc coating is between 7–70 µm [62, 63]. Several alloys are available based on aluminium [64] and magnesium. The alloy with magnesium shows a significantly better corrosion protective property [65] and therefore the thickness of the alloy is only between 7–20 µm [66].

Hot Dip Galvanizing of Parts

Steel parts dipped into hot zinc melt at temperatures between 440–460°C generate a zinc layer with a thickness between 40 and 80 µm [67]. The certain thickness depends on the use of the part and is defined in DIN EN ISO 1461. The surface after cooling shows large or small spangles depending on the process.

At first view the protection is based on a cathodic protection of the steel surface because the zinc layer acts as sacrificial anode (see Table 5.2). Furthermore the corrosion products of the zinc dissolution could influence the passivation process on the uncoated (edges, scratches, etc.) steel surface [68].

Another approach is based on galvanized tin coated steel used for packaging applications. The electrolytic applied metal layer with a thickness between 1 and 6 g/m^2 on steel is used predominantly in Europe for packaging especially of food cans (see Ex. 2.11).

1.4 Basics in Organic Coatings

From a corrosion protection point of view an organic coating increases the diffusion path for water/electrolyte and oxygen to the surface. Low diffusion coefficients (see Eq. 1.38) could be realized with polymeric networks with a high network density and the use of pigments and fillers. Some water vapour permeability of organic coatings are summarized in Table 3.8.

The composition of an organic coating could be summarized as follows:

- Solvent (water, organic solvents or no solvent in powder coatings and 100% systems)
- Resin (organic based or inorganic polymer)
- Cross linking agent (isocyanate, epoxy, amine, melamine or no agent in physical or oxygen drying resins)
- Pigments (Corrosion protective pigments like zinc phosphate, coloured pigments, etc.)
- Filler (calcium carbonate, talcum, barium sulphate, etc.)
- Additives (silicones, waxes, fluorine containing compounds, polymers, catalyst for the cross-linking reaction, etc.)

Some important examples of the components are given below. For a detailed presentation of the formulation of organic coatings see refs. [69–71].

1.4.1 Typical Resins in Organic Coatings

Some important resins especially for corrosion protective coatings are illustrated in Fig. 1.16. All resins have advantages and disadvantages

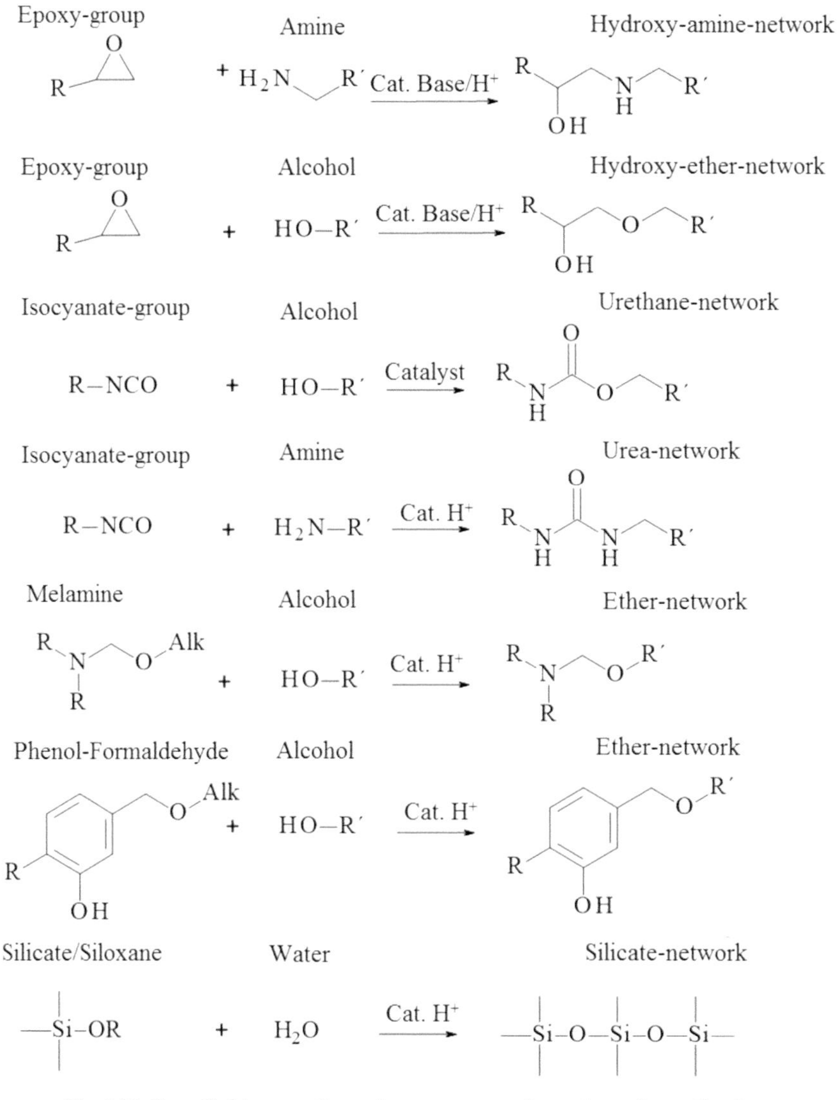

Epoxy-group + Amine (H_2N–R′) —Cat. Base/H^+→ Hydroxy-amine-network

Epoxy-group + Alcohol (HO–R′) —Cat. Base/H^+→ Hydroxy-ether-network

Isocyanate-group (R–NCO) + Alcohol (HO–R′) —Catalyst→ Urethane-network

Isocyanate-group (R–NCO) + Amine (H_2N–R′) —Cat. H^+→ Urea-network

Melamine + Alcohol (HO–R′) —Cat. H^+→ Ether-network

Phenol-Formaldehyde + Alcohol (HO–R′) —Cat. H^+→ Ether-network

Silicate/Siloxane (–Si–OR) + Water (H_2O) —Cat. H^+→ Silicate-network (–Si–O–Si–O–Si–)

Fig. 1.13 Cross-linking reactions of common organic coatings. Cat. = Catalyst.

(see ref. [69–71]) depending on the use of the coating or the use of the coated part. In principle all coatings need a cross-linking-reaction, i.e., the formation of a polymer network during the film formation [72]. All combinations of the resins in Fig. 1.16 and the cross-linking-agents in Fig. 1.17 that are chemically possible are used in industry based on the following chemical reactions (Figs. 1.13 and 1.14).

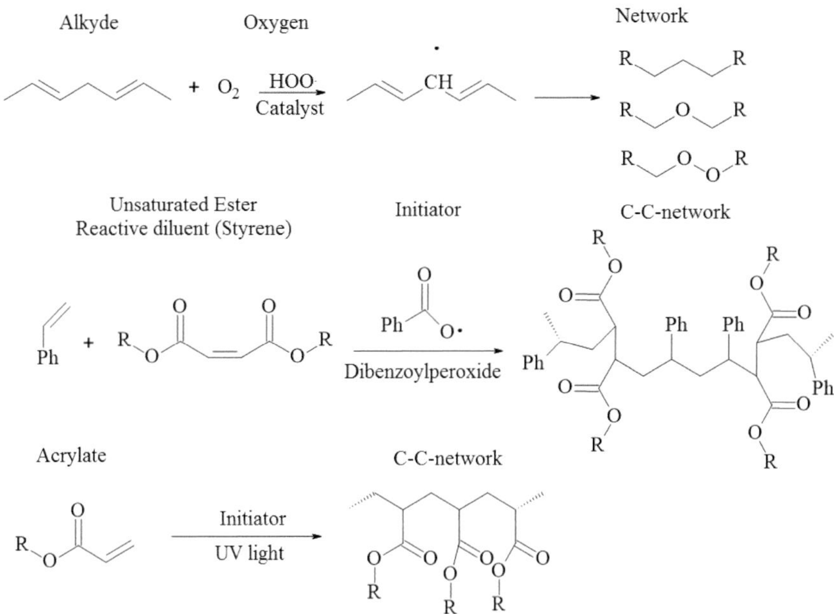

Fig. 1.14 Cross-linking reactions via radicals of common organic coatings.

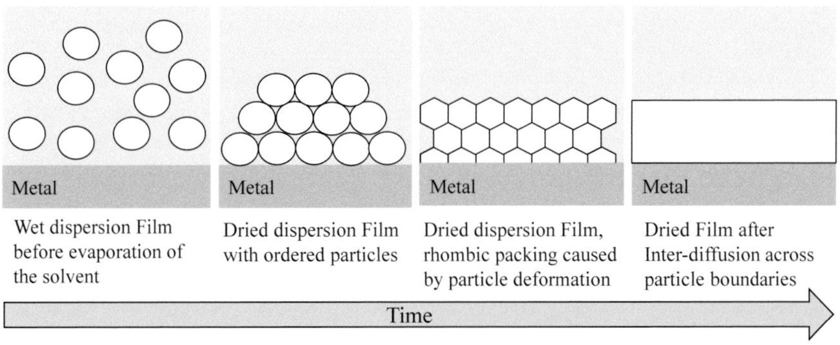

Fig. 1.15 Film formation of physical drying dispersions.

Alkyd resins, unsaturated polyesters and acrylate containing resins will be cross-linked with oxygen, peroxides or UV light with radical intermediates with the following mechanisms.

Especially water based polyacrylic dispersions could be physically dried (Fig. 1.15), i.e., the polymer network will be produced by inter-diffusion of polymer chains of the dispersion particles into each other [see also [72]).

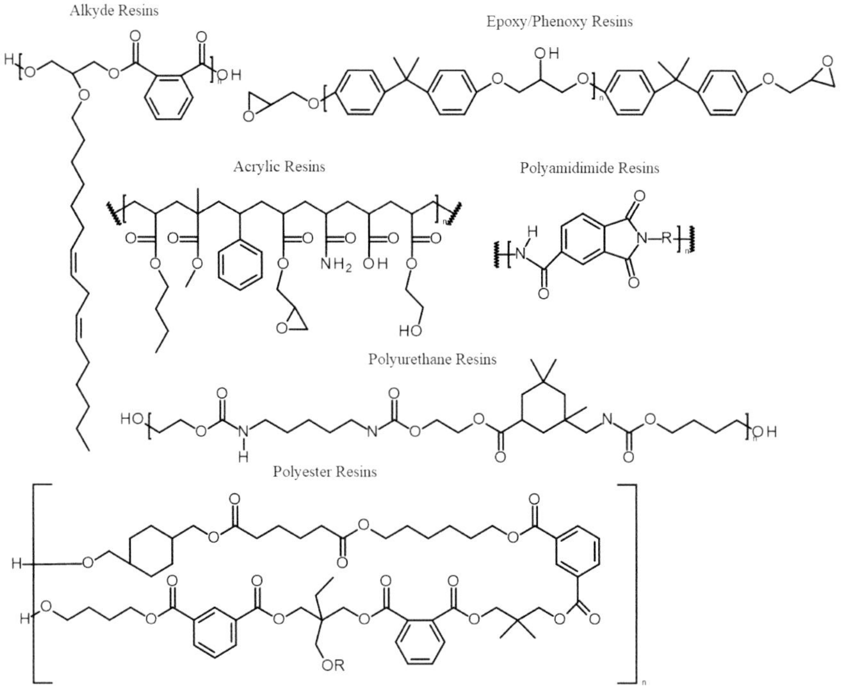

Fig. 1.16 Common resins for organic coatings with some examples for the typical monomers.

At a first look the organic coating is just a barrier from the analytical point of view. To analyze the corrosion process on the surface the coating is often removed from the surface or the measurement could be done through the coating for instance with the scanning Kelvin probe. Actually the transport processes through the coating are very important to analyze the complete process (see above) and at a later state of the corrosion reaction the degradation of the coating has to be analyzed to investigate the corrosion protective properties of the complete system. For the most common organic compounds used in organic coatings spectroscopic data are summarized in Tables 5.8 and 5.9 in the app. The polymeric network generated by the cross-linking reaction could be analyzed by means of infrared and Raman spectroscopy on the applied coating (see Exs. 3.21 and 3.22).

1.4.2 Typical Pigments in Organic Coatings

Hundreds of different pigment types are used in organic coatings [69, 70]. In corrosion protective coatings the pigments could be classified according to the function in the coating as follows:

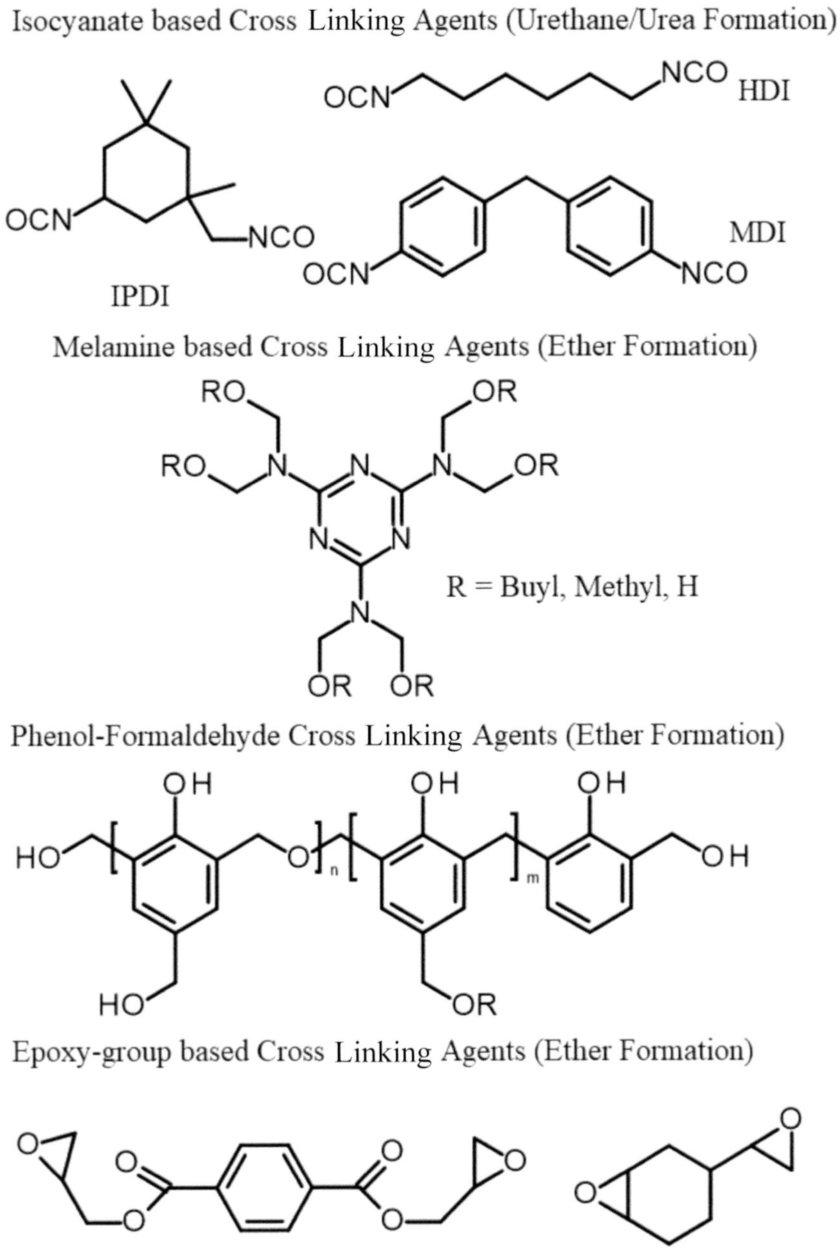

Isocyanate based Cross Linking Agents (Urethane/Urea Formation)

HDI

IPDI

MDI

Melamine based Cross Linking Agents (Ether Formation)

R = Buyl, Methyl, H

Phenol-Formaldehyde Cross Linking Agents (Ether Formation)

Epoxy-group based Cross Linking Agents (Ether Formation)

Fig. 1.17 Common cross-linking-agents for organic coatings. The NCO compounds rarely use as monomers but oligomers (Biuret, Isocyanurate [70]) are common cross-linking agents.

Corrosion Protective Pigments

Zinc

Another way to apply zinc coatings on a steel surface are zinc rich coatings divided into two different systems. Zinc powder coatings use zinc particles with a diameter in the range of 20–40 μm and a zinc content of 90% in the dry film [70, 71]. Zinc flake coatings use thin lamellar particles with a high aspect ratio, i.e., a thickness of some hundred nm and an area of some 20 μm (for details of zinc flakes see [76–78] and for zinc flake coatings see [79–80]). This type of coating is introduced in detail in Chap. 2.2.1.

Zinc and other Phosphates

These type of pigments are called active corrosion protective pigments as they inhibit the corrosion. If the corrosion process is initiated, the low pH values of the anode area dissolute the zinc, calcium, magnesium, aluminium or strontium phosphate pigments followed by a precipitation of phosphates on the surface. Therefore a conversion layer is generated and inhibits or at least decreases the corrosion process [73].

Lamellar Pigments

To increase the distance for the diffusion process of water, ions and oxygen through the coating (see Eq. 1.38) lamellar pigments based on mica, iron oxide, aluminium silicate (Kaolin) or talcum are used to increase the corrosion protective properties [71]. The effect could be investigated by EIS (Chap. 2.2.2).

Pigments for Colouring the Surface

A white coating could be created with titanium dioxide based pigment and a black surface with carbon black particles. Other colours like red (haematite, chinacridone, diketo pyrrolopyrrols, etc.), yellow (goethite, bismutvanadates, etc.), blue (cupper phthalocyanines, etc.) and green (chlorinated cupper phthalocyanines) are of course also possible [71]. In principle all pigments have an influence of the corrosion protective property of the resulting film.

Fillers

The difference between pigments and fillers is possible by the refraction index. Compounds with an index lower than 1.7 are often called fillers [71]. To achieve a dense packaging of the pigments a combination of small and big particles is necessary. Fillers are often the big particles based on calcium carbonates or barium sulphate or other compounds.

1.4.3 Additives

One paragraph is not enough just to classify the enormous amount of additives used in organic coatings. From the analytical point of view additives are often a challenge because they enrich on interfaces and surfaces and disturb the spectroscopic measurement because of the high functionality of the molecules or polymers. On the other hand additives are important for the film forming process and the final properties of the coating and therefore the analysis would be easier without the additives but the results could hardly be used to understand the real coating. For further reading see refs. [69–72]. In the analytical chapters certain additives are mentioned, if they are necessary for the analytical process.

1.4.4 Coating Systems

The final necessary corrosion protective property could be achieved by combination of different coatings. In Fig. 1.18 the coating system used in coil coating application is illustrated. In many applications steel is covered first with zinc followed by a conversion layer. The first organic coating (primer) is often based on epoxy-resins, whereas in the top coat polyacrylates, polyurethane or polyester are commonly used because of the higher UV stability. In Exs. 2.12 and 2.13 the corrosion protective properties and some analytical methods for coil coatings systems are presented.

In the automotive industry a complex coating system for car bodies is used (Fig. 1.19) to achieve a high corrosion protective performance in combination with a good mechanical stability and a mirror like appearance with a broad variety of colours and visual effects [74]. The used primer is cathodic deposition paint (E coat) explained in detail in ref. [75]. The

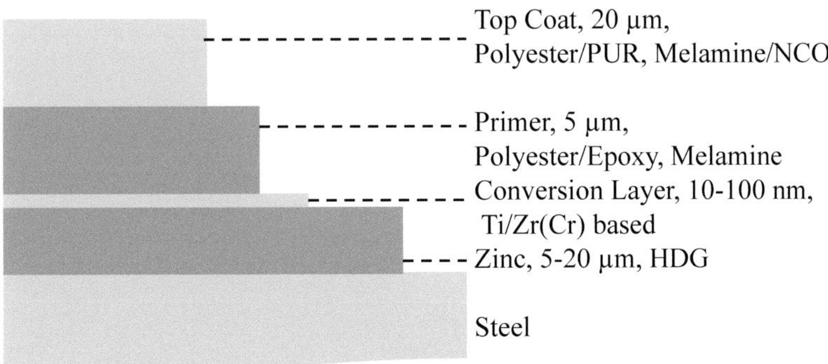

Fig. 1.18 Typical layer system for coil coating.

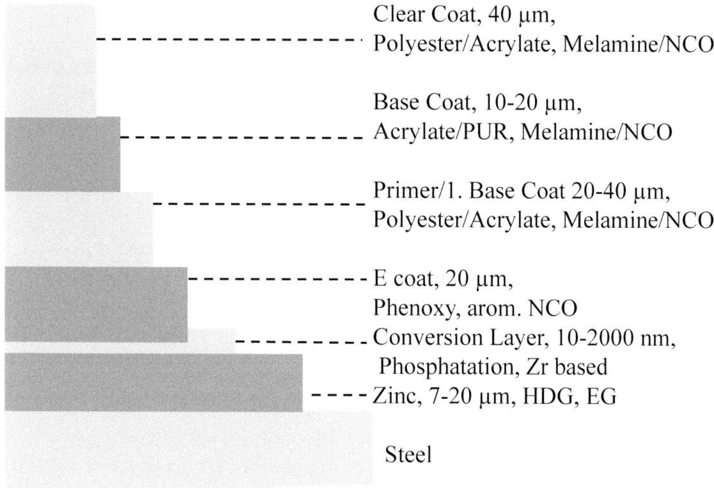

Fig. 1.19 Typical layer system for automotive coatings (compare [74]).

phenoxy resins (Fig. 1.16) with protonated amine groups applied on the cathodic polarized surface by deprotonation in the highly alkaline solution near the cathode (evolution of hydrogen, see Reac. 1.4) and cured catalytic [81–83] with blocked aromatic isocyanates (e.g., MDI based). The resulting coating has a very high network density and a very good adhesion to the pre-treated surface (see Exs. 2.14 and 2.15). The next coating layers are necessary to protect the E coat from UV light. Furthermore every layer fulfilled another property (Fig. 1.19) like stone chip protection (filler), colour (base coat), chemical resistance, UV protection and gloss (clear coat) [74].

To protect the tin surface in food packaging one layer of can coatings is applied (for details see Ex. 2.11) and the same is often used for attaching parts coated with autophoretic coatings (for details see Ex. 3.20) or the so called black E coat. Finally hundreds of special coatings used in industry to optimize corrosion protective properties of the surface. One example is wire enamel used for copper wire presented in Ex. 3.22.

1.5 Literature

1. H. Kaesche, Die Korrosion der Metalle, 3. neubearb. und. erw. Aufl. 1990, Springer Verlag, Heidelberg, 2011, English Edition: H. Kaesche, Corrosion of Metals, Springer Verlag, Berlin-Heidelberg, 2003
2. C.H. Hamann, W, Vielstich, Elektrochemie, Wiley-VCH, Weinheim, 2005; C.H. Hamann, A. Hamnett, W. Vielstich, Electrochemistry, Wiley-VCH, Weinheim, 2007
3. P.W. Atkins, Physical Chemistry, Oxford University Press, 1978

4. M. Pourbaix, Atlas of Electrochemical Equilibria in aqueous solutions, National Association of Corrosion engineers, Houston Texas, 1974
5. M. Pourbaix, Atlas of Electrochemical Equilibria in Aqueous Solutions, Pergamon, New York, 1966
6. R.G. Mortimer, Physical Chemistry, The Benjamin/Cummings Publishing Company Inc., Redwood City, 1993
7. R. Holze, Experimental Electrochemistry, Wiley-VCH, Weinheim, 2009
8. J. Crank, The Mathematics of Diffusion, Clarendon Press, Oxford, 2. Ed., 1975
9. Encyclopaedia of Electrochemistry, Edited by A.J. Bard and M. Stratmann, 2007 Wiley-VCH Verlag GmbH & Co. KGaA, Weinheim, Vol. 4, Corrosion and Oxide Films, Chap. 1.1 G. S. Frankel, Fundamentals of Corrosion
10. E. Bardal, Corrosion and Protection, Springer Verlag, London, 2004
11. R.G. Kelly, J.R. Scully, D.W. Shoesmith, R.G. Buchheit, Electrochemical Techniques in Corrosion Science and Engineering, Marcel Dekker Inc., New York, Basel, 2002
12. E. Barsoukov, J. Ross Macdonald, Impedance Spectroscopy Theory, Experiment, and Applications Second Edition, Wiley-Interscience, Hoboken, 2005
13. D. Meschede, Gerthsen Physik, 23. Auflage, Springer Verlag, Berlin, 2006
14. Encyclopaedia of Electrochemistry, Edited by A.J. Bard and M. Stratmann, 2007 Wiley-VCH Verlag GmbH & Co. KGaA, Weinheim, Vol. 4, Corrosion and Oxide Films
15. Thesis (PhD), S. Toews, Corrosion Protection by Selective Addressing of Polymer Dispersions to Electrochemical Active Sites, University Paderborn, 2010
16. S. Toews, W. Bremser, H. Hintze-Brüning, M. Dornbusch, EUROCORR 2010, Moscow, 2010
17. W. Bremser, M. Dornbusch, H. Hintze-Brüning, S. Toews, Verfahren zur autophoretischen Beschichtung, Beschichtungsmittel und Mehrschichtlackierung, DE 102010019245A1
18. F.M. Song, Electrochim. Acta 56 (2011) 6789–6803
19. F.M. Song, Corros. Sci. 55 (2012) 107–115
20. J. H. Park, G.D. Lee, H. Ooshige, A. Nishikata, T. Tsuru, Corros. Sci. 45 (2003) 1881–1894
21. F.M. Geenen, J.H.W. de Wit, E.P.M. van Westing, Prog. Org. Coat. 18 (1990) 299–312
22. T.-J. Chuang, J. Coat. Technol. 71, 895 (1999) 75–85
23. Thesis (MA), J. Karlsson, Corrosion Mechanisms under Organic Coatings, Chalmers University of Technology, Göteborg, Sweden, 2011
24. J.H.W. de Witt, D.H. van der Weijde, H.J.W. Lenderik, International Corrosion Congress, Proceedings 13th, Nov. 1996, 102/1-102/7
25. M. Stratmann, H. Streckel, Werkst. Korros. 43(1992) 316–320
26. H. S. Wroblowa, S. B. Qaderi, J. Electroanal. Chem. 279 (1990) 231–242
27. H. S. Wroblowa, S. B. Qaderi, J. Electroanal. Chem. 295 (1990) 153–161
28. H. S. Wroblowa, S. B. Qaderi, Poly. Mater. Sci. Eng. 58 (1988) 27–31
29. W. Fürbeth, M. Stratmann, Fresenius J. Anal. Chem. 353 (1995) 337–341
30. G. M. Hoch, International Corrosion Conference series, 1974, NACE-3, Localized Corros., 143–142
31. W. Fürbeth, M. Stratmann, Corros. Sci. 43 (2001) 207–227
32. W. Fürbeth, M. Stratmann, Corros. Sci. 43 (2001) 229–241
33. W. Fürbeth, M. Stratmann, Corros. Sci. 43 (2001) 243–254
34. A. Leng, H. Streckel, M. Stratmann, Corros. Sci. 41 (1999) 547–578
35. F. Fleischhaker, C. Schade, H. Witteler, Preparation for passivating metal surfaces, containing Polymers having acid groups and containing Ti or Zr compounds, WO002013064442
36. M. Dornbusch, R.Wegner, A. Wiesmann, Corrosion-protection agent and method for current-free application thereof, WO002006125499

37. M. Dornbusch, A. Wiesmann, Varnish layer-forming anti-corrosion agent having reduced crack formation, and method for the current less application thereof, WO002008058587
38. W. Rausch, Die Phosphatierung von Metallen, Eugen G. Leuze Verlag, Bad Saulgau, 2005
39. Thesis (PhD), N. Müller, Einfluß von Bad- und Substratkomponenten auf die Bildung von Phosphatschichten, Heinrich-Heine-University Düsseldorf, 2000
40. Thesis (PhD), D. Zimmermann, Der Einfluss von Nickel auf die Phosphatierung von Zink, Heinrich-Heine-University Düsseldorf, 2003
41. T. S. N. Sankara Narayanan, Rev. Adv. Mater. Sci. 9 (2005) 130–177
42. UK3119
43. P. Kuhm, B. Mayer, Nachr. Chem. Tech. Lab. 43, 1995, 9, 956–959
44. http://docslide.net, J. Donofrio, Zinc Phosphating, 57–73
45. O. Nobuyuki, T. Haruyoshi, Method of surface treatment of tin plated cans and tin plated steel sheets, US4339310
46. F. J. Frelin, L.T. Kelly, A. J. Malloy, Coating solution for metal surfaces, US4370177
47. P. Schiefer, K. Wittel, Verfahren zur Oberflächenbehandlung von Aluminium, DE3236247
48. E. M. MUSINGO, W. J. NEILL, L. S. SANDER, Composition and method for non-chromate coating of aluminum, US4921552
49. O. Nobuyuki, T. Haruyoshi, T. Shinich, Verfahren zum Behandeln von Metalloberflächen, DE3408573
50. K. Hackbarth, C. Hirsch, Christina, S. Küpper, V. Lachmann, P. Reeßing, J. Sander, R. Seidel, Chromfreies Korrosionsschutzmittel und Korrosionsschutzverfahren, DE19923084
51. P. Kuhm, S. Sinnwell, JOT, 3, 2016, 46–49
52. P. Taher, K. Lill, J. H. W. de Wit, M. C. Mol, H. Terryn, J. Phys. Chem. C 2012, 116, 8426–8436
53. B. Wilson, N. Fink, G. Grundmeier, Electrochim. Acta 51 (2006) 3066–3075
54. M. A. Smit, J.A. Hunter, J. D. B. Sharman, G. M. Scamans, J. M. Sykes, Corros. Sci. 45 (2003) 1903–1920
55. M. DabalaÁa, L. Armelaob, A. Buchbergera, I. Calliari, Appl. Surf. Sci. 172 (2001) 312–322
56. B. R. W. Hinton, J. Alloys Compd. 180 (1992) 15–25
57. Thesis (PhD), D. E. Walker, Enhanced molybdate conversion coatings, Loughborough University, 2012
58. D.-L. Lee, T. Kang, H.-J. Sohn, H.-J. Kim, Mater. Trans. 43, 1 (2002) 49–54
59. T.W. Jelinek, Galvanisches Verzinken, Eugen G. Leuze Verlag, Saulgau, 1982
60. B. M. Durodola, J. A. O. Olugbuyiro, S. A. Moshood, O. S. Fayomi, A. P. I. Popoola, Int. J. Electrochem. Sci., 6 (2011) 5605–5616
61. N. Fink, B. Wilson, G. Grundmeier, Electrochim. Acta 51 (2006) 2956–2963
62. Stahl Informationszentrum, Broschüre CM094
63. Stahl Informationszentrum, Broschüre CM095
64. A. R. Moreira, Z. Panossian, P. L. Camargo, M. F. Moreira, I. C. da Silva, J. E. R. de Carvalho, Corros. Sci.48 (2006) 564–576
65. Thesis (PhD), R. Hausbrand, Elektrochemische Untersuchungen zur Korrosionsstabilität von polymerbeschichtetem Zink-Magnesiumüberzug auf Stahlband, Ruhr-University Bochum, 2003
66. ThyssenKrupp Steel, Produktinformation ZMg EcoProtect, 2009
67. Stahl Informationszentrum, Broschüre MB329
68. H. Tanaka, T. Kitazawa, N. Hatanaka, T. Ishikawa, T. Nakayama, Ind. Eng. Chem. Res. 51 (2012) 248–254
69. Z. W. Wicks, F. N. Jones, S. P. Pappas, D. A. Wicks, Organic Coatings, Wiley-Interscience, Hoboken, 2007

70. T. Brock, M. Groteklaes, P. Mischke, European Coatings Handbook, Vincentz Network, Hanover, 2010
71. A. Goldschmidt, H.-J. Streitberger, BASF Handbook of Basics Coating Technology, Vincentz Network, Hanover, 2003
72. P. Mischke, Film Formation, Vincentz Network, Hanover, 2010
73. M. Beiro, A. Collazo, M. Izquierdo, X. R. Novoa, C. Perez, Prog. Org. Coat. 46 (2003) 97-106
74. H.-J. Streitberger, Automotive Paints and Coatings, Wiley-VCH, Weinheim, 2. Ed. 2008
75. M. Dornbusch, R. Rasing, Christ, Epoxy Resins, Vincentz Network, Hanover, 2016
76. C. Scramm, T. Voit, G. Wagner, W. Förster, Zweikomponenten-Korrosionsschutzlack, dessen Verwendung und Verfahren zu dessen Herstellung, DE 10 2005 026 523
77. Patent, Zink-Magnesium-Korrosionsschutzpigmente, Korrosionsschutzlack und Verfahren zur Herstellung der Korrosionsschutzpigmente, DE 10 2012 107 634
78. M. Rupprecht, C. Wolfrum, D. Pfammatter, M. Stoll, H. Weiß, Zinkmagnesiumlegierung-Korrosionsschutzpigmente, Korrosionsschutzlack und Verfahren zur Herstellung der Korrosionsschutzpigmente, EP 2 785 806
79. E. G. Maze, G. L. Lelong, E. Dorsett, J. Guhde, I. Nishikawa, Particulate metal alloy coating for providing corrosion protection, US 7678 184
80. Y. Endo, T. Sakai, K. Takasse, Chromium-free water reducible rust inhibitive paint for metals, US 2005/0027056
81. K.-H. Grosse Brinkhaus, M. Neumann, O. Johannpoetter, P. Lux, Use of bismuth subnitrate in electro-dipping paints, WO 2009/021719
82. G. Ott, H. Baumgart, D. Schemschat, S. Przybilla, I. Ross-Lipke, K.-H. Grosse Brinkhaus, Tin catalysts having increased stability for use in electrophoretic surface coatings, WO 2009/095249
83. V. Peters, H. Baumgart, M. Dornbusch, Cathodic electrodeposition paint containing metal-organic compound, WO 2009/106337

2
Electrochemical Methods

2.1 Electrochemical Methods on the Metal Surface

2.1.1 *Current-Density-Potential Diagrams*

Basics

The basics for current-density-potential diagrams are already explained in Chap. 1.

Measurement

Because of the fact, that it is not possible to measure accurately the current and the potential between two electrodes [1, 2], a three-electrode-assembly is used for current-density-potential diagrams (Figs. 2.1 and 2.2) [3, 4] often named polarization diagram and used here. The Working Electrode (WE), i.e., the interesting surface, and the Counter Electrode (CE), often a platinum sheet or wire mesh, have to be of parallel orientation to avoid an inhomogeneous field. The current is investigated between WE and CE. The potential is measured between the WE and the Reference Electrode (RE). As RE are normally electrodes of the second type use, they are easy to handle and have a very constant potential (Table 2.1). The potential of the RE is determined against the normal hydrogen electrode (NHE) (for details see [1, 2]).

The reference electrode is connected with the electrochemical cell by means of a salt bridge, containing often a 3M KCl solution because the same electrolyte is inside the RE. The second reason to use KCl is the fact that the ionic mobility's of K^+ and Cl^- are similar [2]. This is important to minimize the diffusion and polarization of the electrolytes between the RE and the cell. Therefore a Haber-Luggin-Capillary (HLC) is used to connect

Table 2.1 Common reference electrodes for electrochemical analysis of surfaces.

Half cell	Comment	Electrode reaction	Potential [V] relative to the NHE	Lit.
Silver Chloride	Saturated KCl	$AgCl + e^- \Leftrightarrow Ag + Cl^-$	+0.1976	[1]
Electrode	1 M KCl		+0.2368	[1]
	0.1 M KCl		+0.2894, +0.288, +0.2881	[1], [5], [2]
Calomel Electrode	Saturated KCl	$Hg_2Cl_2 + 2e^- \Leftrightarrow 2Hg + 2Cl^-$	+0.2415, +0.2416, +0.2446	[1], [5], [2]
	1 M KCl		+0.2801, +0.2807	[5], [1]
	0.1 M KCl		+0.3337, +0.3338	[1], [5, 2]
Copper Sulphate Electrode	Saturated $CuSO_4$	$CuSO_4 + 2e^- \Leftrightarrow Cu + SO_4^{2-}$	+0.3134, +0.320	[6], [2]
Mercury Sulphate Electrode	0.5 M H_2SO_4	$Hg_2SO_4 + 2e^- \Leftrightarrow 2Hg + SO_4^{2-}$	+0.682	[1]
	Saturated K_2SO_4		+0.650	[1]
	1 M K_2SO_4		+0.660	[2]

the RE with the cell and the very small part at the end of the capillary reduces the diffusion of both electrolytes. Furthermore the distance of the HLC and the WE in the electrolyte has to as small as possible to minimize the so called Ohmic drop, i.e., the voltage drop because of the Ohmic resistance of the electrolyte. There are two often used cell geometries as shown in Figs. 2.1 and 2.2. If the working electrode is in the electrolyte (Fig. 2.1), the edges and the back side of the sample have to be covered (e.g., organic coating) to avoid an inhomogeneous field and an imprecise surface area. Otherwise the transport processes from and to the surface are fast and the stirring occurs a homogenous electrolyte. If the working electrode is flanged on the side of the electrochemical cell (Fig. 2.2), larger samples without covering could be used and the change of the samples is easier. The disadvantage is the possibility of a worse electrolyte exchange on the surface and the challenge to seal the contact between cell and sample.

The selection of the electrolyte is a challenging issue because, if aggressive media like NaCl solutions are used, a dynamic situation on the surface, i.e., corrosive processes, is the consequence. To analyze corrosion processes a lot of conditions are possible, summarized in

Table 2.2. Common standards for corrosive solutions like "sea water" are used in literature in some extent different (Table 2.2) and may cause different data (compare Table 5.3 in the app.).

To analyze the electrochemical properties of the surface without corrosion reactions, buffer solutions achieves a stable situation on the working electrode, but the buffer compounds could interact with the surface. Borate based electrolytes for alkaline conditions are common and should not interact with the surface. A buffer solution could be prepared based on the Hernderson-Hasselbach equation [7]:

$$pH = pK_A + \log \frac{c(Base)}{c(Acid)} \wedge pH = pK_A \pm 1 \qquad \text{Eq. 2.1}$$

Because of the pH value limitation, it is not possible to avoid a change of the buffer solution, if a broad pH range has to be analyzed. Even with phosphoric acid (pK_A values 2, 7 and 12) the pH ranges 4–5 and 9–10 could not be stabilized.

The simulation of different corrosive environments such as marine, soil or tap water could be realized with electrolytes summarized in Table 2.2. The electrolyte and the atmospheric conditions (O_2, CO_2, etc.) define the corrosion mechanism (see Fig. 3.50 for zinc and Fig. 4.1 for steel) and could change the corrosion protective properties illustrated for saturated $CaCl_2$ solutions in the Exs. 2.3, 2.6, and 3.19.

The water itself has to be as pure as possible, because any contamination could interact with the surface and adulterate the measurement. The electrolyte should be prepared fresh because CO_2 will diffuse inside the solution and change the pH value according to the following reactions:

$$CO_2 + H_2O \Leftrightarrow H_2CO_3 \qquad \text{Reac. 2.1}$$

$$H_2CO_3 \Leftrightarrow H^+ + HCO_3^- \wedge pK = 6.37[7] \qquad \text{Reac. 2.2}$$

$$HCO_3^- \Leftrightarrow H^+ + CO_3^{2-} \wedge pK = 10.25[7] \qquad \text{Reac. 2.3}$$

The measurement could be done with a potentiostat in modern devices part of an electrochemical workstation, i.e., a device for all standard experiments like the polarization diagram, cyclic voltammetry and electrochemical impedance spectroscopy.

Table 2.2 Common electrolytes for the analysis of corrosion processes and analysis of metal surfaces. (a) For sea water and brack water see also DIN 50905-4 (2017).

Electrolyte	Comment	pH	Lit.
	Buffers		
Phosphate Buffer	PO_4^{3-}/HPO_4^{2-}	11–13	[8]
Carbonate Buffer	CO_3^{2-}/HCO_3^-	9–11	[8]
Borate Buffer	Sodium Borate/Boric Acid 1:1	9.4	[9]
	0.1 M Boric acid, 0.0375 M $Na_2B_4O_7$*10 H_2O	8.4	[10, 11]
Borate Buffer with NaCl	0.15 M Boric acid, 0.075 M $Na_2B_4O_7$*10 H_2O + 0.03 M NaCl	9.0	[12]
Borate/Sulphate Buffer	0.0194 M Boric acid, 0.01 M $Na_2B_4O_7$*10 H_2O + 0.5 M Na_2SO_4	8.9	[13, 14]
Borate/Nitrate Buffer	0.32 M Boric acid, 0.026 $Na_2B_4O_7$*10 H_2O + 1 M $NaNO_3$	7.1	[13, 14]
Phosphate Buffer	$HPO_4^{2-}/H_2PO_4^-$	6–8	[8]
Sodium sulphate Electrolytes	1% Na_2SO_4	~ 7	[15]
	0.1 M Na_2SO_4	~ 7	[16]
Phthalate Buffer		5.5	[17]
Citrate Buffer	0.145 M Citric acid, 0.035 M Sodium citrate	3.0	[18]
	Corrosive electrolytes		
NaCl Electrolytes	3% NaCl	~ 7	[19, 20]
	3.5% NaCl	~ 7	[21], [22]
	5.0% NaCl	~ 7	[23], [24]
Harrison Solution	3.5% $(NH_4)_2SO_4$, 0.5% NaCl	~ 7	[25]
Tap Water	3.04 mM $CaCl_2$, 0.54 mM $MgSO_4$ in tridest water		[9], [26]
Cooling Water	35 mg/l HCO_3^-, 15 mg/l CO_3^{2-}, 100 mg/l Ca^{2+}, 180 mg/l Cl^-, 200 mg/l SO_4^{2-}, 119 mg/l Na^+	8.2	[27]
Soil Simulating Solution	1.82 g/l K_2SO_4, 37.48 g/l Na_2SO_4, 22.69 g/l NaCl, 0.16 g/l $NaHCO_3$, 29.04 g/l $MgSO_4$, 2 g/l $CaSO_4$	~ 8.1	[28]
Sea Water (a)	315 mg/l Cl^-, 80 mg/l SO_4^{2-}, 60 mg/l Na^+, 48 mg/l K^+, 50 mg/l Mg^{2+}, 70 mg/l NO_3^-		[29]
Sea Water (a)	68.105% NaCl, 14.449% $MgCl_2$, 11.362% Na_2SO_4, 3.194% $CaCl_2$, 1.897% KCl, 0.278% KBr, 0.077% H_3BO_3, 0.068 $SrCl_2$, 0.008 NaF, 0.557% $NaHCO_3$ (weight %)		[30]
Russian Mud Corrosion	Saturated $CaCl_2$ at 0°C		[31]

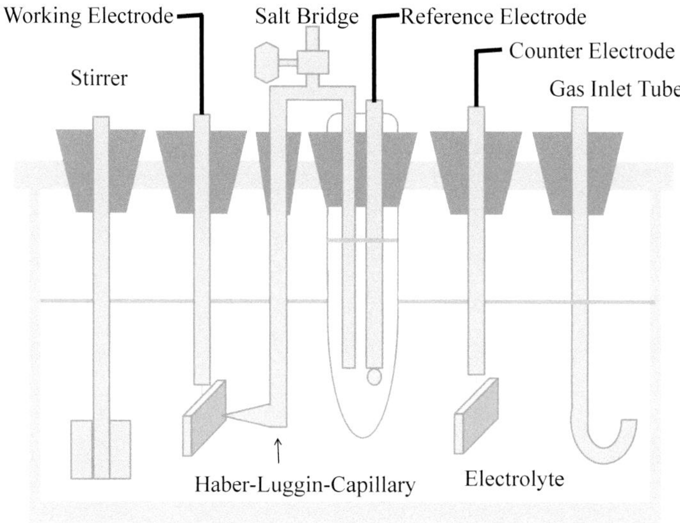

Fig. 2.1 Typical electrochemical cell (compare [3, 4]).

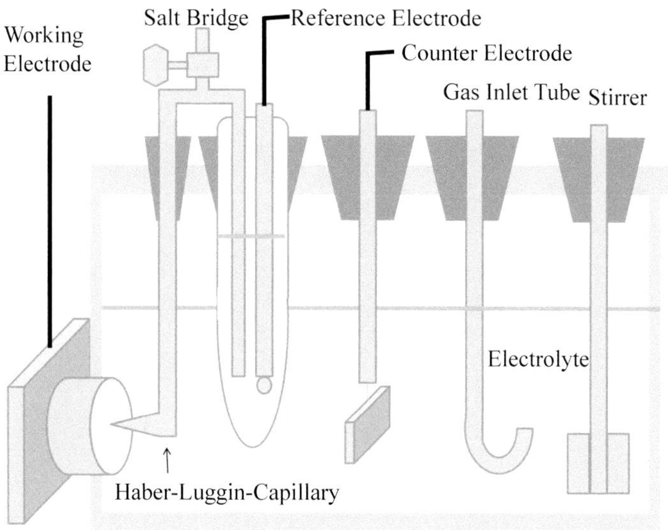

Fig. 2.2 Typical electrochemical cell.

The above mentioned challenges and some more problems to investigate data from the polarization diagrams are illustrated in the following examples.

Example 2.1

The polarization diagram is a powerful tool to investigate i_{cor} and E_{cor} by means of the Tafel equations (Eqs. 1.36 and 1.37). The challenge of this type of analysis is to define the range of data points for the Tafel slope. In theory, if the over-voltage η is $>> \frac{RT}{nF} = \frac{25.7}{n}$ mV at 25°C [1], there should be a linear behaviour in the logarithmic diagram. Actually in literature [32, 33] often higher values like η > 50 mV are used to define the slope. The problem is that near the equilibrium potential both reactions (anodic and cathodic) influence the measurement. Furthermore far from the equilibrium potential transport processes influence the measurement and cause deviations of the linear behaviour, too [4]. Therefore a range in between has to be defined.

In Fig. 2.3 the Tafel slopes have been calculated from η > 50 mV above and below the supposed equilibrium potential. On the diagram of copper and phosphated steel some data in the anodic range has been omitted. The samples have been cleaned with isopropanol and covered with a coating to use the cell in Fig. 2.1. The measurement has been done in a 3% NaCl solution in normal atmosphere and an Ag/AgCl reference electrode in 3 M KCl has been used in the three-electrode-assembly with a platinum sheet as counter electrode. The data have been collected with a scan rate of 20 mV/s.

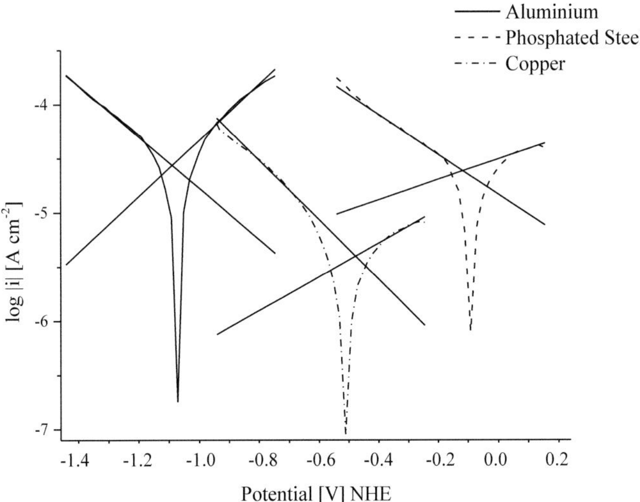

Fig. 2.3 Polarization diagrams of aluminium, tri-cation-phosphated steel and copper in 3% NaCl electrolyte measured with a scan rate of 20 mV/s.

With these data some parameters are available and summarized in Table 2.3. The phosphatation conversion layer on steel cannot change the potential (compare Table 5.3 in the app.) but decreases the current-density because the area for electrochemical reactions is reduced by the phosphate crystals (see also Ex. 2.4). Conversion or in general inhibitors could reduce the corrosion just by covering the surface. Another effect is the change of E_{cor} on the surface, whereas three effects are distinguished:

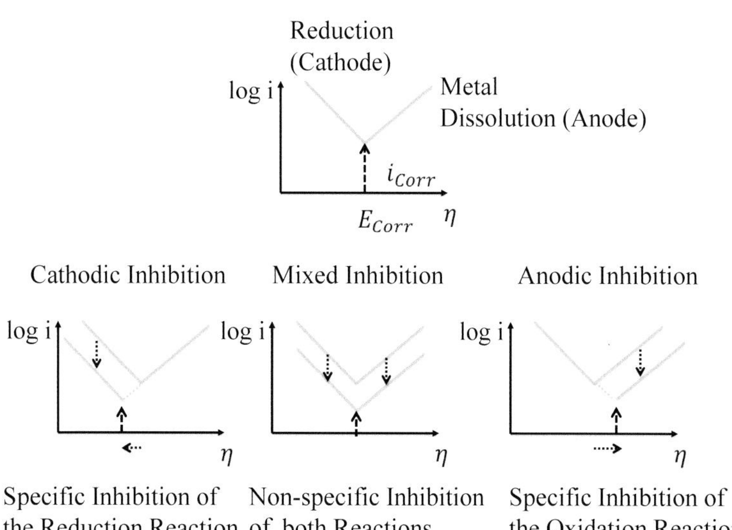

Fig. 2.4 Possible effects of inhibitors or conversion layers on a metal surface.

If the surface is only covered by a layer (molecule, nm or µm scale), a mixed inhibition is the consequence with a reduction of i_{cor}. Inhibitors could act in a specific way on the anodic or the cathodic reaction, visible by a change of E_{cor} in the polarization diagram as shown in Fig. 2.4 that normally increases the corrosion protective effect of the covering. In Table 4.1 some electrochemical data of inhibitors and conversion layers on different metal substrates are summarized.

Table 2.3 Electrochemical data calculated by means of the Tafel equations from Fig. 2.3.

Substrate	E_{Corr} [V]	i_{corr} [mA cm^{-2}]
Aluminium	–1.09	0.0275
Phosphated Steel	–0.48	0.0039
Copper	–0.12	0.0242

The data could be discussed, of course (for copper see [34] and for aluminium [35]). A change of the used data points change E_{cor} and i_{cor}. This might be a reason for the deviating data in literature (see Tables 5.2 and 5.3 in the app.). Another reason is the substrate. From a technical point of view aluminium is always an alloy (in Fig. 2.3 99.5% aluminium has been used) and there are hundreds of aluminium alloys used in industry and of course it is the same for every metal substrate. Yet another effect is based on the treatment before the measurement. In Fig. 2.5 the polarization diagrams illustrated for the metal surfaces cleaned only with isopropanol and cleaned for some seconds in hydrochloric acid followed by cleaning with water and finally with isopropanol before coating the back side and the edges of the samples. All these aspects have to be taken into account to compare polarization diagram data.

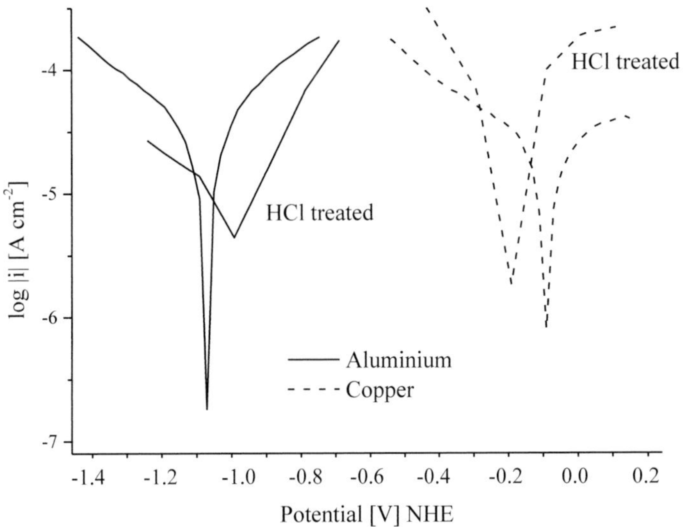

Fig. 2.5 Polarization diagrams of aluminium and copper with different pre-treatments before the measurement in 3% NaCl solution with a scan rate of 20 mV/s.

Example 2.2

The investigation of E_{cor} and i_{cor} of electrolytic zinc coated steel (see Chap. 1.3.2) at different pH values is illustrated in Fig. 2.6. The experiment has been done similar to Ex. 2.1 but different solutions have been used. According to the Pourbaix diagram in Fig. 1.2 changes from active → passive → active should be measured. Starting from pH = 6 the potential is comparable with the theory. The potential at pH = 9.4 is also comparable and the obviously reduced current shows a passive like behaviour. At a pH around 13 a higher potential and a clear active behaviour is not visible

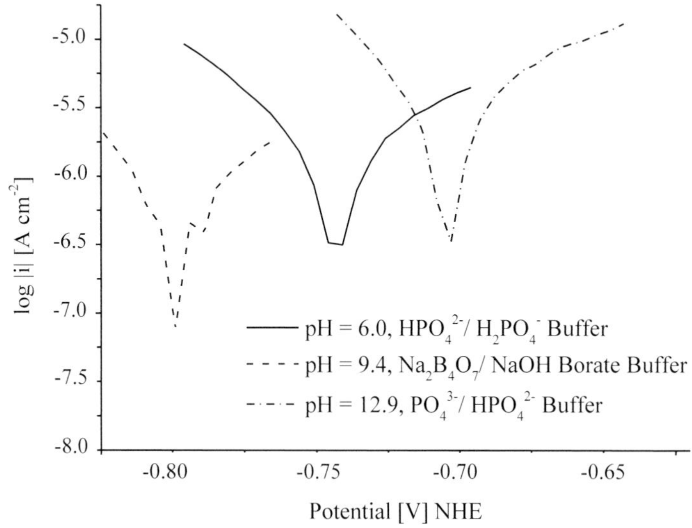

Fig. 2.6 Polarization diagrams of electrolytic zinc coated steel in different buffers measured with a scan rate of 20 mV/s.

in the polarization diagram as assumed in the Pourbaix diagram. This effect was already described in literature [36, 37] and should be caused by a passive layer in this pH range [37]. Therefore the current-density is low in comparison to active aluminium under alkaline conditions (see Fig. 2.8).

The critical fact is the buffer solution in two aspects. First it is questionable, if the phosphate buffer does not interact with the surface and second the borate buffer should not be comparable with the phosphate solutions (compare Eq. 2.1). On the other hand the data are comparable with other data from literature even if different electrolytes are used.

A clearer picture is shown on steel and aluminium. Polarization diagrams on steel and aluminium in phosphate buffer (pH = 6) and carbonate buffer (pH = 11) have been measured in an electrochemical cell similar to Fig. 2.1 with a scan rate of 5 mV/s [8]. Steel shows (Fig. 2.7) a reduction of E_{Ccor} and i_{cor} from pH = 6 to pH = 11 illustrated the change from an active to a passive behaviour. Aluminium (Fig. 2.8) reduces the potential from pH = 6 to pH = 11, too, but the current at pH = 11 increased by two orders of magnitude because aluminium is amphoteric and dissolutes in alkaline conditions. The data correlate with the corresponding Pourbaix diagrams [38] and correlate with observations on technical metal parts therefore even, if different electrolytes are used (compare [39]), reasonable data with polarization diagrams are available.

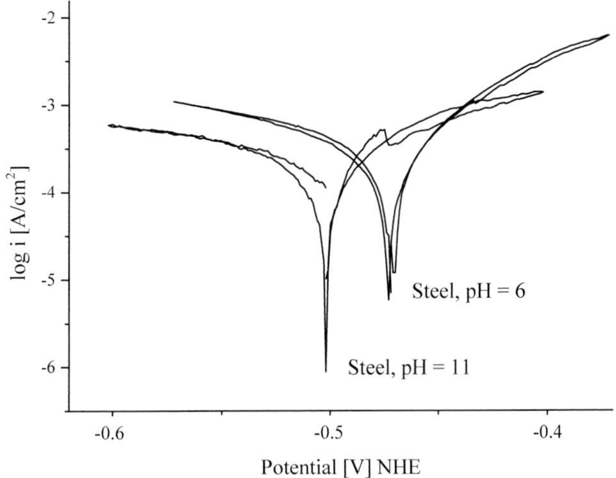

Fig. 2.7 Polarization diagrams of steel in different buffers measured with a scan rate of 5 mV/s (modified from [8, 40]).

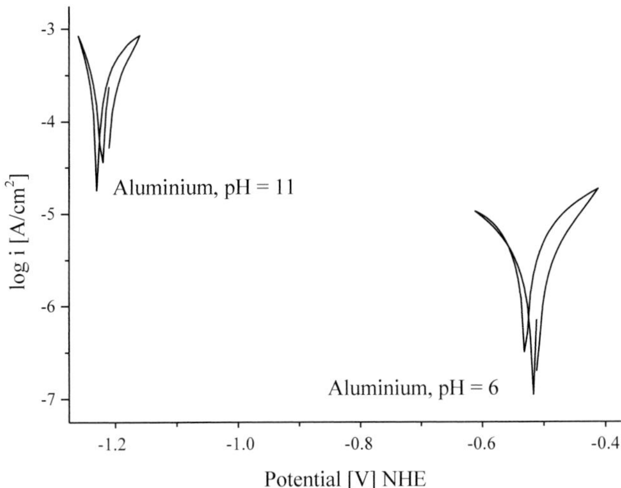

Fig. 2.8 Polarization diagrams of aluminium in different buffers measured with a scan rate of 5 mV/s (modified from [8, 40]).

Example 2.3

Plastic parts on auto car bodies are often coated with chromium layers for optical reasons. The chromium plated plastic parts produced in a galvanic process with an up to five layer system on a copper basis layer on the plastic part (Fig. 2.9). All metal layers applied by a galvanic process which

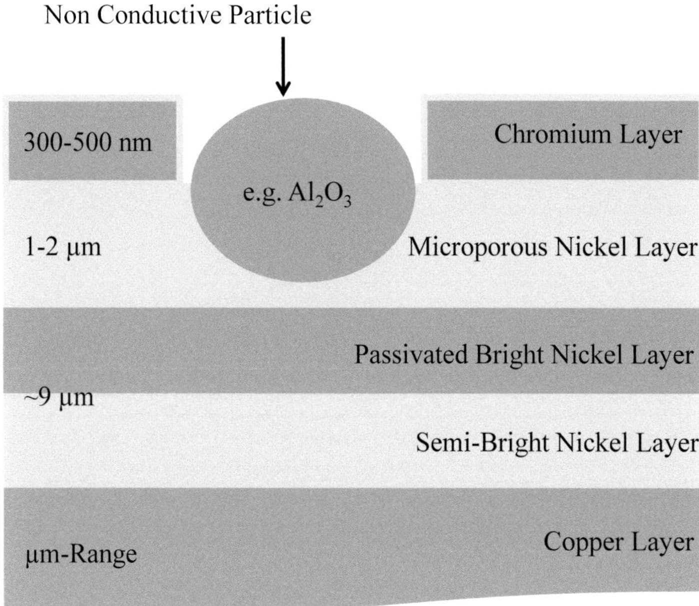

Fig. 2.9 Structure of chromium plated plastic parts in the automotive industry with the micro-pores approach [31].

is able to control the final equilibrium potential of each layer. The bright nickel layer is the most ignoble layer followed by the microporous nickel layer and the chromium layer.

The electrolytic contact between the metal surfaces is achieved with micro cracks or micro pores (e.g., aluminium oxide particles) in the chromium layer which do not reduce the optical effect but allows the electrochemical contact between the layers [31]. The metal coatings applied by galvanic processes and the micro pores are available with non-conducting particles dispersed in the nickel bath and deposit in the nickel layer and inhibit the deposition of chromium in the following galvanic process producing micro pores at this position. Micro cracks could be produced by deposition of chromium hydride in the layer, which destroyed rapidly and generated micro cracks in the chromium layer.

Therefore the bright nickel layer acts as a sacrificial anode for the chromium layer and the corrosive processes occur inside the coating surface to prevent a reduced appearance of the chromium layer [41–44].

This corrosion protective system has worked since decades without any problems in a lot of regions in the world. The increase of chromium plated plastic parts on cars in Russia shows a new phenomenon the so called "Russian mud corrosion" [31, 45, and 46] identified first in Moscow in the last decade. The chromium layer could be destroyed (black or brown spots on the surface or a complete decomposition of the chromium layer) within one winter period and the protection by the nickel layer does not work.

One hypothesis was that this effect is caused by the different electrolyte used on the streets on Moscow during the winter because of the very low temperatures $CaCl_2$ combined with other compounds is used as de-icing salt instead of NaCl. Another reason could be the amount of de-icing salts that generates a mud on the car bodies consisting of dirt, carbon black and $CaCl_2$ and the additives inside. This mud could act as a barrier for the oxygen diffusion through the metal surfaces and could change the electrochemical properties of the surfaces. The first mechanism for that phenomenon was developed by Bauer [45] based on chemical and electrochemical reactions (Fig. 2.10).

The corrosion process starts with the dissolution of nickel as sacrificial anode and protects the chromium layer. Because of the reduced oxygen on the surface (de-icing salt/mud layer on the chromium plated plastic part) the generated new chromium surface, because of the partial destroyed nickel layer, cannot be passivated and the chromium dissolutes because it is more ignoble without a passive layer than the nickel surface.

To prove this mechanism polarization diagrams has been measured in saturated $CaCl_2$ at 0°C with (Fig. 2.11) and without (Fig. 2.12) oxygen in the solution [31, 46]. The experiments were performed in an electrochemical cell similar to Fig. 2.1 and saturated calomel (Table 2.1) as RE with the platinum sheet and CE has been used and the data collected with a scan rate of 5 mV/s [31]. The order of the equilibrium potentials of the layers changed, if the oxygen is reduced on the surface.

Therefore the initiation process according to Bauer could not happen but for the direct dissolution of the chromium layer. The results are reasonable because the industrial galvanic layers in an electrolyte simulating the real situation have been used and proved with several

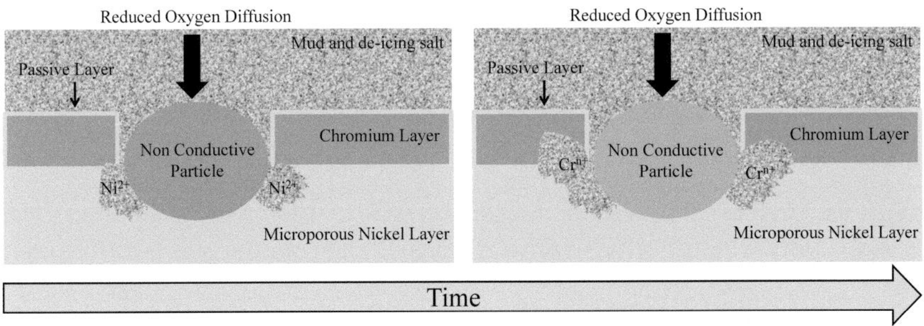

Fig. 2.10 Mechanism of the Russian mud corrosion according to [45].

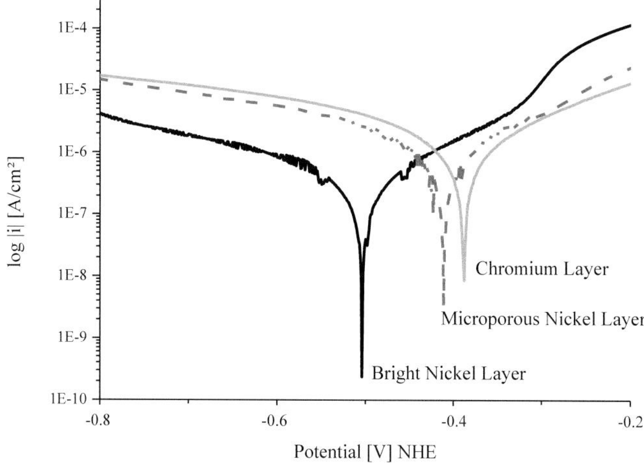

Fig. 2.11 Polarization diagrams of galvanic chromium, microporous and bright nickel layers in aerated saturated $CaCl_2$ electrolyte measured with a scan rate of 5 mV/s [31]. With permission.

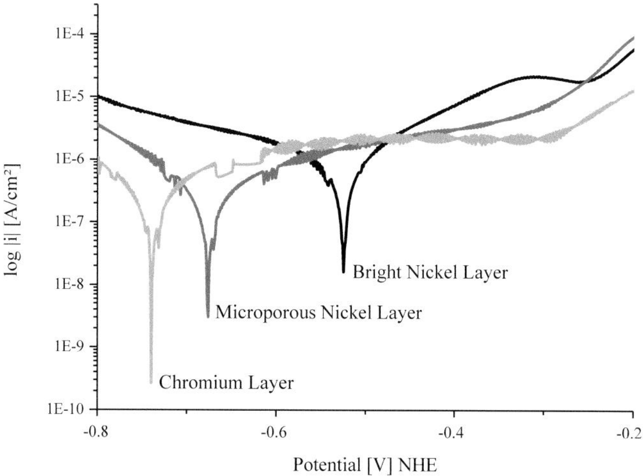

Fig. 2.12 Polarization diagrams of galvanic chromium, microporous and bright nickel layers in deaerated (nitrogen) saturated $CaCl_2$ electrolyte measured with a scan rate of 5 mV/s [31]. With permission.

samples. Furthermore the data correlate with results from real parts in the field [31, 46]. The whole situation is more complex and will be presented in Exs. 2.6 and 3.19 in which the results proved with other independent methods (EIS and UV-VIS, respectively). Why the metal surfaces change the behaviour cannot be clarified with the polarization diagram but with the use of EIS illustrated below.

2.1.2 *Cyclic Voltammetry*

Basics

The Cyclic Voltammetry (CV) is based on a triangular potential-time-sweep (Fig. 2.13) on the WE and detection of the resulting potential-current-diagram shown in Fig. 2.14.

The potential range is between the anodic hydrogen evolution ($E_{initial}$) and the cathodic oxygen evolution (E_{final}). The scan rate v should be in the range of 10–30 mV/s [1] to avoid a deactivation of the electrode. One

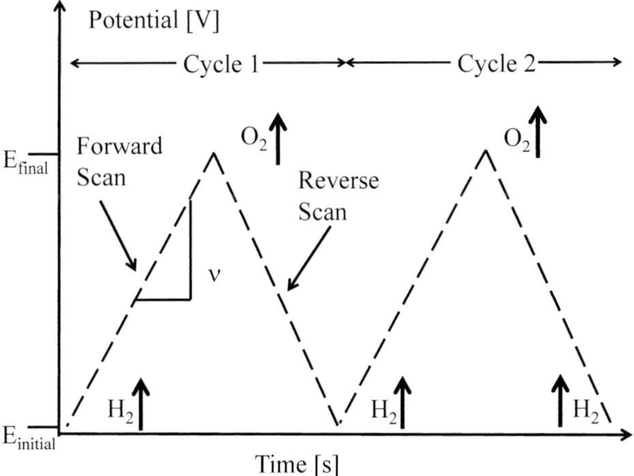

Fig. 2.13 Potential-time excitation signal in cyclic voltammetry experiment (compare [1, 3]).

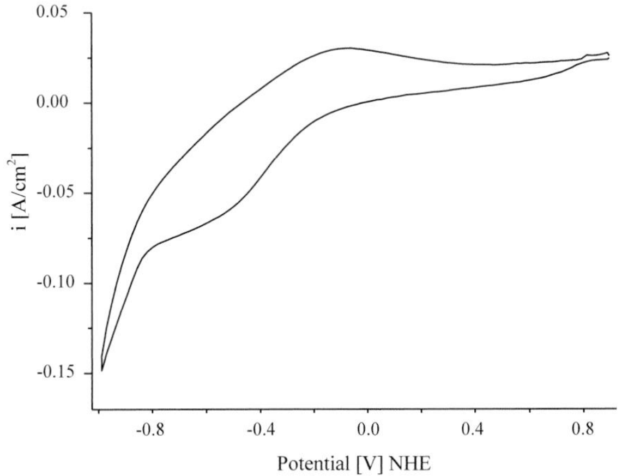

Fig. 2.14 CV diagram of tri-cation-phosphated steel at pH = 10 in carbonate buffer measured with a scan rate of 20 mV/s.

forward scan and one reverse scan is called one cycle and often more than one cycle is necessary to get valid data. The experiment could be performed in electrochemical cell illustrated in Figs. 2.1 and 2.2 and a three-electrode-assembly has to be used to investigate the exact current density and potential of the WE.

The CV method is a powerful tool to analyze electrochemical reactions on conducting surfaces because the peaks in the spectra (Fig. 2.14) are caused by oxidation or reduction reactions of electroactive compounds in solution (faradaic reactions) or absorbed compounds on the electrode and a capacitive current due to the double layer charging (see Fig. 1.1). All processes are dependent on the diffusion from and to the surface (see Fig. 1.6) [1, 3 and 4]. For further reading see ref. [1, 3, 4, 47, 49].

To calculate the current at a certain potential for reversible reactions two situations have to be distinguished:

Uninhibited Charge Transfer Reactions

With the assumption that only one simple redox-reaction:

$$S_{red} \Leftrightarrow S_{Ox} + ne^- \qquad \text{Reac. 2.4}$$

on the WE takes place and the charge transfer through the double layer is uninhibited, the current maximum of a peak in the CV diagram could be calculated with the Randles-Sevcik equation at 25°C as follows [1, 3]:

$$i_{Peak} = nF\left(\frac{nF}{RT}\right)^{\frac{1}{2}} 0.4463\, D_{\mathrm{Red}}^{\frac{1}{2}}\, c_{\mathrm{Red}}^{0}\, \nu^{\frac{1}{2}} = 2.69 \cdot 10^5\, n^{\frac{3}{2}}\, D_{\mathrm{Red}}^{\frac{1}{2}}\, c_{\mathrm{Red}}^{0}\, \nu^{\frac{1}{2}} \qquad \text{Eq. 2.2}$$

i_{Peak} : maximum current [A/cm^2]

D_{Red} : Diffusion constant [cm^2/s] (see Eq. 1.39)

ν : Scan rate [V/s]

n : Number of exchanged electrons (see Eq. 1.2)

$c_{Red}{}^{0}$: Concentration of S_{Red} at t = 0 and c_{Ox} = 0 at t = 0

Therefore the current-density at the peak maximum increases with $\sqrt{\nu}$ and without a change of the peak position.

Inhibited Charge Transfer Reactions

If charge transfer is inhibited, the current-density should be dependent on the scan rate and a shift of the peak maximum occurs, if the scan rate changed [1]. For an inhibited charge transfer reaction with the same assumptions mentioned above Eq. 2.2 at 25°C changed to:

$$i_{Peak} = \pi^{\frac{1}{2}} nF\left(-\frac{\alpha nF}{RT}\upsilon\right)^{\frac{1}{2}}\left(\frac{\alpha n F}{RT}\upsilon t\right)0.282 D_{\mathrm{Red}}^{\frac{1}{2}} c_{\mathrm{Red}}^{0} = 3.01\cdot 10^{5}\, n^{\frac{3}{2}} D_{\mathrm{Red}}^{\frac{1}{2}} c_{\mathrm{Red}}^{0}\, \nu^{\frac{1}{2}} \alpha^{\frac{1}{2}} \quad \text{Eq. 2.3}$$

α: Transition factor (see Eq. 1.33)

The transition factor is implemented in the equation to describe the charge transfer inhibition. Therefore the current-density at the peak maximum will increase with $\sqrt{\nu}$ and with a change of the peak position. Furthermore smaller intensities are expected for inhibited charge transfer reactions. The distinction of the reactions is possible by varying the scan rate because only the inhibited reactions change the positions of the peaks in the CV diagram.

Both equations do not change, if the measurement starts with S_{Ox} and the reaction:

$$S_{Ox} + ne^{-} \Leftrightarrow S_{\mathrm{Red}} \quad \text{Reac. 2.5}$$

takes place on the electrode. The transition factor has to change from α to $(1 - \alpha)$ and the corresponding diffusion constant and concentration for S_{Ox} has to be used for the equation. With the Randles-Sevcik equation the diffusion process on the surface, the number of electrons transferred in an electrochemical reaction and some information about the kinetic of the reaction are available with the measurement of the CV. Therefore the CV is often called the "electrochemical equivalent of spectroscopy" [50] especially for the analysis of the electrochemical behaviour of compounds in solution.

Measurement

The measurement principle has already been mentioned and could be performed in a standard electrochemical cell (see above) with a three-electrode-assembly and the same RE, CE and electrolyte as for the polarization diagrams could be used.

From the corrosion research point of view the CV method allows the description of certain corrosion processes like pitting corrosion on zinc [39] or the stability of tin passive layers [18] always in combination with other electrochemical (EIS) and spectroscopic methods. More often the method is used to describe the protective properties of conversion layers (compare [34, 51, and 52]) and especially the coverage of conversion layers on metal substrate. Therefore the focus of the following examples is on the measurement of the coverage of conversion layers.

Example 2.4

The investigation of the coverage of conversion layers based on the fact that the conversion compounds covers the surface with a nonconductive material therefore all electrochemical reactions and the corresponding current correlates with the free, i.e., electrochemical active, surface [13, 14]. The challenge is to define the surface of the uncoated substrate because a steel surface has different electrochemical properties as a conversion coated one [14]. In literature [13, 14, and 53] the electrochemical methods (CV, EIS) use *in situ*, i.e., during the deposition of a phosphatation layer, and the uncovered surface could be extrapolated with the collected data. To analyze an already coated surface the uncoated surface could only be investigated by removal of the phosphatation layer or an uncoated substrate is used. In Fig. 2.15 both methods are used to define the uncoated surface. The phosphatation layer was removed with hydrochloric acid therefore all measurements have been performed in 3% NaCl electrolytes. A rough determination of the coverage of 90% is possible based on this data by integrating the area in the oxygen evolution range. This resolution is not good enough for the development of conversion layers (but then an *in situ* method is possible, of course) but enough to test the quality of a conversion layer for further coating processes or the analysis of coating systems.

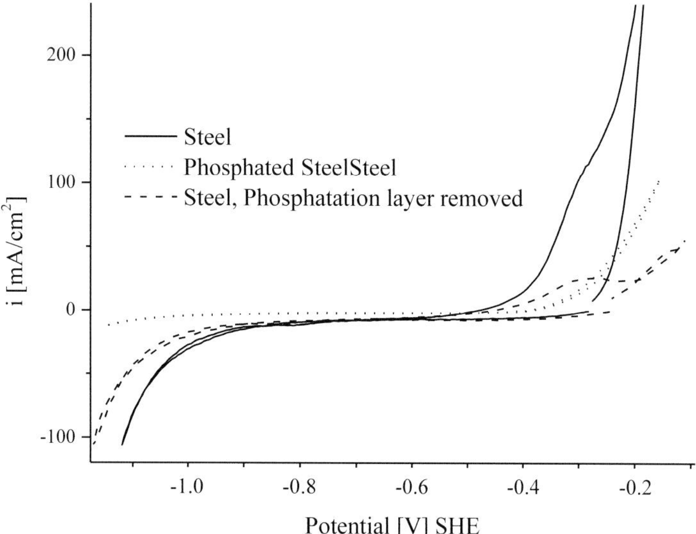

Fig. 2.15 CV diagrams from −1.1 to 0.1 V of phosphated steel, steel and steel after removal of the phosphatation in 3% NaCl solution measured with a scan rate of 20 mV/s.

Example 2.5

The conversion layers based on phosphoric acid have several disadvantages. During the deposition a lot of mud, containing substrate compounds, is produced and without nickel, accelerators and the activation bath dense layers with a high coverage cannot be produced. Therefore alternatives have been developed as already mentioned above. The modern zirconium or titanium based conversion layers reduces the amount of mud and the use of nickel but a acidic fluoride containing bath is necessary and the use of minerals (raw materials for zirconium and titanium) is not sustainable. Therefore conversion layers based on renewable materials should be the optimized solution. One example is conversion layers based on phytic acid (Fig. 2.16) [51]. The structure is based on an inositol ring esterified completely with phosphoric acid. In theory up to 12 hydrogen atoms could be removed to build up salts with the substrate metal.

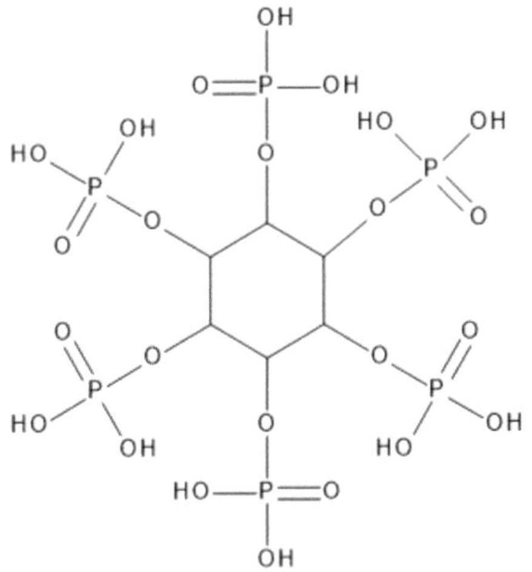

Fig. 2.16 Structure of phytic acid.

Phytic Acid (PA) is a "green" future option for conversion coatings and has been proposed as a sustainable raw material for conversion layers in papers in the last decade [51, 54–57]. Naturally occurring phytic acid is contained in beans, brown rice, corn, sesame seeds and wheat bran [58]. In the last years novel production methods have been developed so that today PA can be easily extracted from rap filter cake [59] as a by-product in the biodiesel production.

To test this approach phytic acid based conversion layers with some additives inside the conversion bath have been applied on steel (C, 0.08%; P, 0.03%; S, 0.03%; Mn 0.4%), zinc coated steel and aluminium at different pH values and immersion times summarized in Table 2.4 [51].

Table 2.4 Conversion solution based on phytic acid. As promoting agent 2 g/l $NaNO_3$ and as levelling agent Sokalan VA64P have been used. The amount of molybdate and tungstate are near the solubility limit of the compounds (compare [51]).

Conversion layer	Ingredients in the solution	pH	Immersion time [min]	Temperature [°C]
Phytic Acid	Phytic acid, KNO_3, Levelling agent	2	6	35
Phytic Acid Mo	Phytic acid, KNO_3, Levelling agent, $(NH_4)_6Mo_7O_{24}$	2	6	35
Phytic Acid W	Phytic acid, KNO_3, Levelling agent, Na_2WO_4	2	6	35

Only on steel a dense homogenous layer is achieved and the electrochemical and spectroscopic methods have been done only on the steel coated surface. The CVs (Fig. 2.17) have been done in a three-electrode-assembly in 3% NaCl solution with a scan rate of 20 mV/s.

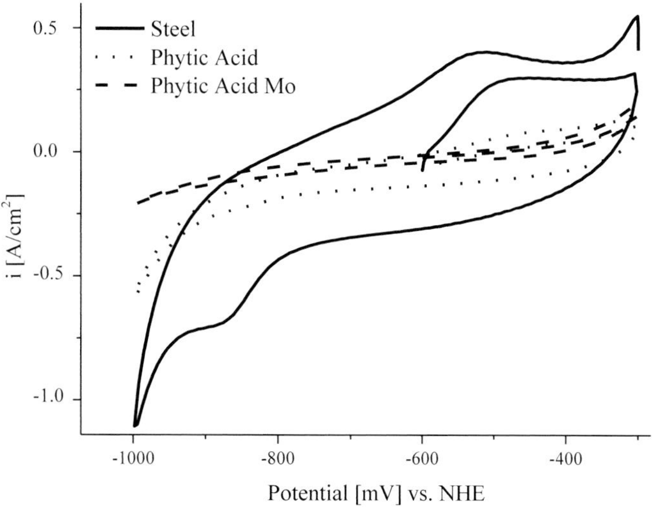

Fig. 2.17 CV diagrams from –0.3 to 1 V of the phytic acid based conversion layer in 3% NaCl solution measured with a scan rate of 20 mV/s. For uncoated steel a scan rate of 100 mV/s has been used.

Unfortunately the dissolution of iron of the uncoated surface was too high with this scan rate and therefore a higher scan rate of 100 mV/s has been used to reduce the iron dissolution (compare Ex. 2.4). The final current-density could be corrected according to Eq. 2.2 with $\sqrt{\nu}$ to calculate the coverage of the phytic acid layer summarized in Table 2.5. In ref. [51]

Table 2.5 Coverage of the conversion layer investigated from the CVs of Fig. 2.16 (tungstate not shown).

Conversion layer	Coverage [%]
Phytic Acid	84
Phytic Acid Mo	92
Phytic Acid W	90

the square root missed in the equation but the ratio of the coverage of the layers does not change.

The high coverages are promising results for a high corrosion protective performance. The performance have been analyzed with EIS (see ref. [51]) and with organic coated samples in the salt spray test (see Ex. 4.4). The chemical structure and morphology has been analyzed by means of infrared spectroscopy (see Ex. 3.13) and SEM/EDX (see Ex. 3.7), respectively.

As already shown on brass surfaces in ref. [57] the phytic acid based conversion layer increases the E_{cor} of the substrate, confirmed on steel in Fig. 2.18.

Based on the polarization diagrams and the CVs, the phytic acid based conversion layer should have a high corrosion protective performance. Based on all data a completely different picture results (see below).

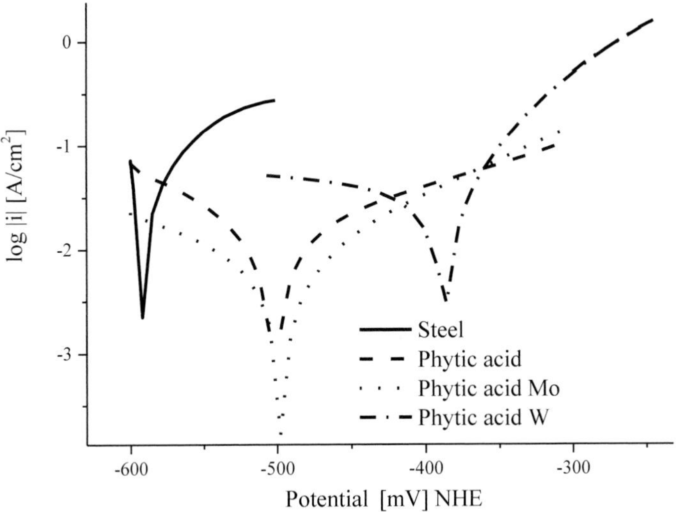

Fig. 2.18 Polarization diagrams of the phytic acid based conversion layers and steel (see Fig. 2.17).

2.1.3 Electrochemical Impedance Spectroscopy

Basics

The electrochemical impedance spectroscopy (EIS) is a powerful but complex tool to analyze processes on a metal and on coated metal surfaces (Chap. 2.2.2). Therefore only a brief introduction with the basics will be presented. For further reading see ref. [4, 60–62].

The EIS is based on the perturbation applied of an electrochemical system, i.e., the WE of a three-electrode-assembly, with a sinusoidal voltage V with low amplitude V_0:

$$V(t) = V_0 \sin \omega t \wedge \omega = 2\pi f \qquad \text{Eq. 2.4}$$

Where V(t) is the voltage at the time t and ω is the radial frequency (rad s^{-1}) easily transformed into the frequency f (in [Hz]) shown in Eq. 2.4. The response of the perturbation is the current I(t) with the amplitude I_0 at the same frequency but shifted in phase with the phase shift γ (Fig. 2.19):

$$I(t) = I_0 \sin(\omega t + \gamma) \qquad \text{Eq. 2.5}$$

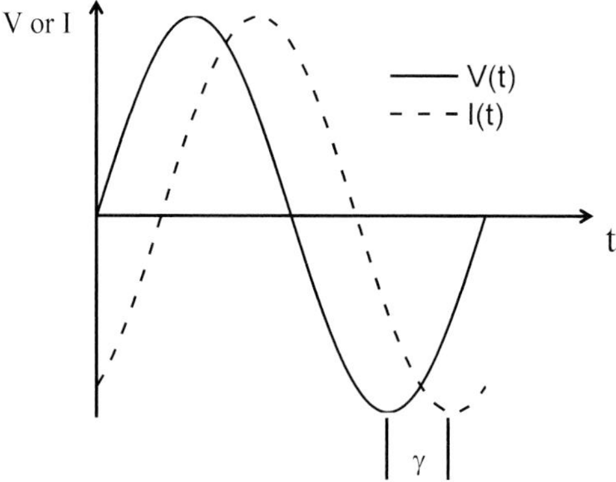

Fig. 2.19 Phase shift of voltage and current.

Analogous to the resistance R in the Ohm's law for a DC circuit the impedance Z is defined as the ratio of voltage and current:

$$Z = \frac{V(t)}{I(t)} \wedge |Z| = \frac{V_0}{I_0} \qquad \text{Eq. 2.6}$$

The impedance has a magnitude |Z| and a phase φ and could be described as a vector in the complex plane as shown in Fig. 2.20. It is therefore convenient to describe impedance in complex notation in three different ways summarized in Eq. 2.7:

$$Z = Z' + iZ'' = |Z|(\cos\varphi + i\sin\varphi) = |Z| \cdot e^{i\varphi} \wedge i = \sqrt{-1} \qquad \text{Eq. 2.7}$$

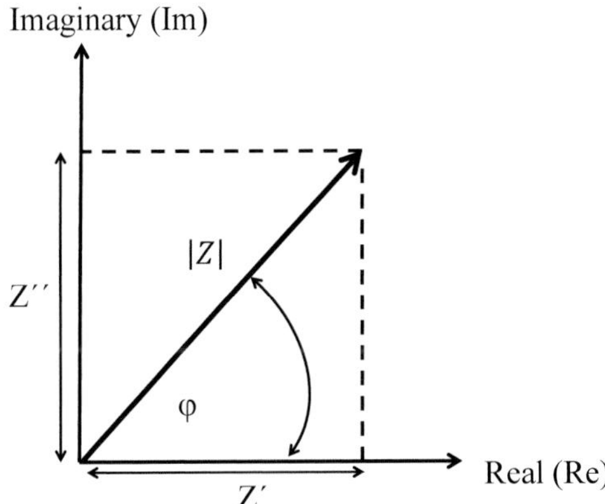

Fig. 2.20 Complex plane to describe the impedance Z and the phase angle φ. Z´ is the real part and Z´´ the imaginary part of the impedance.

The impedance is divided into the real part Z´ and the imaginary part Z´´often referred to as Re and Im. Some useful correlations of complex numbers are summarized in Chap. 5.3 in the app. The magnitude of the impedance could be calculated according to Phytagoras' theorem as follows:

$$|Z| = \sqrt{Z'^2 + Z''^2} = \sqrt{\mathrm{Re}^2 + \mathrm{Im}^2} \qquad \text{Eq. 2.8}$$

The phase angle is available with the simple trigonometric relation:

$$\varphi = \arctan\frac{Z''}{Z'} = \arctan\frac{Im}{Re} \qquad \text{Eq. 2.9}$$

The EIS data is often interpreted with an equivalent circuit, i.e., an electric circuit with electric elements such as resistance and capacitance and elements describing transport/diffusion processes to fit the data and to achieve a physical-chemical model for the processes on the electrode surface. In the following paragraph some important circuit elements are presented and their combinations with the equations to calculate the impedance, magnitude and phase.

Resistance

A pure resistance R does not change with the frequency of the voltage and is therefore constant without any phase shift (see Table 2.6).

Capacity

The impedance of a pure capacitor with the capacitance C could be calculated according to the relationship:

$$Z = \frac{1}{i\omega C} = -\frac{i}{\omega C} \wedge i^2 = -1 \quad \text{Eq. 2.10}$$

The phase shift is $-\frac{\pi}{2}$ and therefore the capacitance depends on the frequency and is entirely imaginary [60].

Combinations of R and C

In general the impedance of an electrical circuit with more than one element could be defined based on Kirchhoff´s law [63] as follows:

Serial Connection

The impedance is the sum of the impedances in the circuit:

$$Z = \sum Z_i \quad \text{Eq. 2.11}$$

Parallel Connection

The reciprocal impedance is the sum of the reciprocal impedances in the circuit:

$$\frac{1}{Z} = \sum \frac{1}{Z_i} \quad \text{Eq. 2.12}$$

The impedance of a serial connection of a capacitance and a resistance could be calculated with the use of Eqs. 2.11 and 2.10:

$$Z = Z_R + Z_C = R - \frac{i}{\omega C} \quad \text{Eq. 2.13}$$

The impedance of a parallel connection of a capacitance and a resistance has to be calculated based on Eqs. 2.12 and 2.10:

$$\frac{1}{Z}=\frac{1}{Z_R}+\frac{1}{Z_C}=\frac{1}{R}+i\omega C=\frac{1+i\omega RC}{R} \Rightarrow Z=\frac{R}{1+i\omega RC} \qquad \text{Eq. 2.14}$$

The impedance spectra illustrated with the so called Nyquist and Bode plots (see below) and for the Nyquist plot the calculation of the real and imaginary parts is necessary. Furthermore for calculations the separation of the real and the imaginary parts of a complex number is very useful (see Chap. 5.3) therefore Eq. 2.14 could be extended with the conjugated complex number (see Eq. 5.18) of the denominator to separate both parts as follows:

$$Z=\frac{R(1-i\omega RC)}{(1+i\omega RC)(1-i\omega RC)}=\frac{R-i\omega R^2 C}{1+\omega^2 R^2 C^2}=\frac{R}{1+(\omega RC)^2}-i\frac{\omega R^2 C}{1+(\omega RC)^2} \qquad \text{Eq. 2.15}$$

With this strategy all impedance relations could be separated in real and imaginary parts even, if the equivalent circuit is complex as illustrated in Chap. 2.2.2. In Table 2.6 some equivalent circuits with the corresponding impedance, magnitude and phase equations are summarized. The corresponding Nyquist plots of the equivalent circuits, i.e., the graph in the complex plane but with the negative imaginary part above the real part are shown in Fig. 2.21.

Another way to present the data of an EIS measurement is the so called Bode plot, i.e., the plot of the logarithm magnitude (Eq. 2.8) and the phase (Eq. 2.9) on another axis versus the logarithm frequency. The Bode plots of a resistance and a capacitance are summarized in Fig. 2.22.

Table 2.6 Basic equivalent circuits and elements.

Circuit element	Equivalent circuit	Impedance Z (Z = Z′+ i Z″)	Magnitude |Z|	Phase φ
Resistance	R	$Z=R$	$\|Z\|=R$	0
Capacitance	C	$Z=\frac{1}{i\omega C}$	$\|Z\|=\frac{1}{\omega C}$	$-\frac{\pi}{2}$
Resistance and Capacitance in Series	R C	$Z=R-i\frac{1}{\omega C}$	$\|Z\|=\sqrt{R^2+\frac{1}{\omega^2 C^2}}$	$\arctan\left(-\frac{1}{\omega RC}\right)$
Resistance and Capacitance in Parallel	C R	$Z=\frac{R}{1+(\omega RC)^2}-i\frac{\omega R^2 C}{1+(\omega RC)^2}$	$\|Z\|=\sqrt{\frac{R^2}{1+(\omega RC)^2}}$	$\arctan(-\omega RC)$
Randles Equivalent Circuit	R_{El} C R	$Z=R_{El}+\frac{R}{1+(\omega RC)^2}-i\frac{\omega R^2 C}{1+(\omega RC)}$	$Z=\sqrt{R_{El}^2+\frac{2R_{El}R+R^2}{1+(\omega RC)^2}}$	$\arctan\frac{Z''}{Z'}$

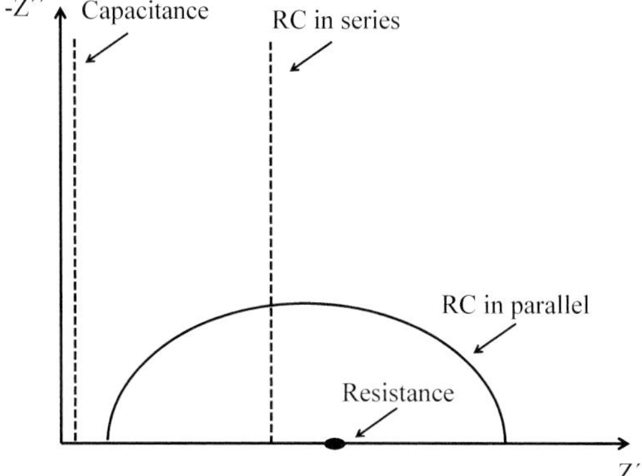

Fig. 2.21 Principal Nyquist diagrams of the equivalent circuits in Table 2.6.

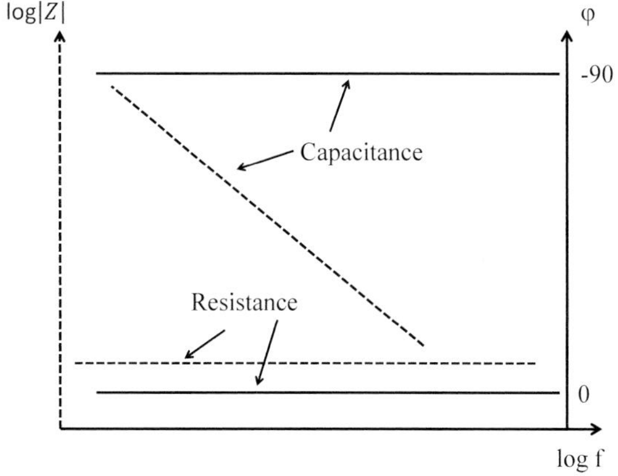

Fig. 2.22 Principal Bode plots of a resistance and a capacitance.

Interface Metal/Electrolyte with Transfer Reactions

If a metal surface is in contact with an electrolyte, a double layer results on the surface (Fig. 1.1). The double layer acts like a capacitor with the double layer capacity C_{DL} and, if transfer reactions are possible, with the charge transfer resistance R_{CT} in parallel to the capacity in the equivalent circuit. Because of the fact that the electrolyte resistance R_{El} has be considered it is added in series to the double layer elements, shown in Table 2.6 (Randles equivalent circuit). With the relation Z = Re + i Im the corresponding impedance equation could be transformed as follows [61]:

$$\left(\mathrm{Re}-\frac{2R+R_{El}}{2}\right)^2+(\mathrm{Im})^2=\left(\frac{R}{2}\right)^2 \qquad \text{Eq. 2.16}$$

Eq. 2.16 is the function of a semicircle in the complex plane. The corresponding plot is illustrated in Fig. 2.23 with the investigation of the values of the elements in the Randles equivalent circuit on both the Nyquist and the Bode plot.

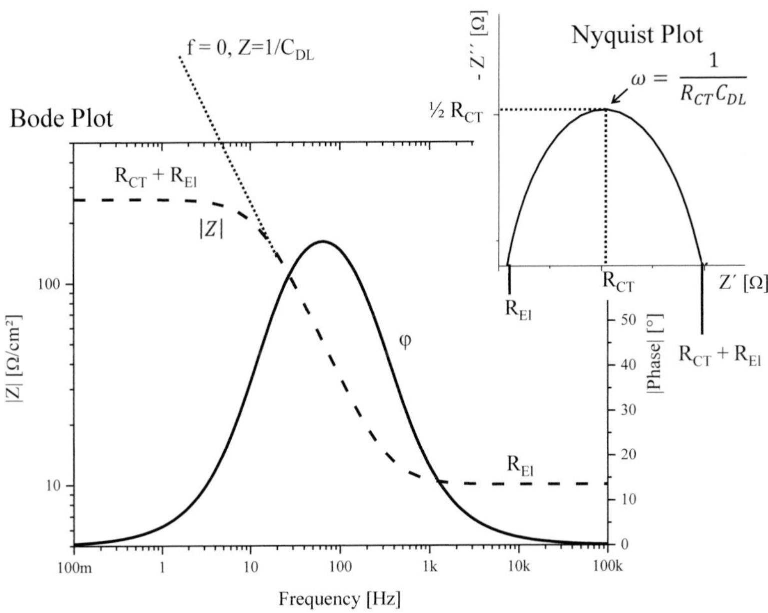

Fig. 2.23 Principal Bode and Nyquist plot of the Randles equivalent circuit.

In principle all data are available with one plot but it is often useful to generate both to get the complete picture. At high frequencies the impedance of the double layer capacity C_{DL} is small and the electrolyte resistance R_{El} could be investigated. With decreasing frequency C_{DL} increases (Eq. 2.10) and the capacity dominates the spectrum with a linear behaviour in the Bode plot. At very low frequencies the impedance of C_{DL} is higher than the charge transfer resistance and therefore the Ohmic resistances could be investigated at very low frequencies. The diameter of the semi-circle in the Nyquist plot gives R_{CT} and the maximum allows the calculation of time constant τ and therefore the value of C_{DL} (see Fig. 2.23) [60]:

$$\tau = R_{CT} C_{DL} \qquad \text{Eq. 2.17}$$

The Randles approach neglected the diffusion processes or in general the transport phenomenon from and to the surface. To describe the complete

Faradaic impedance, i.e., the impedance occurs by electrochemical reactions on the metal surface, at least two different situations have to be distinguished:

Semi Infinite Diffusion

Warburg [74] established the equation for a semi-infinite diffusion process at the electrode based on a simple reaction similar to Reac. 2.4. The Warburg impedance Z_W for this reaction with the assumption that the diffusion coefficients of both species are equal $D_{Ox} = D_{Red}$ (compare Eq. 1.38) is defined as (some derivations for this equation illustrated in ref. [1, 60, and 61]):

$$Z_W = \frac{W}{\sqrt{\omega}} - i\frac{W}{\sqrt{\omega}} \quad \wedge \quad W = \frac{2RT}{n^2\, F^2\, c^0\, \sqrt{2}\, \sqrt{D}} \qquad \text{Eq. 2.18}$$

The Warburg constant W contains the Faradaic constant F, the ideal gas constant R, the number of transferred electrons n, the concentration of the reactants in solution c^0 and the diffusion coefficient D.

The real and the imaginary part have the same value and $Z_W = 0$ for $\omega \rightarrow \infty$. The Warburg impedance in combination with other elements is visible for diffusion controlled reactions at small frequencies in the Nyquist plot as a straight line with the phase angle $\varphi = 45°$ (Fig. 2.24). In fast stirring solutions or/and at a high metal ion concentration c_{Ox} the Warburg impedance is not visible [61].

Finite Diffusion

In some cases a finite diffusion layer on the metal surface has to be assumed for the calculation of the Faradaic impedance especially in stirred solutions. In this situation there is a certain distance x (compare Fig. 1.1) from the surface with a constant concentration and therefore a finite diffusion has to be based for the calculation. The equation for the finite Warburg impedance Z_W^{finite} is shown in Eq. 2.19 [60, 60a]:

$$Z_W^{finite} = R_0 \frac{\tanh\left(x\sqrt{\frac{i\omega}{D}}\right)}{x\sqrt{\frac{i\omega}{D}}} \quad \wedge \quad R_0 = \text{Diffusion resistance for } \omega \rightarrow 0 \qquad \text{Eq. 2.19}$$

In the Nyquist spectrum the impedance will bend over the real axis at low frequencies rise to a distorted semi-circle [60, 60a] shown in Fig. 2.24. Another possibility for a finite diffusion is that there is no transport of electroactive species through a layer, whatever structure. The impedance Z_D^{finite} in that case could be described with the following equation [60, 60a]:

$$Z_D^{finite} = \frac{x^2}{DC_0} \frac{\coth\left(x\sqrt{\frac{i\omega}{D}}\right)}{x\sqrt{\frac{i\omega}{D}}} \wedge C_0 = \text{Limiting Capacity for } \omega \to 0 \qquad \text{Eq. 2.20}$$

At low frequencies the impedance behaves like a capacitance shown in Fig. 2.24. If diffusion processes are relevant at the interface, the equivalent circuit in Table 2.6 has to be expanded to the also named Randles equivalent circuit shown in Fig. 2.25.

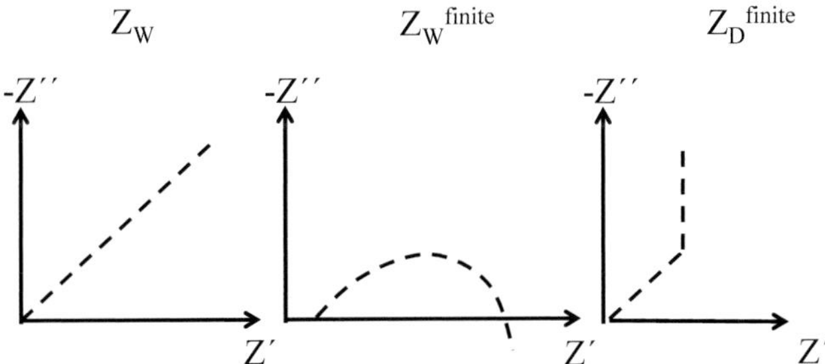

Fig. 2.24 Sketches of the possible diffusion controlled impedance behaviour in the Nyquist plot at low frequencies according to the Eq. 2.18–2.20.

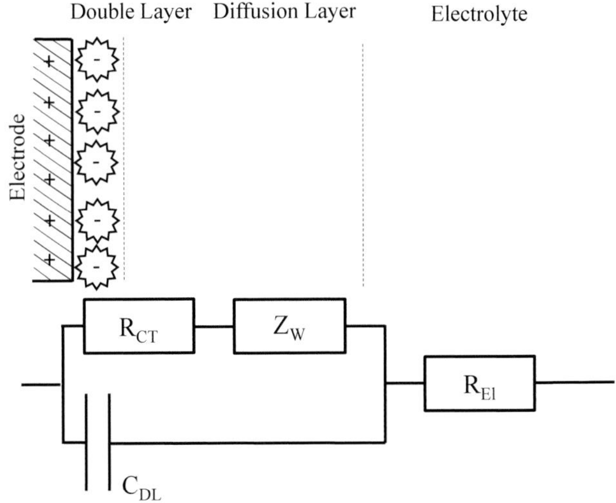

Fig. 2.25 Randles equivalent circuit with a diffusion controlled process according to [1] (compare Fig. 1.1).

In reference [60a] a complete overview of all relevant diffusion situations (finite, semi-finite, cylindrical and spherical electrodes, etc.) is presented.

Measurement

The measurement is performed with the electrochemical cells in Figs. 2.1 and 2.2 but often the cell is similar to Fig. 2.2 as a small device clamped on the test sheet sealed with a rubber O-ring based on a Teflon tube is used with a three-electrode-assembly and often graphite electrodes used as CE. The EIS measurements could be analyzed as mentioned above, if a linear behaviour of the signal could be achieved. This is the fact, if the applied voltage is low, i.e., 1–10 mV [61], and the system is stable during the measurement. This is a challenge, if impedances in aggressive electrolytes at very low frequencies have to be investigated because this measurement is time consuming and the surface of the WE will change, of course. Because of that often a frequency range between 10–100 mHz and 100 kHz–1 MHz is used to achieve a testing time of some minutes. If very low frequencies (10^{-4} Hz and lower) are used, the testing time increases to some hours. Finally the linearity of the data has to be proved and every supplier of EIS devices provides mathematical methods to test the impedance data. One method available with the software SIM (Zahner Elektrik GmbH & Co KG) is based on the transfer function $H_0(\omega)$ [64] to calculate the magnitude of the impedance with the data of the phase with the following equation:

$$\ln\left|H_0(\omega)\right| \approx const. + \frac{\pi}{2}\int_{\omega_0}^{\omega_i} \varphi(\omega)\, d\ln\omega + \gamma \frac{d\varphi(\omega)}{d\ln\omega} \wedge \gamma = -0.52 \qquad \text{Eq. 2.21}$$

For details of the transfer function and the mathematical methods in EIS see ref. [64]. The test procedure shows, if the calculated data are similar to the experimental data, i.e., if the system was in a quasi-stationary state or in the worst case it defines the number of data points in the spectra, which are useful for the interpretation of the experiment. To avoid artefacts because of an in-stationary state of the surface the measurement starts with a cycle at 1 kHz to 100 kHz or 1 MHz and then down to 10–100 mHz, i.e., the first run to the higher frequency allows the system to be stationary but a test according to Eq. 2.21 is always necessary.

Example 2.6

In Ex. 2.3 the Russian mud corrosion phenomenon has already been illustrated. The fact that chromium depending on the amount of oxygen and the electrolyte changing from the noblest to ignoble according to the nickel layer, a change of the passive behaviour should be able to be investigated by EIS.

Figures 2.26 and 2.27 illustrate the Bode and the Nyquist plots of galvanic deposited nickel and chromium (based on a Cr(VI) solution) layer in a nearly saturated $CaCl_2$ and N_2 flushed solution [46, 65]. The measurement was conducted starting with a cycle from 1 kHz to 100 kHz followed with a decrease frequency to 10 mHz in a three-electrode-assembly with a graphite CE and calomel RE. Above 66 Hz 20 periods for a data point have been collected and below 66 Hz three data points with a total testing time of 20 minutes [65].

The nickel surface in Fig. 2.26 shows only a one time constant (Eq. 2.17) and the data could be simulated with the simple Randles equivalent circuit (Table 2.6) and with the simulation a R_{CT} value of 100 $k\Omega/cm^2$ has been calculated. The time constant describes the double layer on the nickel surface. Therefore nickel shows one semicircle in the corresponding Nyquist diagram (Fig. 2.27). The chromium surface applied with a Cr(VI) galvanic bath shows two time constants (Fig. 2.26) and the situation could not be described with the Randles equivalent circuit. In constrast to nickel, the chromium surface have a passive layer, which could be described with the model based on Young [66]. Young developed the model based on measurements on niobium passive layers

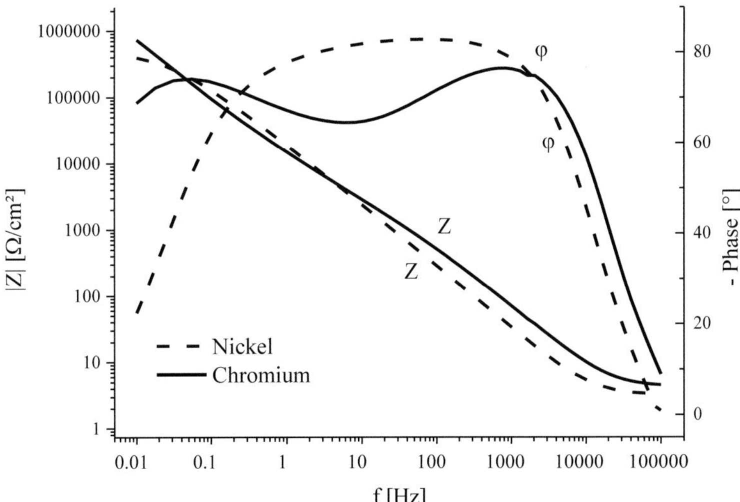

Fig. 2.26 Bode plots of galvanic nickel and chromium layers in nearly saturated $CaCl_2$ and N_2 flushed solution [31, 65].

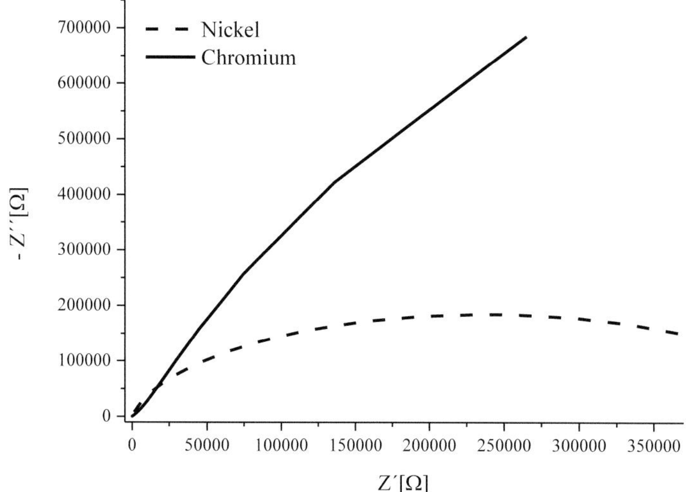

Fig. 2.27 Nyquist plots of galvanic nickel and chromium layers in nearly saturated $CaCl_2$ and N_2 flushed solution [31, 65].

with the result that the conductivity κ of the passive layer decreases with decreasing distance x relative to a characteristic length δ to the metal surface as follows:

$$\kappa(x)=\kappa(0)\cdot e^{-\frac{x}{\delta}} \qquad \text{Eq. 2.22}$$

The corresponding impedance Z_Y is defined as [67]:

$$Z_Y=\frac{p}{i\omega C}\ln\frac{1+i\omega\tau e^{\frac{1}{p}}}{1+i\omega\tau} \wedge p=\frac{\delta}{d} \;;\tau=\frac{\varepsilon_0\,\varepsilon_r}{\kappa(0)} \qquad \text{Eq. 2.23}$$

p : penetration depth [m]

τ : time constant [s]

d : thickness of the passive layer [m]

Furthermore the capacity of an equivalent circuit often does not act like a pure capacity and then a constant phase element CPE according to the Kramer-Kronig-relation [68] is a useful element to describe the surface [69, 70]:

$$Z_{CPE}=\frac{1}{(i\omega)^{\alpha}C} \wedge \alpha=0-1 \qquad \text{Eq. 2.24}$$

This behaviour is caused by the roughness of the surface or an inhomogeneous allocation of the current but other unknown reasons are also possible [71]. With a pure capacity $\alpha = 1$ (compare Eq. 2.10) and a Warburg impedance results with $\alpha = 0.5$ (compare Eq. 2.18). For $\alpha = 0$ an Ohmic resistance is described with the CPE.

With these elements an equivalent circuit of chromium layers could be generated shown in Fig. 2.28.

The Nyquist plot (Fig. 2.27) shows no Warburg impedance, i.e., a straight line in the low frequency range with a phase angle of 45° but the fit of the data with the Warburg impedance in the equivalent circuit is better (error of 2.3%) as without it (error 5.9%) but this is not a sufficient reason for a complex equivalent circuit. The reason for Z_W in the equivalent circuit is based on the point defect model [65, 72, and 73] that causes the existence of the Warburg impedance with the migration of cations and anion vacancies in the passive layer.

The combination of the nickel surface and the chromium layer in a chromium plated plastic part with micro pores (see Fig. 2.9) or micro cracks (see Ex. 2.3) has been analyzed in saturated $CaCl_2$ flushed with CO_2 resulting a pH = 3.5 [65]. The comparison of the industrial defect containing surfaces with the defect free chromium layer used in Figs. 2.26 and 2.27 is shown in Fig. 2.29 and prove the same behaviour because the electrochemical active surface is dominated by the chromium surface and the small surface of the nickel layer is more or less not visible in the diagram. Therefore a chromium surface with a passive layer could be assumed on the industrial surfaces.

The impedance spectra of the chromium layer with micro pores are illustrated in Figs. 2.30 and 2.31.

In the Nyquist plot a loop is visible after 20 hours in the low frequency range indicating a change from the passive into the active state of the surface [62]. With longer treatment Warburg diffusion is visible indicating the corrosion of the chromium layer. The corresponding Bode plots shows

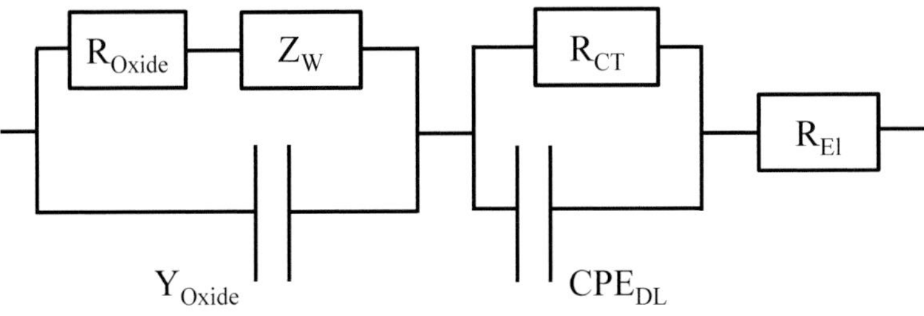

Fig. 2.28 Equivalent circuit for chromium layers in nearly saturated $CaCl_2$ and N_2 flushed solution.

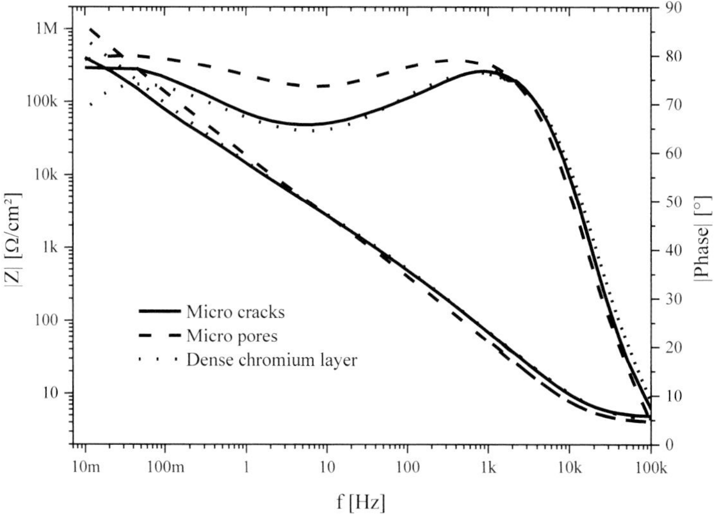

Fig. 2.29 Bode plots of the comparison of different chromium layers in nearly saturated $CaCl_2$ and CO_2 flushed solution [31, 65].

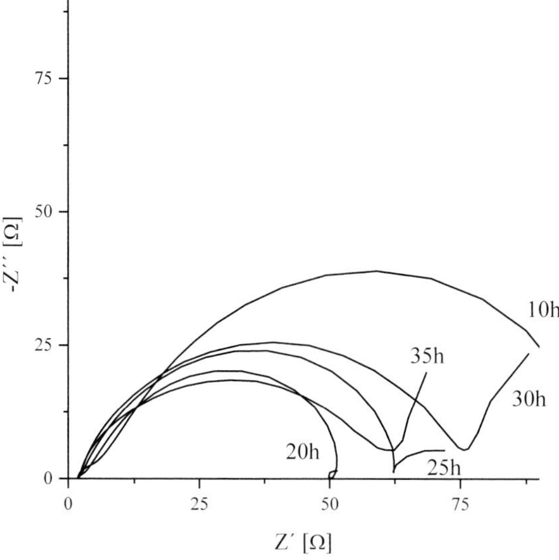

Fig. 2.30 Nyquist plots of the chromium layers with micro pores in nearly saturated $CaCl_2$ and CO_2 flushed solution at different treatment times in hours [31, 65].

the decrease of the passive layer with the decrease of the first maximum in the phase. Finally corrosion products precipitate on the surface and generating a layer with the corresponding maximum in the Bode plot.

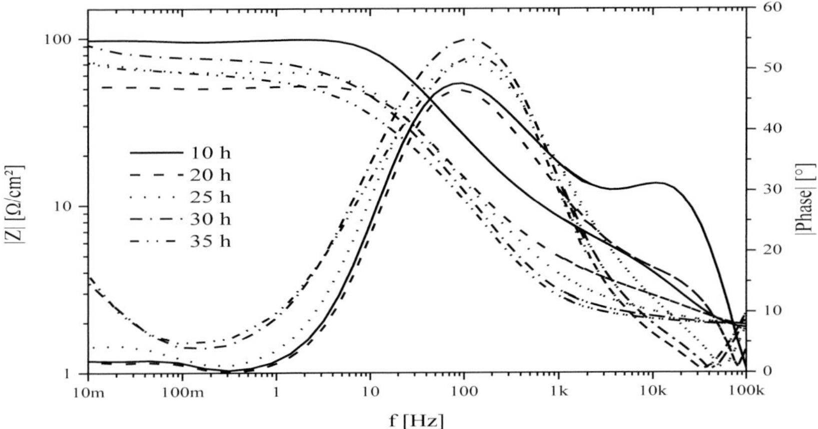

Fig. 2.31 Bode plots of the chromium layers with micro pores in nearly saturated $CaCl_2$ and CO_2 flushed solution at different treatment times in hours [31, 65].

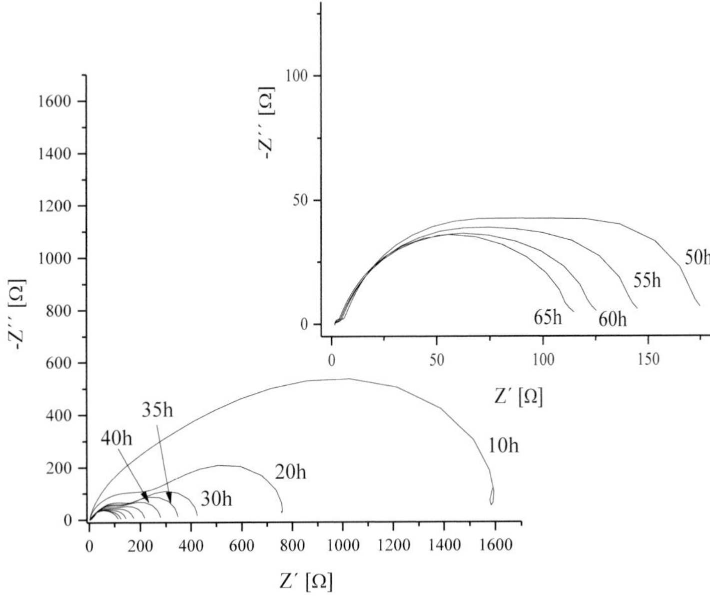

Fig. 2.32 Nyquist plots of the chromium layers with micro cracks in nearly saturated $CaCl_2$ and CO_2 flushed solution at different treatment times in hours [31, 65].

The impedance spectra of the chromium layer with micro cracks are illustrated in Figs. 2.32 and 2.33.

There is a loop in the Nyquist plot but no Warburg impedance is visible but a third semicircle appears in the diagram after some time. The

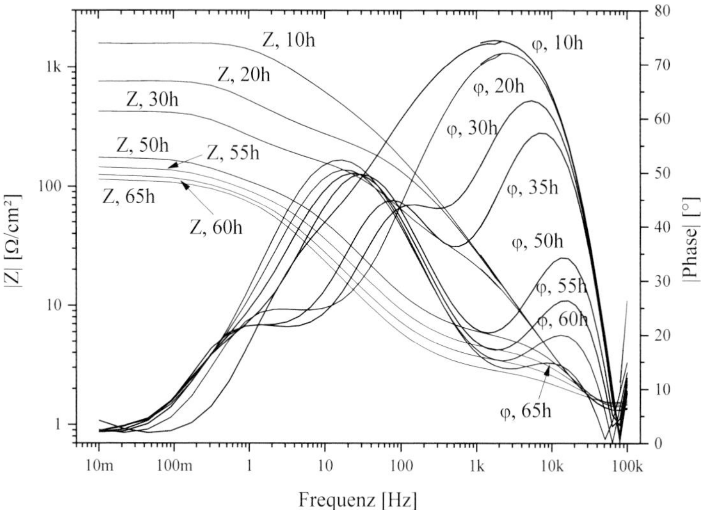

Fig. 2.33 Bode plots of the chromium layers with micro cracks in nearly saturated $CaCl_2$ and CO_2 flushed solution at different treatment times in hours [31, 65].

total impedance in the Bode plot is significantly higher as in the micro pores containing system. Therefore the chromium layer should be passive during the treatment and the nickel layer under the chromium layer dissolutes and the surface increases and causes the visible semicircle in the Nyquist plot. This is the desired process, i.e., the chromium surface is passive and the nickel surface protects the chromium layer by means of contact corrosion. For further reading see refs. [31, 46].

Both mechanism investigated with EIS are similar to the phenomenon visible in parts from the field test, i.e., real parts from car bodies showed that the chromium layer with micro cracks corroded significantly slower than the surface with micro pores. Therefore the EIS in combination with the polarization diagrams (Ex. 2.3) are able to describe the corrosion mechanism in the field because similar surfaces have been used for the electrochemical methods.

Example 2.7

To get an impression of the behaviour of other metal surfaces the EIS spectra of aluminium, copper and phosphated steel are shown in the following paragraph.

Phosphated Steel

The interesting fact regarding the EIS spectra of conversion layer is the barrier effect, i.e., if the conversion coating protects the surface with an insulating coating. Therefore a useful conversion coating behaves like a pure capacitor, i.e., in the Bode plot (Fig. 2.34) visible with a straight line in the graph and in the Nyquist plot (Fig. 2.35) with one semi-circle. Actually a conversion layer applied with an acid corrosion process (see Chap. 1.3.1)

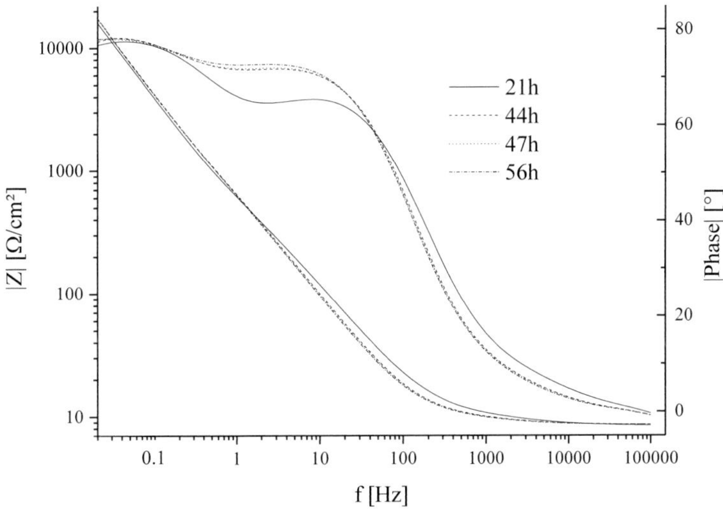

Fig. 2.34 Bode plots of phosphated steel in aerated 3% NaCl solution at different treatment times in hours.

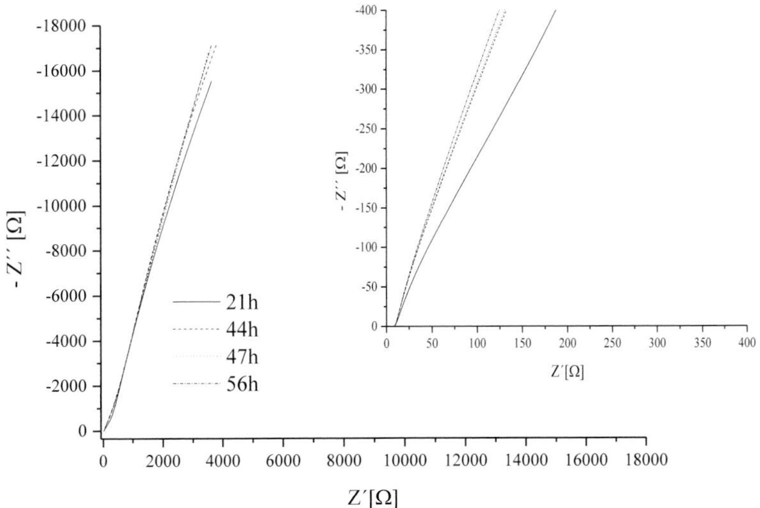

Fig. 2.35 Nyquist plots of phosphated steel in aerated 3% NaCl solution at different treatment times in hours.

cannot cover the surface 100% and therefore the spectra shows the barrier effect of the conversion layer and the electrochemical processes on the uncovered surface area.

The phase angle in the Bode plot changes one time and therefore the Nyquist plot shows two semi-circles visible, if the scale on both axes is similar as in Fig. 2.35.

Aluminium and Copper

On aluminium and copper surfaces in 3% NaCl solution corrosion might happen through the passive layer of the metals. Therefore the corrosion mechanism is controlled by the charge transfer process and therefore diffusion processes (Warburg impedance at 45°) are not visible in the Nyquist plots (Figs. 2.38 and 2.39) but the Bode plots show that the phase angle in the low frequency range is near to 45° at the beginning of the measurement and increases during testing time significant on copper (Fig. 2.36) and slightly on aluminium (Fig. 2.37).

A detailed analysis of the processes is quite complex as shown on chromium in Ex. 2.6. For instance the course of the first spectra of the copper surface in the Nyquist plot is quite similar to the finite diffusion according to Eq. 2.20 (see Fig. 2.24) but of course there are other reasons

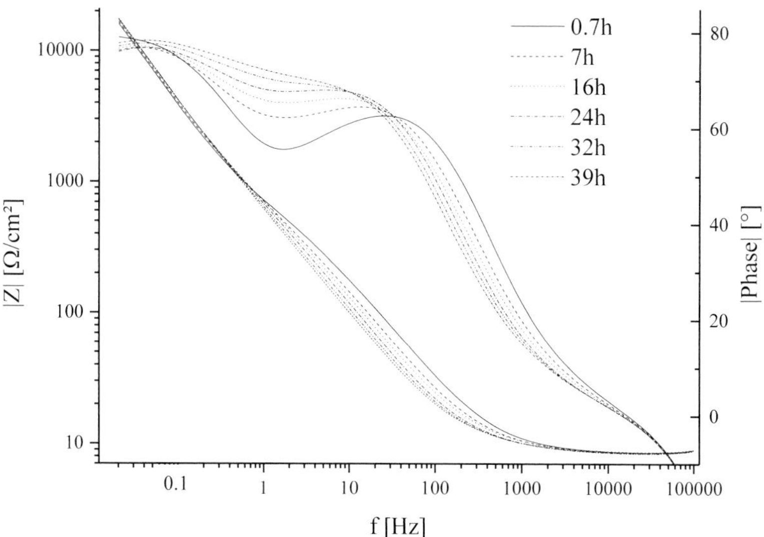

Fig. 2.36 Bode plots of copper in aerated 3% NaCl solution at different treatment times in hours.

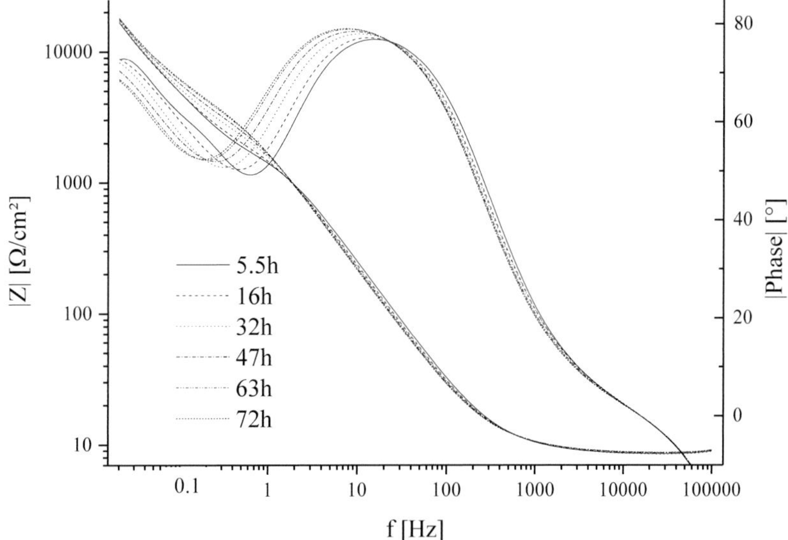

Fig. 2.37 Bode plots of aluminium in aerated 3% NaCl solution at different treatment times in hours.

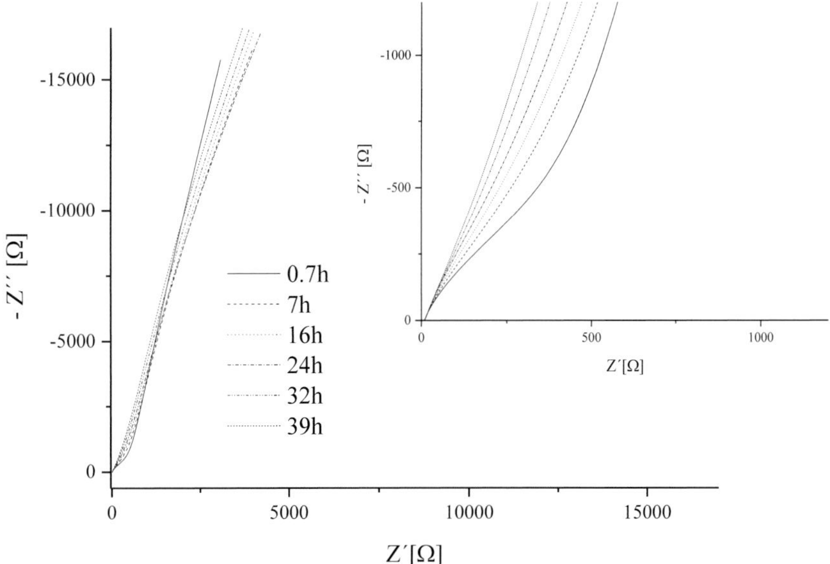

Fig. 2.38 Nyquist plots of copper in aerated 3% NaCl solution at different treatment times in hours.

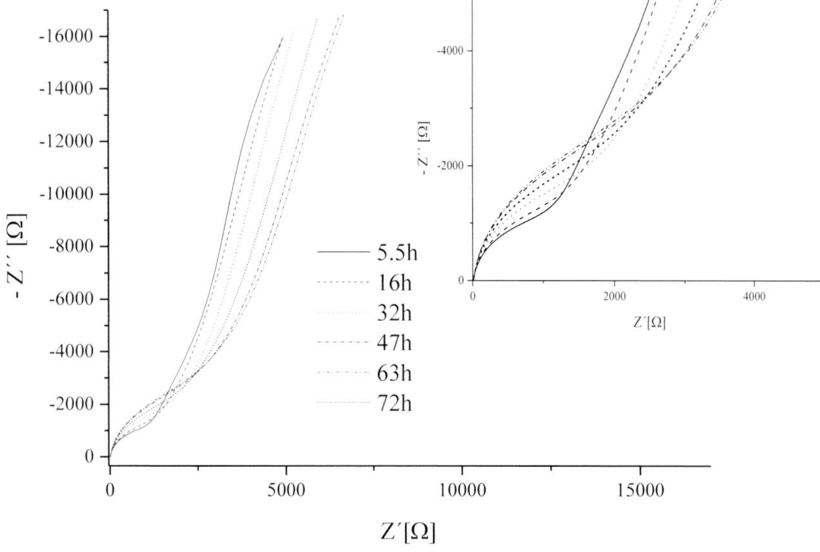

Fig. 2.39 Nyquist plots of aluminium in aerated 3% NaCl solution at different treatment times in hours.

possible and the fact that a Warburg impedance is not visible in the spectra did not confirm that the process is not diffusion controlled.

To estimate the stability of a passive or conversion layer the absolute impedances in the spectra and investigation of diffusion processes give a first hint to compare surfaces regarding the corrosion protective properties. In Table 2.7 the magnitudes of the impedance of the spectra in this example are summarized.

Table 2.7 Magnitudes of the impedance of the spectra in Figs. 2.34, 2.36 and 2.37 in 3%NaCl.

Surface	|Z| [Ω] at 20 mHz	Time [h]
Phosphated Steel	17554	44
Aluminium	17691	47
Copper	17590	39

The data show that a conversion layer is able to increase the barrier properties of a steel surface to the level of noble or valve metals.

2.2 Electrochemical Methods on the Coating Surface

2.2.1 *Cyclic Voltammetry*

Basics

The basics for CVs are already explained in Chap. 2.1.2.

Measurement

The basics for the measurement of CVs are already explained in Chap. 2.1.2.

The electrochemical methods such as the polarization diagram and CV are only useful on conducting surfaces and organic coatings are insulating normally (see Chap. 2.2.2). The zinc rich primers based on zinc powder or zinc flakes with an organic or inorganic resin are an exception. These surfaces are between an organic coating and a metal surface illustrated in the following example.

Example 2.8

Fastener (screws) coated with zinc in all mentioned methods (see Chap. 1.3.2). The lacquer application is based on a dip-spin-process, i.e., a bulk of fasteners dipped into a coating bath followed by a spin process to remove the coating to a certain thickness. Finally the bulk is dried in an oven. The challenge in this application process is to generate a constant coating thickness in the thread of the fastener to achieve a constant corrosion protection performance. To apply the coating true to gauge a thickness above 20 μm cannot be exceeded. Therefore zinc dust coatings are not useful because the zinc particles are too big with a diameter in the range of 20–40 μm. A solution realized in the 80's of the last century are zinc flake coatings containing zinc flakes with a high aspect ratio illustrated in Fig. 2.40 in a resin (for details see DIN EN 13858).

The stability of the coatings against corrosion is very high with a durability of some hundred to thousands hours in the salt spray test (see Chap. 1.4.2 and Chap. 4.2.2). To reduce the quality test time electrochemical methods has been used to describe the performance of zinc flake coatings illustrated in the following example.

CV measurements have been performed on fasteners coated with a zinc flake system in a three-electrode-assembly shown in Fig. 2.41. No homogeneous field is possible because of the thread and because of the application the thickness of the coating could vary.

Measurements on zinc rich coatings on smooth surfaces have already been done [75, 15] and the surfaces behave like a zinc surface but as a

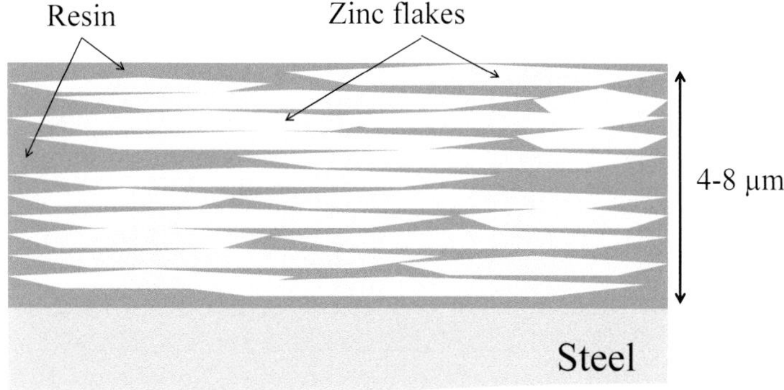

Fig. 2.40 Scheme of a zinc flake coating on steel.

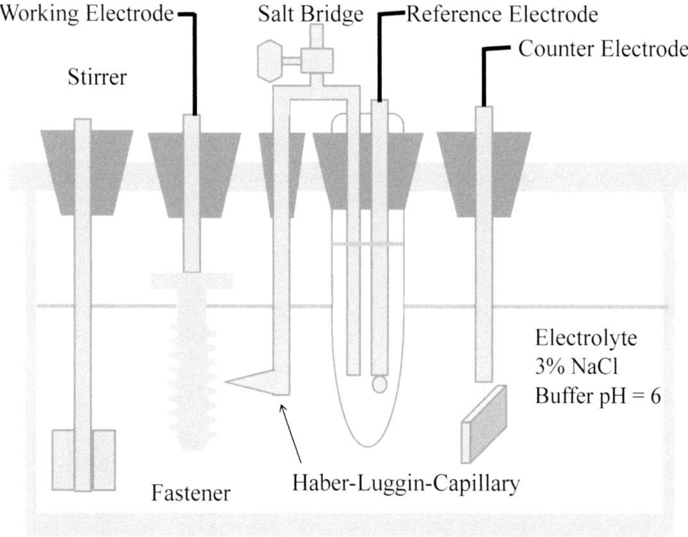

Fig. 2.41 Electrochemical cell for CV measurement on fastener surfaces.

coating. Therefore the first question was, if the CV measurements on fasteners are possible and valid data are achievable.

In Fig. 2.42 some CVs of zinc flake coatings in 3% NaCl solution buffered at pH = 6 from different fastener batches are illustrated [76].

The expected deviations are small enough to try to analyze the time dependent measurements. In Fig. 2.43 the CVs of one batch of a zinc flake coated fastener at different immersion times are collected.

With some more data it was possible to show [76] that the surface passivates during the treatment in NaCl solution. This result is ambivalent

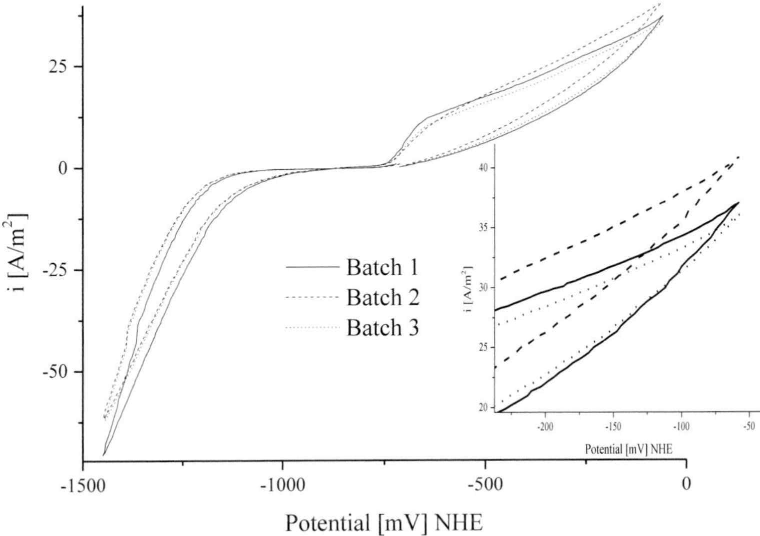

Fig. 2.42 CVs of different batches of a zinc flake coating on one type of steel fastener in 3% NaCl solution buffered at pH = 6, measured in a three-electrode-assembly with a calomel electrode as RE and a platinum sheet as CE according to [76]. With permission.

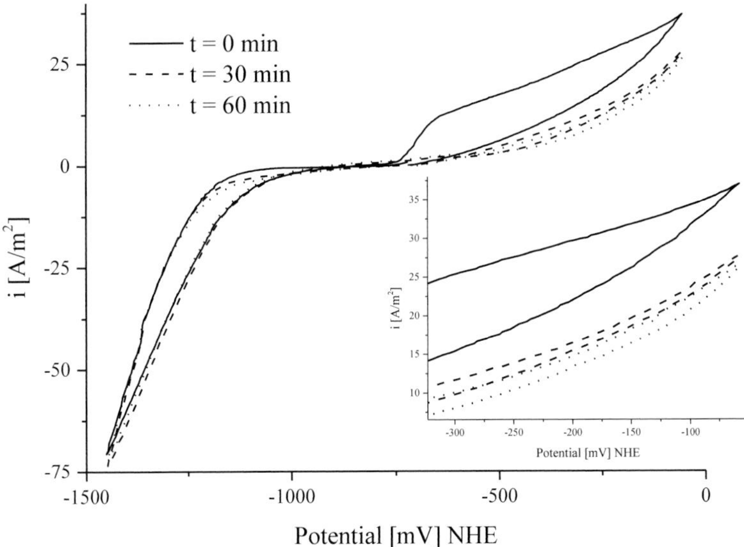

Fig. 2.43 CVs of one batch of a zinc flake coating on one type of steel fastener for different immersion times in 3% NaCl solution buffered at pH = 6, measured in a three-electrode-assembly with a calomel electrode as RE and a platinum sheet as CE according to [76]. With permission.

because the passivating reduces the corrosion rate on the zinc surface but it decreases the cathodic protection of the steel fastener. Therefore the process has to be analyzed in detail by means of EIS illustrated in Ex. 2.9.

2.2.2 *Electrochemical Impedance Spectroscopy*

Basics

The basics for EIS are already explained in Chap. 2.1.3.

Diffusion Processes on Organic Coatings

Organic coatings are normally insulated and dense and could be described regarding EIS at a first look as a pure capacitor C_{Coat}. Furthermore all coatings have defects or pores generating an Ohmic resistance R_{Coat} similar to R_{CT}. Therefore the Randles equivalent circuit (Table 2.6) is often used to analyze EIS data of organic coatings. Based on this model the amount of water in a coating during immersion of the surface in an electrolyte could be investigated with the Brasher-Kingsbury-equation (BK-equation) [77]. The derivation of the equation starts with an assumption for the total permittivity [78] as follows:

$$\varepsilon_{measurement} = \varepsilon_{Coating}^{\frac{V_{coating}}{V_{Sum}}} \varepsilon_{Water}^{\frac{V_{Water}}{V_{Sum}}} \varepsilon_{Air}^{\frac{V_{Air}}{V_{Sum}}} \wedge V_{Sum} = V_{Coating} + V_{Water} + V_{Air} \wedge \varepsilon_{Air} = 1 \quad \text{Eq. 2.25}$$

The volumes could be separated, if there is no interaction of the water/electrolyte with the organic network of the coating

$$\varepsilon_{measurement} = \varepsilon_{Coating}^{\frac{V_{coating}}{V_{Sum}}} \varepsilon_{Water}^{\frac{V_{Water}}{V_{Sum}}} \quad \text{Eq. 2.26}$$

The time rate of change follows, if no water is in the coating before the measurement starts:

$$\frac{\varepsilon_{measurement}}{\varepsilon_{meas.,t=0}} = \frac{\varepsilon_{Water}^{\frac{V_{Water}}{V_{Sum}}}}{\varepsilon_{Water}^{\frac{V_{Water,t=0}}{V_{Sum}}}} \wedge V_{Water,t=0} = 0 \Rightarrow \varepsilon_{Water}^{\frac{V_{Water,t=0}}{V_{Sum}}} = 1 \quad \text{Eq. 2.27}$$

$$\frac{\varepsilon_{measurement}}{\varepsilon_{meas.,t=0}} = \varepsilon_{Water}^{\frac{V_{Water}}{V_{Sum}}} = 80^{x} \wedge \varepsilon(Water) = 80 \wedge C = \varepsilon\,\varepsilon_0\,\frac{A}{d} \quad \text{Eq. 2.28}$$

In Eq. 2.28 two assumptions are implemented. First the permittivity of water does not change into the coating during the water/electrolyte uptake and second the thickness of the coating and the roughness of the surface of the coating do not change during the water/electrolyte uptake because the correlation of the capacity and the permittivity for a parallel plate capacitance is only given, if the surface area A and the distance d of the capacitance, i.e., the coating thickness, are constant during immersion.

$$\frac{C_{measurement}}{C_{meas.,t=0}} = 80^{x} \qquad \text{Eq. 2.29}$$

$$x = \frac{\log\left(\frac{C_{measurement}}{C_{meas.,t=0}}\right)}{\log 80} \qquad \text{Eq. 2.30}$$

There are a lot of assumptions necessary for Eq. 2.30 but often especially on highly cross-linked and thick (> 50 µm) organic coatings the BK-equation is a useful tool to estimate the amount of water uptake in a coating. The assumptions of the BK-equation are usually not fulfilled because an interaction of water or electrolyte with the polymer network, i.e., a swelling of the polymer, takes place therefore the thickness of the coating is changed during the water uptake and the transport through the coating could occur in different ways summarized in Fig. 2.44.

The swelling of the polymer network could cause an increase of the coating thickness and the transport of water or electrolyte (see Exs. 2.10 and 3.18 (UV-VIS)) into the coating occurs at defects or interfaces between

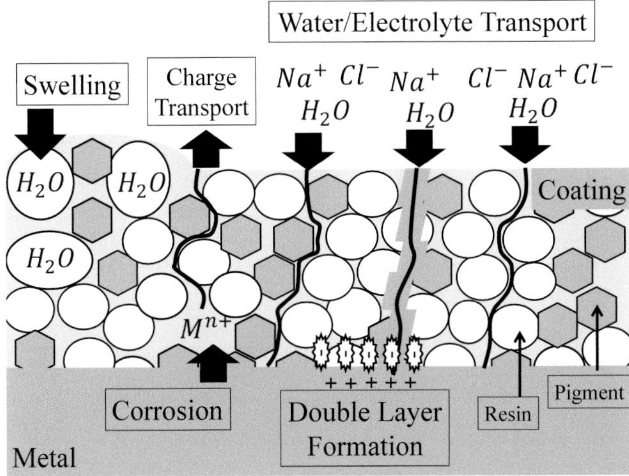

Fig. 2.44 Possible processes in an organic coating during immersion in an electrolyte.

resin particles or at pigment surfaces. Finally at the coating/metal interface corrosion could happen or a double layer could arise. Therefore the processes have to be divided into diffusion processes, polymer processes and corrosion processes.

The transport process by diffusion could be described with the second Fickian law [7]:

$$\left(\frac{dc}{dt}\right) = D\left(\frac{d^2c}{dx^2}\right) \qquad \text{Eq. 2.31}$$

With the necessary boundary conditions for a transport process through a polymer network Crank [79] found a solution for the differential equation as follows:

$$\frac{M_t}{M_s} = \frac{\log\frac{C_{Coat,t}}{C_{Coat,0}}}{\log\frac{C_{Coat,s}}{C_{Coat,0}}} = 1 - \frac{8}{\pi^2}\sum_{n=0}^{\infty}\frac{1}{2n+1}\cdot e^{\frac{-(2n+1)^2 D\pi^2 t}{4d^2}} \qquad \text{Eq. 2.32}$$

M_t : Amount of water at time t

$C_{Coat'0}$: Capacity at time t = 0

d : Thickness of the coating

M_S : Amount of water at saturation

$C_{Coat,S}$: Capacity at saturation

$C_{Coat,t}$: Capacity at time t

D : Diffusion constant

If water reacts with the polymer network, the pure diffusion process has to be expanded. In literature [8, 80, 81] an established way for the description of a non-ideal diffusion process, i.e., a non Fickian transport process, is the addition of a linear term with the factor SC in Eq. 2.33 as follows (compare Eq. 1.38):

$$\frac{M_t}{M_s} = \frac{\log\frac{C_{Coat,t}}{C_{Coat,0}}}{\log\frac{C_{Coat,s}}{C_{Coat,0}}} = 1 - \frac{8}{\pi^2}\sum_{n=0}^{\infty}\frac{1}{2n+1}\cdot e^{\frac{-(2n+1)^2 D\pi^2 t}{4d^2}} + SC\cdot t \qquad \text{Eq. 2.33}$$

At a first look Eq. 2.33 is a reasonable compromise to describe a diffusion controlled transport into an organic coating considering chemical reactions of water or electrolyte with the organic network or pigments in the coating.

The use of the Fickian law to describe processes in organic coatings is based on an equivalent circuit. The disadvantages of this procedure are that there is a risk to select an unsuitable equivalent circuit and the uncertainty, if the mathematical simulation of the data achieves the right capacity and resistance values. To prove the equivalent circuit a second independent method may have to be used. One possibility is the measurement of the resistance R_{Coat} of the organic coating by means of a highly resistive electrode according to DIN IE 60092 [8] before and after the treatment with the electrolyte to test both the equivalent circuit and the mathematical simulation.

Phase Model for Organic Coatings

Another method avoids these disadvantages by an interpretation of the shape of the impedance spectra without the use of an equivalent circuit and the simulation of the experimental data. In Fig. 2.45 the principal situations of an organic coating on a metal substrate treated with a corrosive electrolyte are illustrated. The EIS spectra especially for long term measurements could be assigning to a phase and the change of the phases could be correlated against the treatment time (see Ex. 2.10). The correct interpretation of the spectra could be supported by means of simple derivatives of the spectra to define the number of semi-circles (the first derivative shows the number of maxima in the graph) and to distinguish between a semi-circle and a linear behaviour (the first derivative is constant in a linear behaviour) of the spectra. Furthermore test methods (Eqs. 2.35, 2.37 and 2.39) based on the impedance equations (Eqs. 2.34, 2.36 and 2.38) of the corresponding equivalent circuits could be used to prove the interpretation of the data as follows:

Phase I–II:

$$Z = R_{El} + \frac{R_{Coat}}{1+\left(\omega R_{Coat} C_{Coat}\right)^2} - i\frac{\omega C_{Coat} R_{Coat}^2}{1+\left(\omega R_{Coat} C_{Coat}\right)^2} \qquad \text{Eq. 2.34}$$

For the whole frequency range follows:

$$Test\,(without\, R_{El}): \frac{Z'}{Z''} = -\frac{R_{Coat}}{\omega C_{Coat} R_{Coat}^2} = -\frac{1}{\omega C_{Coat} R_{Coat}} = -\frac{1}{\omega \tau} \qquad \text{Eq. 2.35}$$

Phase III [83]:

$$Z = R_{El} + \frac{N - \omega^2 MN + \omega^2 PO}{\left(1-\omega^2 M\right)^2 + \omega^2 P^2} + i\,\omega \frac{O - PN - \omega^2 MO}{\left(1-\omega^2 M\right)^2 + \omega^2 P^2} \qquad \text{Eq. 2.36}$$

$M = C_{Coat} R_{Coat} C_{DL} R_{CT}$; $N = R_{Coat} + R_{DL}$; $O = C_{DL} R_{CT} R_{Coat}$; $P = C_{Coat} R_{Coat} + C_{DL} R_{CT} + C_{Coat} R_{CT}$

Test (without R_{El}): For $0 < \omega < \infty$ R could be ignored

For $0 < \omega < \infty \Rightarrow N = 0$ it follows:

$$\frac{Z'}{Z''} = \frac{\omega^2 PO}{\omega\left(O - \omega^2 MO\right)} = \frac{\omega P}{1 - \omega^2 M} = \frac{\omega\left(\tau_{Coat} + \tau_{DL} + C_{Coat} R_{CT}\right)}{1 - \omega^2 \tau_{Coat} \tau_{DL}} \qquad \text{Eq. 2.37}$$

Phase IV–V:

$$Z = R_{El} + \frac{\left(\frac{a^2}{R} - a\sqrt{b}\sqrt{\frac{\omega}{2}}\right) + \left(b\sqrt{b}\sqrt{\frac{\omega}{2}}\right)}{\left(Cb\omega + \frac{a}{R} - \sqrt{b}\sqrt{\frac{\omega}{2}}\right)^2 + \left(\omega C a + \sqrt{b}\sqrt{\frac{\omega}{2}}\right)^2} + i \frac{-\omega C\left(b^2 + a^2\right) - \sqrt{b}\sqrt{\frac{\omega}{2}}\left(a+b\right) + \frac{ab}{R}}{\left(Cb\omega + \frac{a}{R} - \sqrt{b}\sqrt{\frac{\omega}{2}}\right)^2 + \left(\omega C a + \sqrt{b}\sqrt{\frac{\omega}{2}}\right)^2}$$

Eq. 2.38

$$a = R^2\omega; b = W^2; c = -CW^2\omega + \omega R - W\sqrt{\frac{\omega}{2}}; d = \omega^2 C R^2 + W\sqrt{\frac{\omega}{2}}$$

Tests (without R_{El}):

For $0 < \omega < \infty$ R could be ignored and for $\omega \to \infty \Rightarrow W = 0$

$$0 < \omega < \infty \Rightarrow a = 0 \Rightarrow \frac{Z'}{Z''} = \frac{b\sqrt{b}\sqrt{\frac{\omega}{2}}}{-Cb^2\omega - \sqrt{b}\sqrt{\frac{\omega}{2}}\,b} = -\frac{1}{CW\sqrt{2}\sqrt{\omega} - 1}$$

$$\omega \to \infty \Rightarrow b = 0 \Rightarrow \frac{Z'}{Z''} = \frac{\frac{a^2}{R}}{-\omega C a^2} = -\frac{1}{\omega\tau}$$

Eq. 2.39

The graph of the quotient Z'/Z'' against the reciprocal frequency gives a hint which phase corresponds to the impedance spectrum. In combination with the Nyquist and the Bode plots a correlation of the experimental data to the phase is possible.

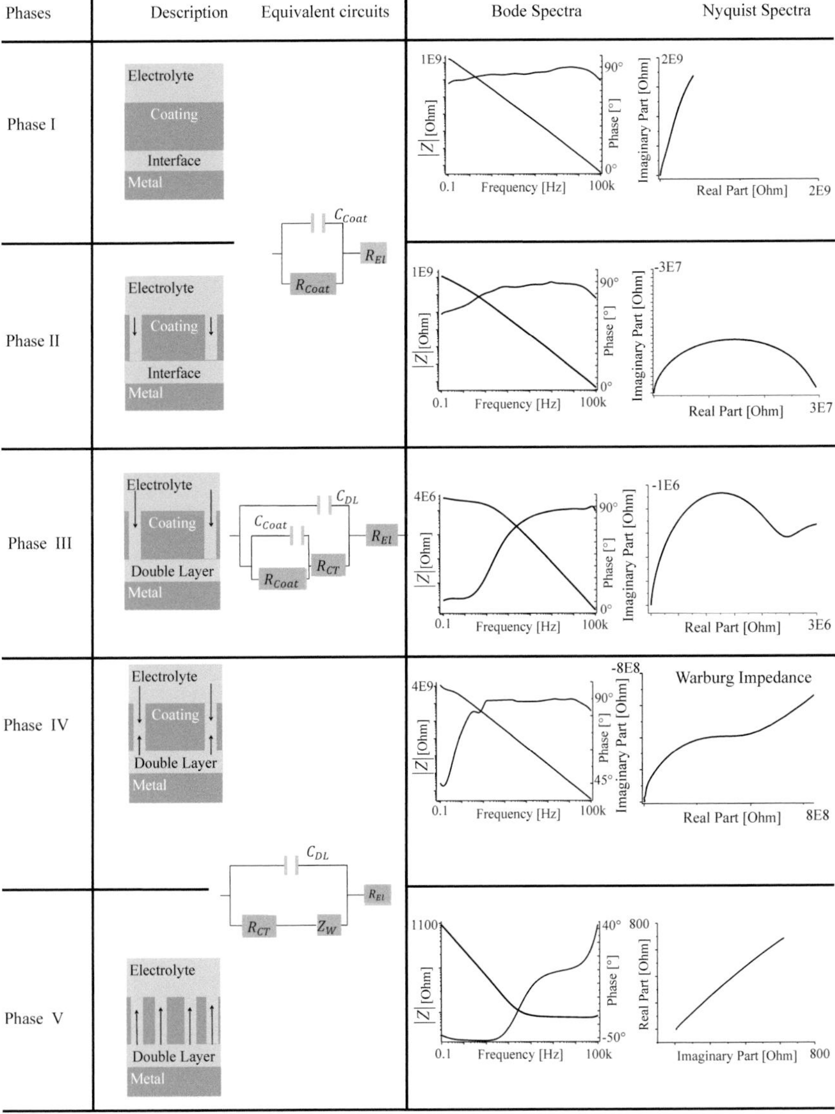

Fig. 2.45 Phases of organic coatings on metal substrates. Examples for the phase models are illustrated in [82].

The phase model has the advantage to measure the performance of an organic coating without time consuming simulation of the EIS data. Finally, if a coating shows a transport process (Phase IV or V), the barrier properties are low and it does not matter what kind of diffusion process occurred and what is the reasonable equivalent circuit for the electrochemical situation.

Measurement

The basics for the measurement of EIS are already explained in Chap. 2.1.3.

In contrast to the EIS on metal substrates often [84–89] a two-electrode-assembly is used. There are as many arguments for a two- as for the three-electrode-assembly for EIS measurements on organic coatings. There is often no difference in the resulting data and with a two-electrode-assembly more samples could be measured in simple electrochemical cells shown in Fig. 2.46. In our laboratory we use cut waste-pipes (PVC) pressed or glued [82] on the coating surface with a two-electrode-assembly, whereas the CE is platinum mesh wire or a stainless steel sheet.

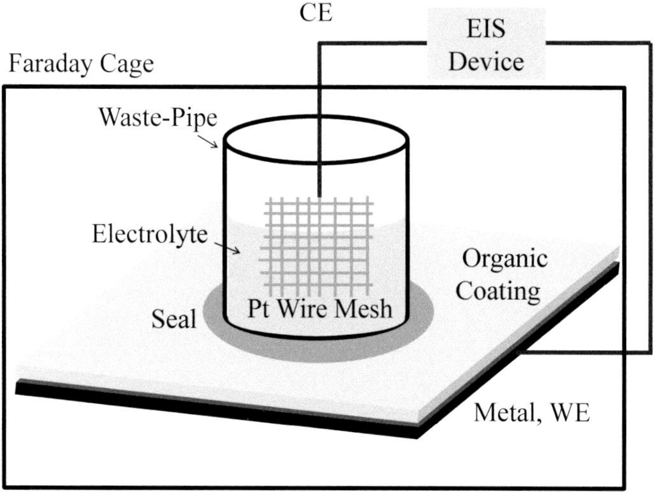

Fig. 2.46 Electrochemical cell for a two-electrode-assembly on organic coatings. The setup is standardized in DIN EN ISO 16773-2 and ASTM G106-89.

Example 2.9

In Ex. 2.8 the CVs of zinc flake coated fasteners are discussed. With the same samples EIS spectra have been performed in a three-electrode-assembly. The passivation detected with CV should decrease the possibility of the zinc flake containing coating to protect the steel substrate but increases the stability of the coating. The EIS spectra after 24 hours (Fig. 2.47) and 48 hours (Fig. 2.48) immersion in NaCl solution show a distorted semi-circle in the high and a linear behaviour in the low frequency range, whereas the slope of the linear range is too high for Warburg diffusion. The distorted

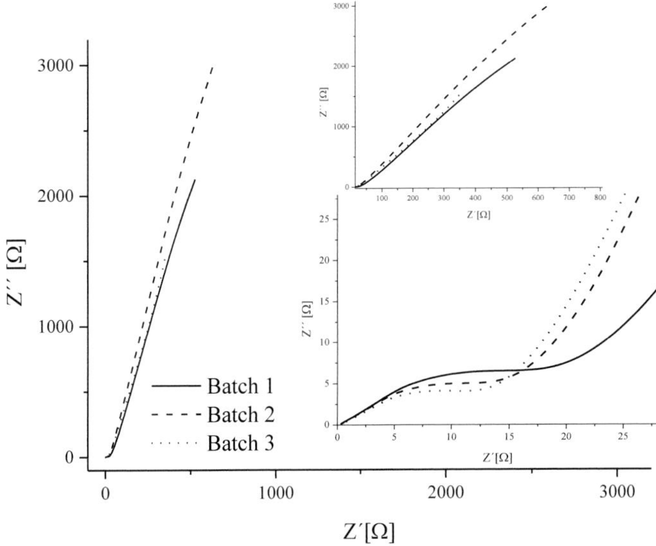

Fig. 2.47 EIS spectra after 24 hours immersion of zinc flake containing coating on a fastener analogue to Ex. 2.8 (compare [76]). With permission.

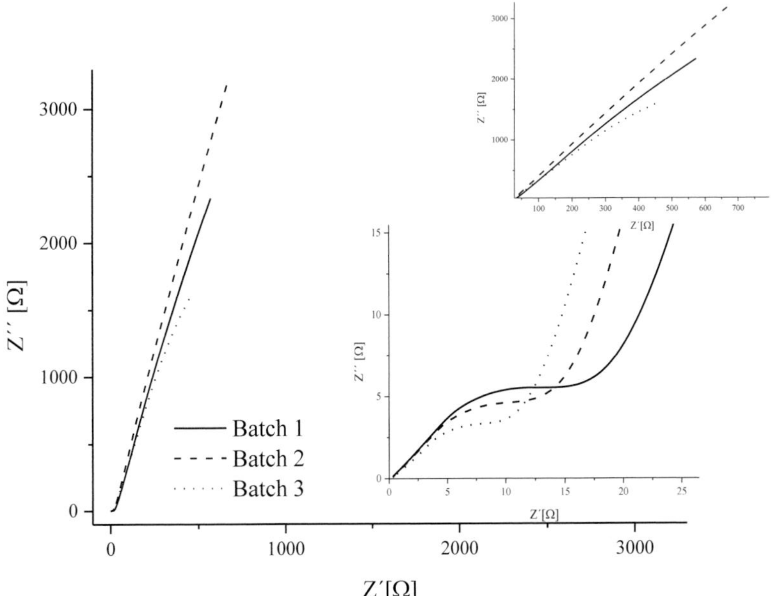

Fig. 2.48 EIS spectra after 48 hours immersion of zinc flake containing coating on a fastener analogue to Ex. 2.8 (compare [76]). With permission.

semi-circle is comparable with the finite diffusion according to Eq. 2.19 (see Fig. 2.24) and the linear behaviour with the finite diffusion according to Eq. 2.20.

In [90] a complex equivalent circuit has been used to simulate EIS spectra on zinc rich primer but based on Eqs. 2.19 and 2.20 a qualitative description of the situation is possible without any simulation. The zinc in the coating dissolutes in the aggressive solution and passivates because of the precipitation of corrosion products. A passive layer (see Fig. 2.43) is formed and generating a finite diffusion controlled process on the surface. Therefore the cathodic protection is still possible but the dissolution of zinc is decreased. The difference of the batches in Figs. 2.47 and 2.48 is the rate of the formation of a second semi-circle (batch 3 after 48 hours), i.e., the acceleration of the corrosion process.

The reduced diffusion of the zinc cations should be based on the resin of the coating and the precipitation of corrosion products, i.e., the resin reduces the transport process but does not insulate the surface. The EIS data explain the high corrosion protective performance of zinc flake coatings and by analysis of the linear range in the low frequency range a fast test method could be developed [76]. Finally the data show that zinc rich coatings behave more than a metal surface than an organic coating.

Example 2.10

Water based coatings dried often at ambient conditions and dried according to the process illustrated in Fig. 1.15. A weaker barrier property in comparison to a solvent based chemical cross-linked coating is often assumed in literature. Actually the high end performance of an epoxy resin based organic coating cured by an amine cross-linking-agent could not be achieved with water based dispersion resins with the same coating thickness until now. On the other hand the performance of polyacrylic based dispersions could vary in a broad range.

In the following example different polyacrylic based dispersions have been applied on steel samples and have been analyzed by means of long term EIS. The clear coats applied with a dry thickness of 35–40 μm and the measurements have been performed with the setup in Fig. 2.46 for one week.

The water uptake has been calculated with Eq. 2.30 of different polyacrylic dispersions varies not only in the different amount of water from 3–25% but additional in the kinetic of the process.

In Fig. 2.49 the time dependent water uptake of two coatings with a low water uptake of 6% with saturation after 25 hours and one coating with a water uptake of more than 19% without saturation but a continuous increase of the water uptake according to Eq. 2.33 are illustrated.

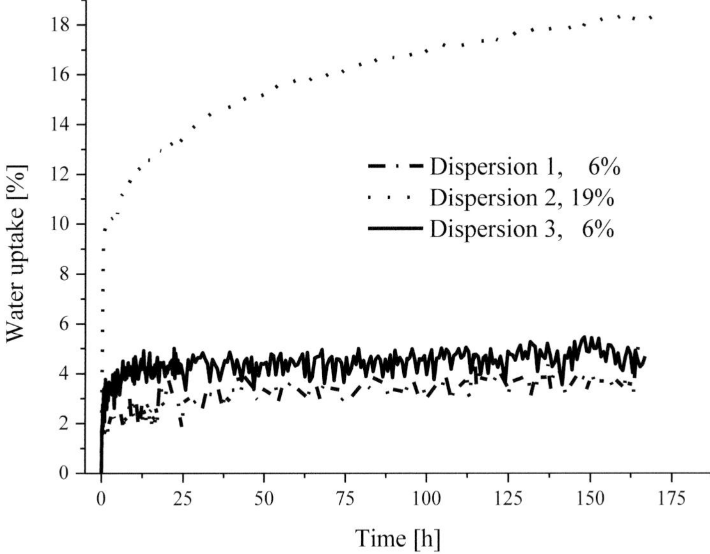

Fig. 2.49 Time dependent behaviour of the water/electrolyte uptake of coatings calculated by the data from the simulated EIS with the Randles equivalent circuit from Table 2.6. The final water uptake in [%] is calculated with Eq. 2.30 shown in the legend (modified from [82]).

Table 2.8 Water/electrolyte uptake after 175 hours of dispersion 4 in different NaCl electrolytes calculated by the data from the simulated EIS with the Randles equivalent circuit from Table 2.6 and with the use of Eq. 2.30.

NaCl concentration [%]	Water/Electrolyte uptake [%]
1	21 ± 1
3	13 ± 4
5	12 ± 5

Besides the effect of the chemical or structural composition of the dispersion the amount of ions, i.e., the ion strength, in the solution affected the total water uptake (Table 2.8) and the electrochemical reactions at the metal/coating interface (Fig. 2.50). The water uptake increases with decreasing ion strength in solution. This effect was already described in refs. [91, 92].

Furthermore the intense noise of the data in Table 2.8 of the higher ion strengths shows that the model, i.e., the Randles equivalent circuit, is able to calculate the capacity for the Brasher-Kingsbury equation, but is not suitable to fit the data. This effect is caused by corrosive processes visible

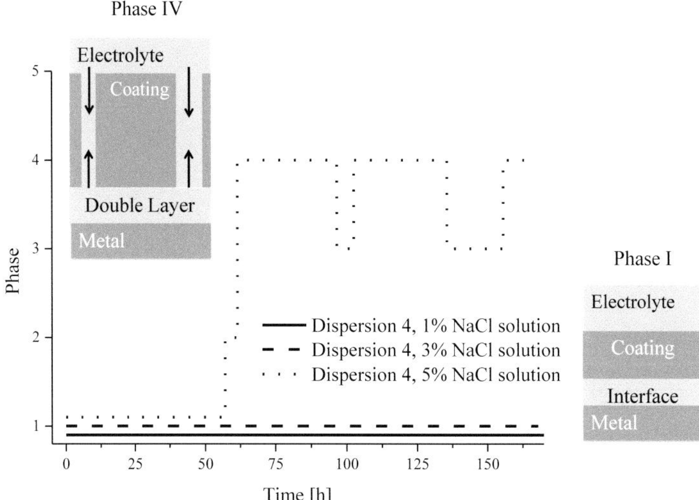

Fig. 2.50 Time dependent phase analysis of the coating samples from dispersion 4 (modified from [82]).

in Fig. 2.50. The phase analysis of the different ion strengths in solution shows that at least at 5% NaCl solution corrosive processes occurs at the interface because a phase 4 is achieved after 60 hours and corrosion is visible with the naked eye on the coating surface after the measurement. Therefore the water uptake, i.e., the total amount of water, is not relevant for the barrier properties but the amount of ions in the electrolyte above the coating. Therefore it is necessary to measure the amount of electrolyte in the coating. The investigation of the electrolyte uptake has been done by means of UV-VIS spectroscopy (see Ex. 3.18).

A first hint of the behaviour is given by the measurement after a constant climate test (see Chap. 4.2.3). In the test the coating is treated with water vapour for 48 hours and directly after this procedure the EIS measurement in a 3% NaCl solution starts in the continuous line in Fig. 2.51. The sample has been dried for one week and a second EIS measurement has been done shown with the dotted line in the same figure. The negative water uptake shows at a first look that water diffuses inside the coating in the constant climate chamber and is transported outside the coating during the EIS measurement because of the osmotic pressure against the NaCl solution. An open question is the different behaviour before and after drying of the surface in the phase analysis illustrated in Fig. 2.51 because there is no reason for a different resistance of the coating, if just water diffuses in- and outside the coating but a phase shift from 1 to 2 is only visible in the direct measurement and not after drying.

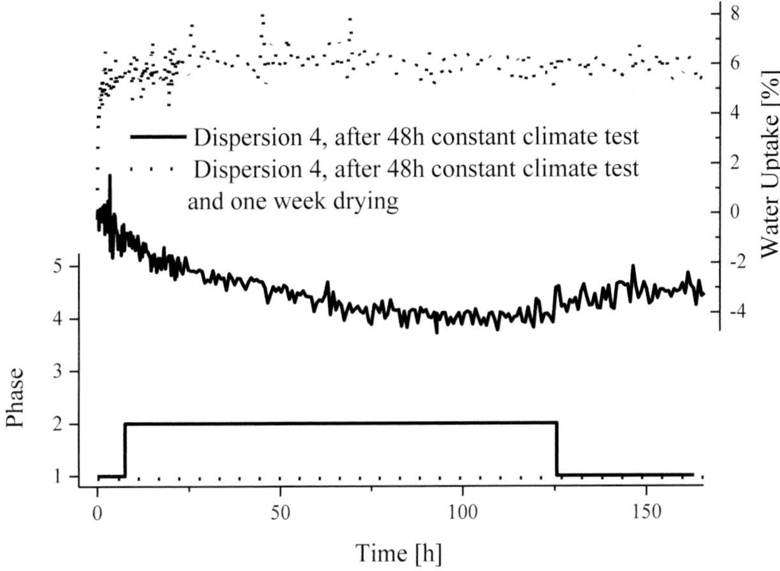

Fig. 2.51 Time dependent phase analysis of the coating samples from dispersion 4 after 48 hours constant climate test and after an additional one week drying at ambient conditions in comparison with the corresponding water uptake calculated with Eq. 2.30 based on the capacity data calculated with the Randles equivalent circuit in Table 2.6 (modified from [82]).

Example 2.11

Can coatings are often based on phenoxy resins combined with phenol-formaldehyde resins as cross-linking-agents (compare Figs. 1.16 and 1.17). The coating is necessary because the dissolution of tin ions from the metal surface into the food has to be prohibited (see Chap. 1.3.2). The disadvantage of the so called gold varnish is the use of bisphenol A (BPA) as monomer for the phenoxy resin diffuses from the coating into the food. Therefore BPA should be avoided in can coating resins (for details of the BPA challenge and alternative resins see [93]) for food packaging.

In the following example BPA has been replaced in a typical gold varnish (phenoxy resins cured with phenol-formaldehyde-resin) in different concentrations in the resin illustrated in Figs. 2.52 and 2.53. The clear coatings have been applied on tin coated steel (2.8 g/m^2 tin) with a dry film thickness of 5–6 µm without any pre-treatment. The spectra after 5 (Fig. 2.52) and 24 (Fig. 2.53) hours shows the high resistance of BPA based resins and the thin clear coatings allow to see the diffusion processes in the low frequency range. The replacement of BPA reduces the resistance, i.e., the barrier properties, of the coating but the complete

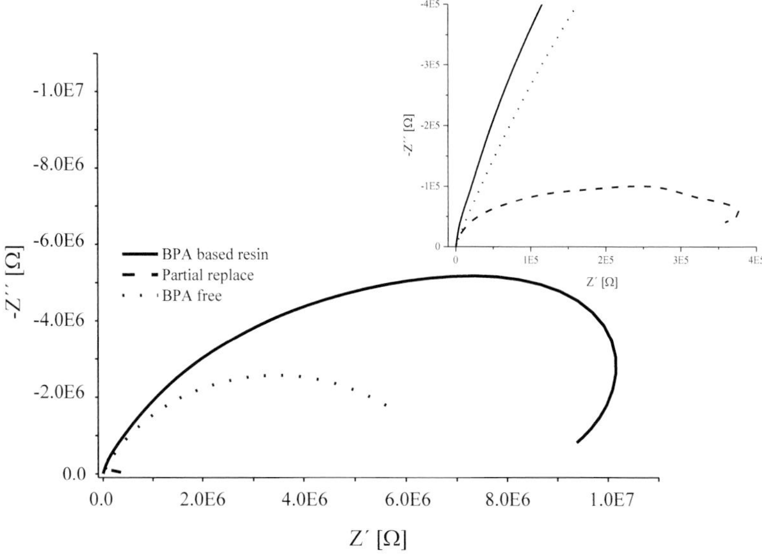

Fig. 2.52 Nyquist spectra of different can coatings immersed in 3% NaCl solution for 5 hours.

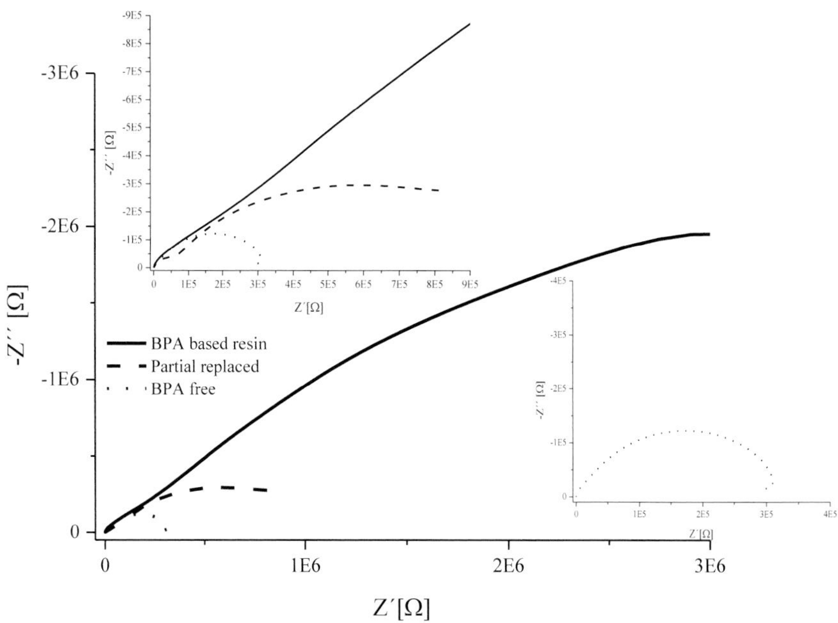

Fig. 2.53 Nyquist spectra of different can coatings immersed in 3% NaCl solution for 24 hours.

BPA free coating shows a stable semi-circle, i.e., phase 2, after 5 hours and diffusion processes after 24 hours. The coating with the partially replaced BPA behaves like the standard with visible diffusion processes (distorted semi-circle in Fig. 2.24) and both BPA containing coatings show phase 3 after 24 hours, i.e., two semi-circles despite the higher coating resistance in comparison to the BPA free system. It is still an open question why the BPA free system inhibits the diffusion process and, if this property inhibits corrosion for a long term treatment, but the data show that the coating resistance is only one aspect for the corrosion protective properties of a coating and the transport processes through the coating and the processes at the interface give only the complete picture. Furthermore the diffusion process interpreted as a corrosion process is reasonable because often the corrosion is visible with the naked eye but in this example the diffusion could be also caused by the transport of electrolyte through the coating. The EIS is not able to differentiate between these processes therefore a second independent method is necessary (see Chap. 3.2.2).

Example 2.12

Another aspect of corrosion protective properties is the use of active pigments. In the following example coil coating primers (see Chap. 1.4.4) have been analyzed with EIS in Harrison solution (see Table 2.2). A chromate containing primer applied on a chromate containing pre-treatment have been compared with a chromate free primer on a chromate free pre-treatment (for details see [94]). Both systems are represented after 500 hours salt spray test (see Ex. 4.5) the same good performance with more or less no delamination at the scratch. The EIS data in Figs. 2.54 and 2.55 show a completely different picture. The chromate free system shows a water uptake of approximately 3% with slight increase during time whereas the chromate containing system shows a continuous increase of water uptake after short incubation time perfectly described by Eq. 2.33. The curve fitting in the figures are necessary to calculate the SC factor according to Eq. 2.33. Therefore the chromate containing system shows an intense polymer swelling that decreases the barrier properties of the coating.

The corrosion protective property of an organic coating could not be described by the coating resistance and the EIS data do not give the complete picture because the rate of transport processes, active pigments, i.e., electrochemical and chemical reactions at the interface, have to be taken into account and could be investigated with other methods (see below).

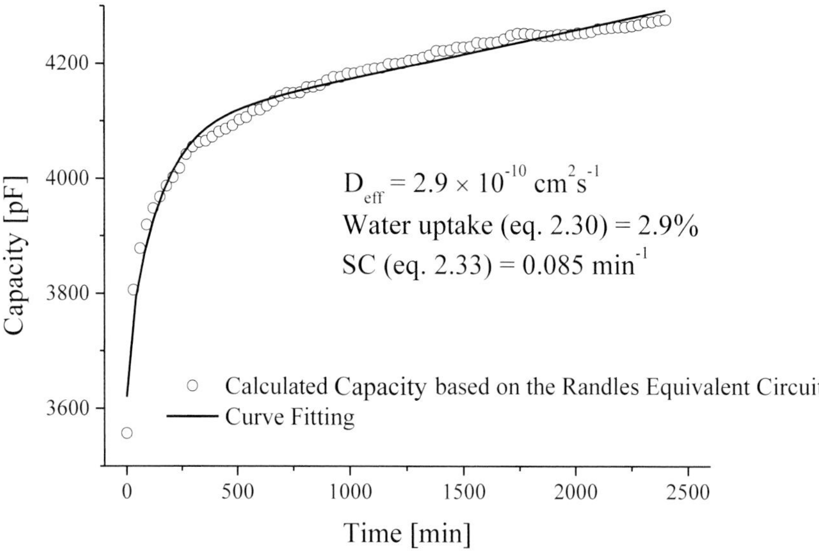

Fig. 2.54 Time dependent behaviour of the water/electrolyte uptake of the chromate free coil coating primer on a chromate free pre-treatment calculated by the data from the simulated EIS with the Randles equivalent circuit from Table 2.6 from [94]. With permission.

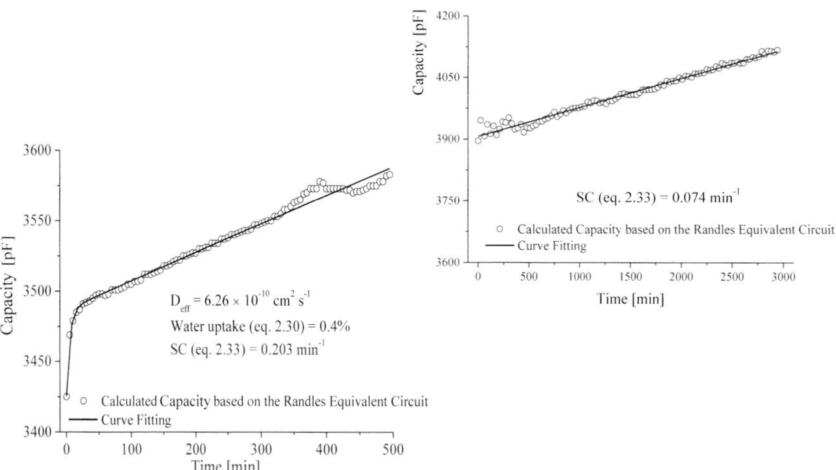

Fig. 2.55 Time dependent behaviour of the water/electrolyte uptake of the chromate containing coil coating primer on a chromate based pre-treatment calculated by the data from the simulated EIS with the Randles equivalent circuit from Table 2.6 from [94]. With permission.

2.2.3 *Scanning Kelvin Probe*

Basics

A very detailed presentation of the Scanning Kelvin Probe (SKP) with the historical background and the use in electrochemistry is given in ref. [95, 96]. In the following paragraph the necessary basics for the use in corrosion research and especially on organic coated substrates are presented.

Measuring Signal of the SKP

In Fig. 2.56 the principal setup of a SKP is illustrated. The measurement is based on the investigation of the potential between the probe with a known, i.e., calibrated, potential and the sample surface. Lord Kelvin [97] developed a method to investigate the potential difference $\Delta\Psi$ between two electrodes by applying an external potential U_{ext} and varying the distance between the electrodes and therefore changing the capacity ΔC of the system by measuring the resulting charge ΔQ as follows:

$$\Delta Q = (\Delta\Psi + U_{ext})\cdot \Delta C \qquad \text{Eq. 2.40}$$

After compensation the applied potential is equal to the potential difference of the electrodes:

$$U_{ext} = -\Delta\Psi \qquad \text{Eq. 2.41}$$

As it is a challenge to measure the charge accurately, Zisman [98] optimized the method by vibrating one electrode with a sinusoidal frequency ω with the amplitude Δd (see Fig. 2.56) and measuring the induced current i(t):

$$i(t) = \frac{dQ(t)}{dt} = (\Delta\Psi + U_{ext})\cdot\frac{dC(t)}{dt} \qquad \text{Eq. 2.42}$$

For the periodic change of the distance follows:

$$d(t) = d_0 + \Delta d \cos \omega t \qquad \text{Eq. 2.43}$$

From this it follows that the capacity of a plate capacitor (compare Eq. 2.28) with the permittivity ε, the electric field constant ε_0 and the surface A:

$$C(t) = \frac{\varepsilon\varepsilon_0 A}{d(t)} = \frac{\varepsilon\varepsilon_0 A}{d_0 + \Delta d\cos\omega t} \qquad \text{Eq. 2.44}$$

and furthermore for the current:

$$i(t)=\left(\Delta\Psi+U_{ext}\right)\varepsilon\,\varepsilon_0\,A\frac{d}{dt}\left(\frac{1}{d_0+\Delta d\cos\omega t}\right) \qquad \text{Eq. 2.45}$$

The derivation results the equation for the current:

$$i(t)=\left(\Delta\Psi+U_{ext}\right)\frac{\varepsilon\,\varepsilon_0\,\omega\,A\,\Delta d\sin\omega t}{\left(d_0+\Delta d\cos\omega t\right)^2}\ \wedge\ if\ \Delta d\ \ll d_0 \Rightarrow \left(d_0+\Delta d\cos\omega t\right)^2\approx d_0^2 \qquad \text{Eq. 2.46}$$

The approximation in Eq. 2.46 is suitable for $\Delta d < 0.15\ d_0$ [96] and from this it follows for the current:

$$i(t)=\left(\Delta\Psi+U_{ext}\right)\frac{\varepsilon\,\varepsilon_0\,\omega\,A\,\Delta d\sin\omega t}{d_0^{\ 2}} \qquad \text{Eq. 2.47}$$

With the modulation of U_{ext} to the situation i = 0 than Eq. 2.41 could be used to investigate the potential difference.

The potential is independent from the distance but the signal to noise ratio of the current increases rapidly with the increase of d_0 especially, if the surface of the probe is small, normally around 100 µm. Therefore d_0 has to be as small as possible (some µm) and constant during the measurement. If corrosion takes place under an organic coating, the thickness of the coating could increase (see Figs. 1.10 and 1.11) and therefore a height control is necessary to analyze corrosion processes under organic coatings. The height control could be realized with a second sinusoidal signal explained in detail in refs. [96, 99] and the principal setup is illustrated in ref. [100].

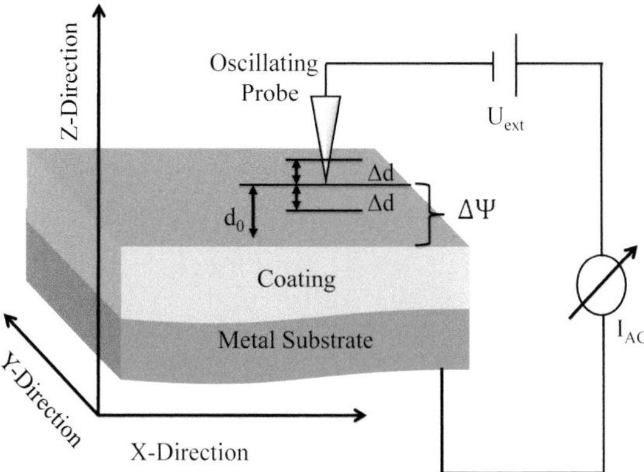

Fig. 2.56 Schematic setup of a Scanning Kelvin Probe (SKP).

Interpretation of the Measuring Signal of the SKP

With the SKP the delamination (compare Figs. 1.11 and 1.12) of an organic coating could be analyzed and the different areas and processes summarized in Fig. 2.57 could be investigated by measuring the potential between the probe and the sample (see above). The interpretation of the measuring signal is explained in the following paragraph.

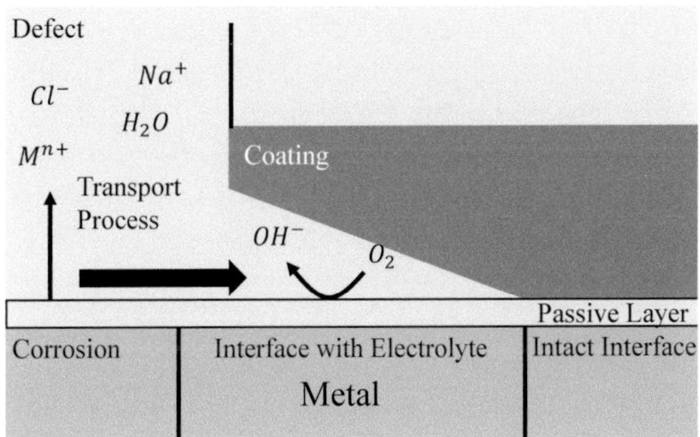

Fig. 2.57 Scheme of the processes on an organic coated metal substrate with a defect (e.g., scratch) in contact with an electrolyte and oxygen.

The potential between the probe and the sample is based on the work function w of the surfaces (for details see ref. [101]) allows investigating the electrode potential of a surface.

On an electrolyte covered sample Stratmann et al. [102, 103] showed that there is a correlation between the Volta potential difference $\Delta\Psi_{Probe/Electrolyte}$, i.e., the potential between the probe and the electrolyte surface, and the corrosion potential E_{Corr} of the sample. The potential E_{Corr} is defined in relation to reference electrode, the probe, as follows:

$$E_{Corr} = \varphi + \varphi_{Probe} \qquad \text{Eq. 2.48}$$

Thus the potential could be described as follows:

$$E_{Corr} = \left(\frac{w_{Probe}}{F} - \chi_{Electrolyte} - \varphi_{Probe} \right) + \Delta\Psi_{Probe/Electrolyte} \qquad \text{Eq. 2.49}$$

If the surface properties of the probe during the measurement are constant, i.e., the work function w_{Probe}, the potential φ_{Probe} and the dipole moment of the electrolyte $\chi_{Electrolyte}$, the corrosion potential of the surface could be investigated, if the probe is calibrated against a known potential (see Ex. 2.13).

On an organic coated sample in contact with an electrolyte a similar relation is given for the corrosion potential [96]:

$$E_{Corr} = \left(\frac{w_{Probe}}{F} - \chi_{Coating} - \varphi_{Probe} \right) + \Delta\varphi_{Donnan} + \Delta\Psi_{Probe/Coating} \qquad \text{Eq. 2.50}$$

The Donnan potential [101, 104–106] is caused by fixed charged groups in the coating relevant for a high amount of charges in the coating, e.g., ion exchange resins. Leng et al. [107] showed on an epoxy resin cured with polyamidoamines (compare Figs. 1.13 and 1.16) that the Donnan potential could be ignored. The Volta potential difference is caused by every interface between the sample and the probe and therefore on metal substrates the resulting potential is the sum of all interface potentials as follows (see Fig. 2.57):

$$\Delta\Psi_{Probe/Coating} = \Delta\varphi_{Passive}^{Substrate} + \Delta\varphi_{Passive} + \Delta\varphi_{Coating}^{Passive} - \frac{1}{F}\mu_{Electrons}^{Substrate} - \frac{w_{Probe}}{F} + \chi_{Coating} \qquad \text{Eq. 2.51}$$

$\Delta\varphi_{Passive}^{Substrate}$: Galvani potential between the substrate and the passive layer (often oxides)

$\Delta\varphi_{Passive}$: Potential drop in the passive layer

F : Faraday constant

$\Delta\varphi_{Coating}^{Passive}$: Potential drop between the passive layer and the coating

$\mu_{Electrons}^{Substrate}$: Chemical potential of the electrons in the substrate

$\chi_{Coating}$: Dipole moment of the coating

Therefore the measurement signal changes from the dry interface to a wet (electrolyte) interface and additionally corrosion processes occur at the interface.

Transport of Water/Electrolyte at the Interface Metal/Polymer

The transport of ions with the charge z along the distance x in an electrolyte is based in principle on two driving forces, the diffusion based on the concentration c difference (see Eq. 1.38) and the migration based on the potential E difference summarized as the total flux j_{Ion} in the Nernst-Planck equation as follows [108]:

$$j_{Ion} = j_{Diff} + j_{Mig} = -D\frac{dc(x)}{dx} + \left(-\frac{F}{RT} D c z \frac{dE(x)}{dx} \right) \qquad \text{Eq. 2.52}$$

If both driving forces have to be taken into account or only one, is discussed below in detail.

Fürbeth and Stratmann showed [108] that in oxygen containing atmosphere, i.e., under corrosive conditions, the diffusion force dominates only in the beginning of the delamination and then the migration is the dominant force for the transport.

With the assumption that only diffusion causes the transport of ions with the concentration c_0 through the interface along the distance x for the time t and with the following boundary conditions [109]:

$$c = c_0 \, for \; x = 0 \; and \; t > 0$$
$$c = 0 \, for \; x > 0 \; and \; t = 0 \qquad \text{Eq. 2.53}$$

the second Fickian law (Eq. 2.31) could be solved with the following equation [79]:

$$c = c_0 \left[1 - erf \left(\frac{x}{2\sqrt{Dt}} \right) \right] \qquad \text{Eq. 2.54}$$

The formal diffusion coefficient could be investigated with the following equation:

$$\bar{x} = 2\sqrt{D}\sqrt{t} \qquad \text{Eq. 2.55}$$

The main diffusion length $\bar{x}$ [109] has to be measured and the slope against $\sqrt{t}$ results the diffusion coefficient.

Corrosion Processes at the Interface Metal/Polymer

If corrosive processes occurs under the coating, Stratmann et al. showed that the ion transport at the interface between anode and cathode is the rate determining step [110, 111] and the alkaline pH value under the coating in a cathodic delamination process (see Fig. 1.11) causes the delamination of the coating. Furthermore the delamination rate behaves linear against time (see below) on zinc coated steel [111] and on steel surfaces [112]. For anodic delamination processes, i.e., often filiform corrosion, the assumed mechanism (see Fig. 1.11) could be proved [113] and the delamination of the coating (epoxy based system in ref. [113]) could also be caused by the cathodic side processes.

Measurement and Analysis

The collected data from the surface could be illustrated in a 3-dimensional diagram as shown in Fig. 2.58. The potential in the intact and the

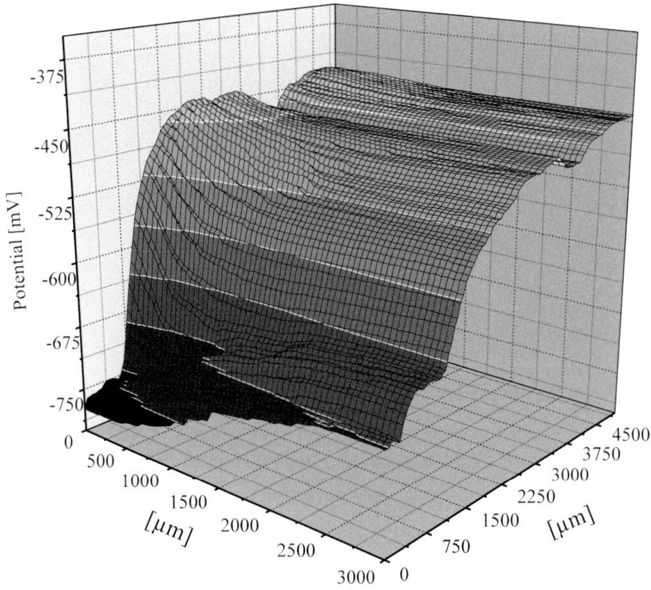

Fig. 2.58 3-dimensional illustration of the potential change from the defect under the coating of an organic coating on a zinc coated steel substrate.

delaminated area shows in corrosive processes the kind of delamination (see Fig. 1.11 for anodic and cathodic delamination) and for transport processes the driving force for the migration. In Fig. 2.58 the corrosion process under an organic coating on zinc coated steel is illustrated. The measurement starts at the defect with the low potential caused by the anodic dissolution of zinc. The potential increased rapidly under the coating to a passive state (see Chap. 1) of zinc, i.e., the cathodic area for the oxygen reduction is located under the coating.

From the diagrams collected over one or several weeks one line scan will be selected (Fig. 2.59) and the inflection point of the graphs are calculated with the second derivative of the data.

Especially if the potential difference is low or several processes happen at the same time, the inflection point cannot be investigated with the second derivative of the data. Other methods like a constant potential or a fit of the sigmoidal graph and comparison of the fit data should be used to investigate the delamination rate. In principle on amphoteric metals more than one corrosive processes are possible because in the alkaline conditions of the cathode an anodic dissolution of the metal is also possible (for zinc see Fig. 1.2).

Finally the graph of inflection points versus time t results the delamination rate under the coating (Fig. 2.60). The delamination rate is

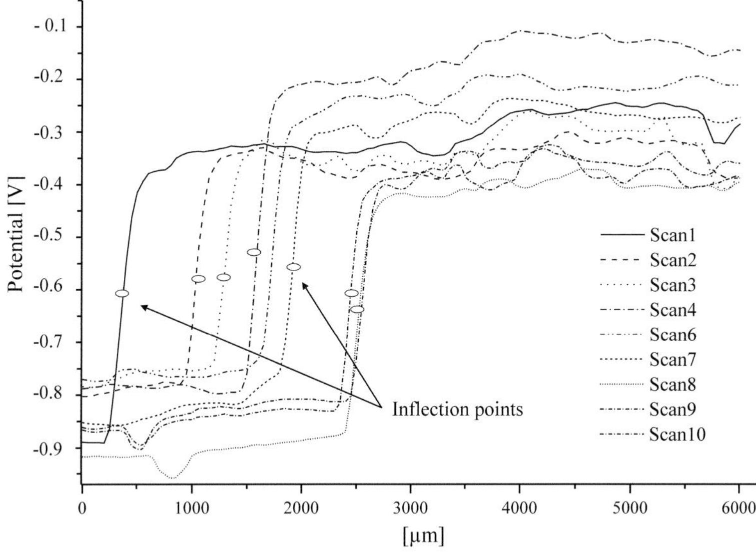

Fig. 2.59 Line scans from the 3-dimensional data (Fig. 2.58) at different times and the investigation of the inflection points.

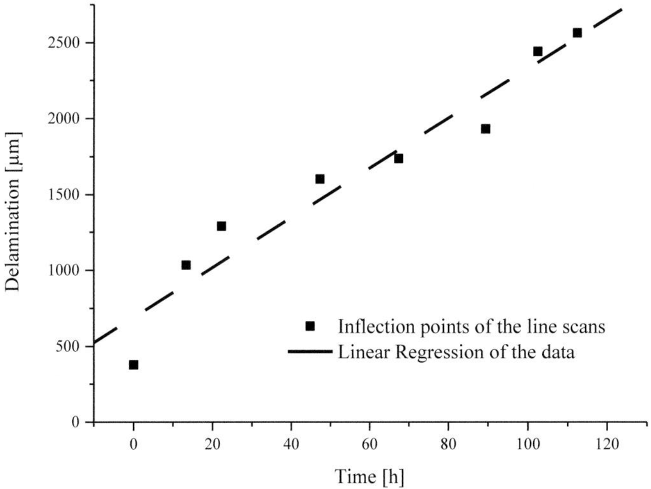

Fig. 2.60 Investigation of the delamination rate from the inflection points of Fig. 2.59.

the relevant value for corrosive processes whereas the diffusion constant available with Eq. 2.55 is the interesting value for transport processes (with the graph of the inflection points versus $\sqrt{t}$).

Example 2.13

The delamination rate of coil coating primer (see Exs. 4.5 and 2.12) is low. Model coatings, i.e., often clear coats not optimized for corrosion protection show a constant and fast delamination, easily measureable by means of the SKP illustrated in Fig. 2.61 [94]. Directly after the defect is filled with electrolyte the potential decrease in the defect and delamination occurs through the coating surface within some hours. Because of the homogenous delamination front the analysis with a line scan similar to Fig. 2.59 achieves reasonable results.

The industrially used coil coating primer based on a chromate free pre-treatment and without chromate containing pigments shows after thousand hours the first hint of a delamination process shown in Fig. 2.62. This example illustrates the limitation of the SKP method because the real time analysis of the delamination process is time consuming on optimized coatings. Furthermore the slow process causes often an inhomogeneous delamination front therefore the measurement has to be proceeded for a long time to get reasonable results.

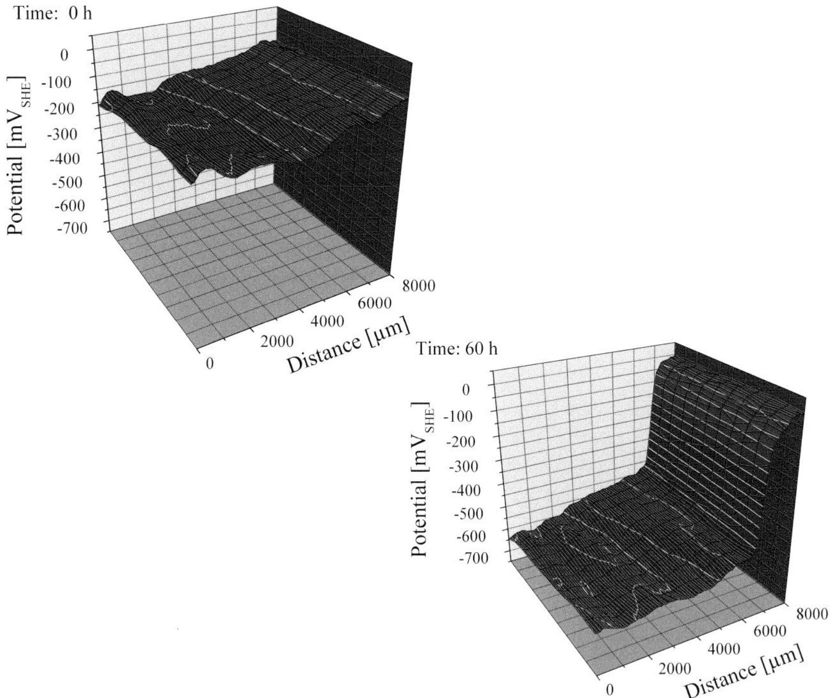

Fig. 2.61 SKP graphs of the delamination of a model coating on zinc coated steel from [94]. With permission.

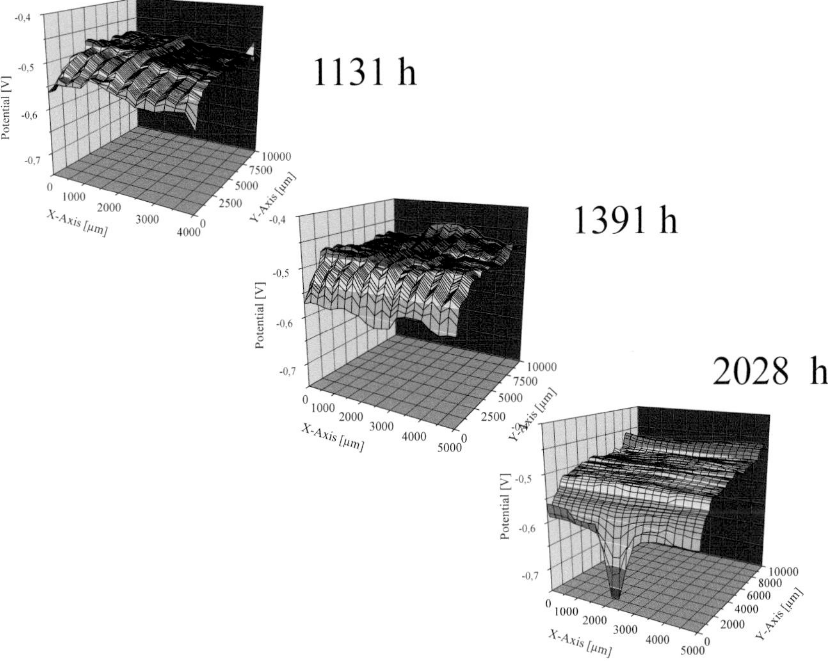

Fig. 2.62 SKP graphs of the delamination of coil coating primer on zinc coated steel from [94]. With permission.

Example 2.14

In Fig. 2.63 the experimental setup for the investigation of the corrosive delamination of an E coat on a zinc substrate is shown [114]. The coated sample is fixed in a device to apply the NaCl containing electrolyte at the edge of the sample and in addition to this a scratch on the sample filled with electrolyte also is applied. At a first look the delamination rate of a scratch and at the edge of a sample should be compared. Because of the fact that the edge defect was not dense a small amount of electrolyte could penetrate in a thin (around 500 µm) crevice at the other edge of the sample. Therefore three areas of corrosive processes have been applied on the sample and could be analyzed at the same time with the Scanning Kelvin Probe (SKP) with height control in a climate chamber in humid air (> 95% relative humidity).

The 3-dimensional data have been analyzed as illustrated in Fig. 2.59 and the final delamination rates are illustrated in Fig. 2.64. First of all there is no difference in the scratch and the edge delamination and the corrosion rate of 5–6 µm/h correlates with the industrially used alternating climate test with 2 mm delamination after 10 weeks (see Chap. 4.2.5).

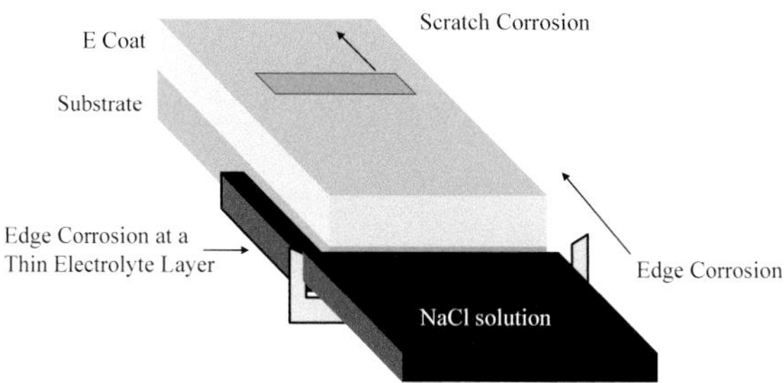

Fig. 2.63 Setup for the SKP measurement of an E coated sample (compare [114, 115]).

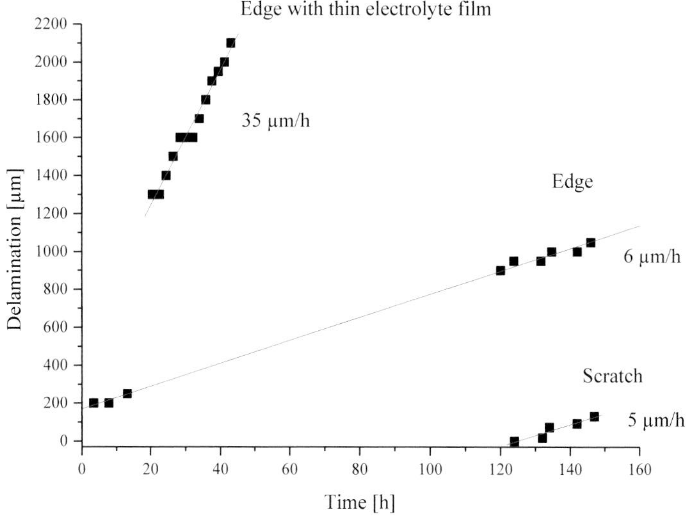

Fig. 2.64 Delamination rates of the different defects from Fig. 2.63 [115].

The delamination rate of the edge with a thin electrolyte layer is seven times higher. This is not a surprising effect. The cathodic process, i.e., the reduction of oxygen of the surface, is a diffusion controlled process (see Chap. 1.1.2). Therefore if the diffusion length (see Eq. 1.38) through the electrolyte is reduced from some millimetres (scratch and edge) to 500 µm, the corrosion rate should increase. This is the reason why the SKP experiments have to be done in a climate chamber with a very high humidity to avoid any change of the electrolyte defect. For the consequences in industrial tests see Chap. 4.2.5.

Example 2.15

As already mentioned the transport processes at the interface are relevant because they should control the corrosion rate. To investigate the transport processes the experimental setup is the same as for the corrosive process but in an oxygen free atmosphere normally realized with a nitrogen flushed climate chamber. The line scans in Fig. 2.65 have been measured on a pigmented and a clear E coat (see Fig. 1.19) on steel (without conversion layer) in nitrogen atmosphere (O_2 content < 3%) with a relative humidity between 84–86%. Because of the fact that the corrosion rate on E coated steel samples is very low, the samples have been stored in a nitrogen flushed climate chamber in NaCl solution for some hundred hours before the experiment in the SKP with height control started [8]. Before the SKP chamber is flushed with nitrogen, there is enough time for corrosive processes at the defects of the samples and that causes the low potential in the defect. After a certain time corrosion processes stops in the oxygen free atmosphere and only transport processes are visible with the SKP.

The data prove the results from Fürbeth and Stratmann [108] on model clear coats that migration is the most important driving force for the transport process at the interface (see Eq. 2.52). The clear E coat generates a significant higher potential drop as the pigment E coat results a faster transport process (Fig. 2.65). On the other hand the interpretation of the data with the assumption of a diffusion process according to Eqs. 2.54 and 2.55 shows reasonable results illustrated in Fig. 2.66. Furthermore, if the concentration of NaCl in the electrolyte decreases, the potential drop increases but the delamination rate decreases as shown in Table 2.9 indicating a diffusion controlled process. Therefore the driving

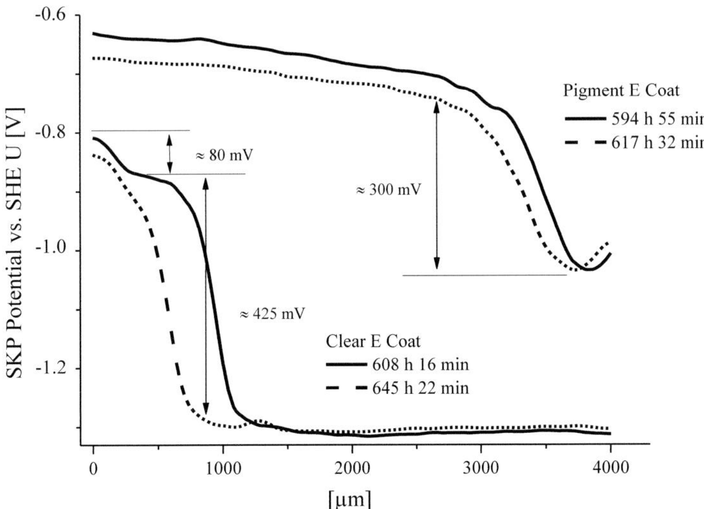

Fig. 2.65 Line scans of pigmented and clear E coats on steel (without pre-treatment) in nitrogen atmosphere (defect at 4000 μm) modified from [8].

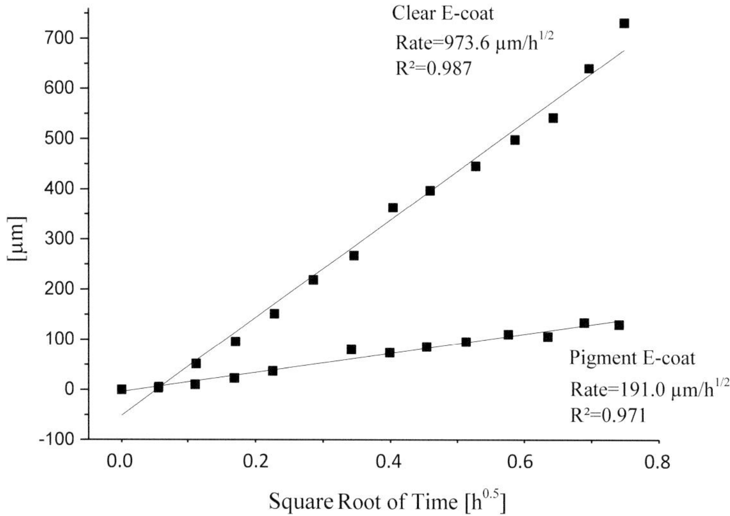

Fig. 2.66 Transport rate of clear and pigment E coat on steel (without pre-treatment) with a 0.5 M NaCl electrolyte in nitrogen atmosphere (modified from [8]).

Table 2.9 Potential drop and delamination rate of pigment E coat on steel (without pre-treatment) in nitrogen atmosphere.

NaCl concentration [mol/l]	Potential drop [mV]	Delamination rate [μm/h$^{0.5}$]
1	270	73.6
0.5	300	26.0
0.1	350	16.3

force depends at least on the chemical situation of the interface and the organic coating and the dominance of one driving force is in principal an open question.

The fact that the transport process depends on the ion concentration could be caused by two effects. First the diffusion depends on the amount of ions, if high concentration are used (the Fickian law is based on diluted solutions) and the potential also depends on the ion concentration, i.e., the migration is a function of the ion concentration (Table 2.9).

2.3 Literature

1. C.H. Hamann, W, Vielstich, Elektrochemie, Wiley-VCH, Weinheim, 2005; C.H. Hamann, A. Hamnett, W. Vielstich, Electrochemistry, Wiley-VCH, Weinheim, 2007
2. G. Wranglen, An Introduction to Corrosion and Protection of Metals, Institut för Metallskydd, Stockholm, 1972

3. J. Wang, Analytical Electrochemistry, Second Edition, Wiley-VCH, 2000
4. C. M. A. Brett, A. M. O. Brett, Electrochemistry, Principles, Methods and Applications, Oxford University Press, New York, 1993
5. R. W. Revie, H. H. Uhlig, Corrosion and Corrosion Control, John Wiley & Sons, Hoboken New Jersey, 2008
6. H. A.G. Stern, D. R. Sadoway, J. W. Tester, J. Electroanal. Chem. 659, 2 (2011) 143–150
7. P.W. Atkins, Physical Chemistry, Oxford University Press, 1978
8. M. Reichinger, W. Bremser, M. Dornbusch, Electrochim. Acta 231 (2017) 135–152
9. K. Saurbier, J.W. Schultze, J. Geke, Mater. Sci. Forum 11-112 (1992) 73–84
10. W. S. Li, J. L. Luo, Int. J. Electrochem. Sci., 2 (2007) 627–665
11. Y.-J. Li, B. Wu, X.-P. Zeng, Y.-F. Liu, Y.-M. Ni, G.-D. Zhou, H.-H. Ge, Thin solid films 405 (2002) 153–161
12. J. Morales, F. Diaz, J. Hernandez-Borges, S. Gonzalez, Corros. Sci. 48 (2006) 361–371
13. J. W. Schultze, N. Müller, Reinigen und Vorbehandeln 53 (1999) 17–22
14. E. Klusmann, U. König, J.W. Schultze, Mater. Corr. 46 (1995) 83–91
15. Thesis (PhD), F. Theiler, Untersuchung über den Mechanismus des Korrosionsschutzes von Stahl durch Zinkstaubanstriche, Eidgenössische Technische Hochschule Zürich, Schweiz, 1974
16. Thesis (PhD), T. Lostak, Elektrochemische und Grenflächenchemische Untersuchungen an Zr-basierenden Konversionsschichten auf verzinktem Bandstahl, University Duisburg-Essen, 2013
17. Thesis (PhD), P. Keller, Elektrochemische und oberflächenanalytische Untersuchungen zur anodischen Deckschichtbildung auf Zinn und Kupfer/Zinn-Legierungen, Heinrich-Heine-University Düsseldorf, 2006
18. C. A. Gervasi, P. A. Palacios, Ind. Eng. Chem. Res. 52 (2013) 9115–9120
19. A. Brinkmann, A. Zockoll, Farbe&Lack 02 (2010) 1ff
20. M. DabalaÁa, L. Armelaob, A. Buchbergera, I. Calliari, Appl. Surf. Sci. 172 (2001) 312–322
21. W. Guixiang, Z. Milin, W. Ruizhi, Appl. Surf. Sci. 258 (2012) 2648–2654
22. N. N. Voevodin, N. T. Grebasch, W. S. Soto, L. S. Kasten, J. T. Grant, F. E. Arnold, M. S. Donley, Prog. Org. Coat. 41 (2001) 287–293
23. V. Moutarlier, M. P. Gigandet, L. Ricq, J. Pagetti, Appl. Surf. Sci. 183 (2001) 1–9
24. M. Hernandez, J. Genesca, J. Uruchurtu, F. Galliano, D. Landolt, Prog. Org. Coat. 56 (2006) 199–206
25. D. J. Mills, S. Mabbutt, R. Akid, JCSE, 2, 2000, 18
26. DIN 51360/2
27. M. Saremi, C. Dehghanian, M. Mohammadi Sabet, Corros. Sci. 48 (2006) 1404–1412
28. K. Belmokre, N. Azzouz, F. Kermiche, M. Wery, J. Pagetti, Mater. Corr. 49 (1998) 108–113
29. B. M. Durodola, J. A. O. Olugbuyiro, S. A. Moshood, O. S. Fayomi, A. P. I. Popoola, Int. J. Electrochem. Sci. 6 (2011) 5605–5616
30. V. Di Noto, M. Mecozzi, Appl. Spectrosc. 51, 9 (1997) 1294–1302
31. C. Langer, W. Wendland, K. Honold, L. Schmidt, J. Gutmann, M. Dornbusch, Corrosion Analysis of Decorative Microporous Chromium Plating Systems in Concentrated Aqueous Electrolytes, Engineering Failure Analysis 91 (2018) 255–274 and Thesis (PhD), C. Langer, Klärung des spezifischen Korrosionsmechanismus an dekorativen Chromüberzugssystemen, University Duisburg-Essen, 2018
32. Encyclopaedia of Electrochemistry, Edited by A.J. Bard and M. Stratmann, 2007 Wiley-VCH Verlag GmbH & Co. KGaA, Weinheim, Vol. 4, Corrosion and Oxide Films, Chap. 1.3 G. S. Frankel, D. Landolt, Kinetics of Electrolytic Corrosion Reactions, 25–49
33. M. A. Ameer, A. M. Fekry, A. A. Ghoneim, F. A. Attaby, Int. J. Electrochem. Sci. 5 (2010) 1847–1861
34. I. Rotariu, G. L. Turdean, F. Kormos, D. Macaovici, G. Tolnai, I. Felhosi, P. Nagy, L. Trif, E. Kalman, Mat. Sci. Forum, Vols. 537-538 (2007) 247–254

35. B. O. Oni, N. O. Egiebor, N. J. Ekekwe, A. Chuku, JMMCE, 7, 4 (2008) 331–346
36. S. Bonk, M. Wicinski, A. W. Hassel, M. Stratmann, Electrochem. Commun. 6 (2004) 800
37. Thesis (PhD), J. C. Fenster, Zinkkorrosion in alkalisch wässrigen Lösungen, Heinrich-Heine-University Düsseldorf, 2009
38. M. Pourbaix, Atlas of Electrochemical Equilibria in aqueous solutions, National Association of Corrosion engineers, Houston Texas, 1974
39. F. H. Assaf, S.S. Abd El-Rehiem, A. M. Zaky, Mater. Chem. Phys. 58 (1999) 58–63
40. Thesis (PhD), M. Reichinger, Investigations of the direction-driven water and ion transport along interfaces and through polymer networks, University Paderborn, 2017
41. T. Pearson, S. Handy, L. Forst, *Galvanotechnik* (2009) 1730–1739
42. K.-H. Tostmann, Korrosion: Ursachen und Vermeidung, Wiley-VCH, Weinheim, 2001
43. E. Kunze, Korrosion und Korrosionsschutz in verschiedenen Gebieten: Teil 1, Wiley-VCH, Berlin, 2001
44. R. Kiefer, R. Stilke, E. Boese, A. Heyn, M. Engelking, R. Hillert, *Galvanotechnik* (2012) 1904–1914.
45. G. Bauer, C. Donner, P. Hartmann, P. Wachter, *Galvanotechnik* (2010) 1960–1969
46. C. Langer, M. Dornbusch, Corrosion Behavior of Decorative Chromium Layer Systems in Concentrated Aqueous Electrolytes, NACE Corrosion 2017, New Orleans, 2017
47. C. M. A. Brett, A.M. O. Brett, Electrochemistry, Principles, Methods and Applications, Oxford University Press, New York, 1993
48. P. H. Rieger, Electrochemistry, Second Edition, Chapman & Hall, New York, London, 1994
49. R. Holze, Experimental Electrochemistry, Wiley-VCH, Weinheim, 2009
50. W. R. Heinemann, P.T. Kissinger, Current Separations, West Lafayette, 9, 12 (1989) 15–18
51. M. Dornbusch, T. Biehler, M. Conrad, A. Greiwe, D. Momper, L. Schmidt, M. Wiedow, JUnQ, 6, 2 (2016) 1–7
52. A. Losch, J. W. Schultze, Appl. Surf. Sci 52 (1991) 29–38
53. A. Losch, J. W. Schultze, J. Electroanal. Chem. 359 (1993) 39–61
54. F. Pan, X. Yang, D. Zhang, Appl. Surf. Sci 255 (2009) 8363–8371
55. R. K. Gupta, K. Mensah-Darkwa, J. Sankar, D. Kumar, Trans. Nonferrous Met. Soc. China 23 (2013) 1237–1244
56. G. Meng, F. Sun, Y. Shaoa, T. Zhang, F. Wang, C. Dong, X. Li, Electrochim. Acta 55 (2010) 5990–5995
57. X. Hongyin, Lili, Adv. Mater. Res. 399-401 (2012) 1967–1971
58. S. Shen, X.-Y. Guo, P. Song, Y.-C. Pan, H.-Q. Wang, Y. Wen, H.-F. Yang, Appl. Surf. Sci. 276 (2013) 167–173
59. U. Sohlig, F. Ruf, K. Suck, E. Neitmann, K. Müller, R. Fischl, C. Pickardt, P. Eisner, Process, for Obtaining Phytic Acid from Rape Press Cake, WO2013001043
60. Encyclopedia of Electrochemistry, Edited by A.J. Bard and M. Stratmann, 2007 Wiley-VCH Verlag GmbH & Co. KGaA, Weinheim, Vol. 3, Instrumentation and Electroanalytical Chemistry, Chap. 2.1 B. Speiser, Linear Sweep and Cyclic Voltammetry, 81–104
60a. A. Lasia, Electrochemical Impedance Spectroscopy and Its Applications, Modern Aspects of Electrochemistry, B. E. Conway, J. Bockris, and R.E. White, Edts., Kluwer Academic/Plenum Publishers, New York, 1999, Vol. 32, p. 143–248.
61. H. Kaesche, Die Korrosion der Metalle, 3. neubearb. u. erw. Aufl. 1990, Springer Verlag, Heidelberg, 2011, English Edition: H. Kaesche, Corrosion of Metals, Springer Verlag, Berlin-Heidelberg, 2003, S. 35ff
62. E. Barsoukov, J. Ross Macdonald, Impedance Spectroscopy Theory, Experiment, and Applications Second Edition, Wiley-Interscience, Hoboken, 2005
63. D. Meschede, Gerthsen Physik, 23. Auflage, Springer Verlag, Berlin, 2006

64. Thesis (PhD), K.-A. Schiller, Optimierung der dynamischenTransferfunktionsanalyse für die Impedanzspektroskopie und die intensitätsmodulierte Photospektroskopie zur Anwendung an instationären und verteilten elektrochemischen Systemen, University Erlangen-Nürnberg, 2012
65. Thesis (MA), L. Schmidt, Elektrochemische Untersuchung des Korrosionsmechanismus an galvanischen Chrom- und Nickelschichten, Niederrhein University of Applied Sciences, Krefeld, 2016
66. L. Young, Anodic Oxide Films, Academic Press, New York, 1961
67. C. A. Schiller, Electrochim. Acta 46 (2001) 3619–3625
68. R. de L. Kronig: *On the theory of dispersion of X-rays*. In: *Journal of the Optical Society of America*. Band 12, Nr. 6, 1926, S. 547–556 H. A. Kramers: *La diffusion de la lumiere par les atomes*. In: *Atti Cong. Intern. Fisici, (Transactions of Volta Centenary Congress) Como*. Bd. 2, 1927, S. 545–557
69. A. Hassanzadeh, Corros. Sci. 49 (2007) 1895–1906
70. E. E. Oguzie, K. L. Oguzie, C. O. Akalezi, I. O. Udeze, J. N. Ogbulie, V. O. Njoku, ACS Sustainable Chem. Eng. 1 (2013) 214–225
71. Encyclopaedia of Electrochemistry, Edited by A. J. Bard and M. Stratmann, 2007 Wiley-VCH Verlag GmbH & Co. KGaA, Weinheim, Vol. 3, Instrumentation and Electroanalytical Chemistry, Chap. 3.5 P. A. Christensen, *In situ* Infrared Spectroelectrochemistry, 530–571
72. Yu, Haobo, J. Phys. Chem. 116 (2012) 25478–25484
73. D. D. Macdonald, Electrochim. Acta 56 (2011) 1761–1772
74. E. Warburg, Ann. Phys. Chem. 303, 3 (1899) 493–499
75. N. Hammouda, H. Chadli, G. Guillemot, K. Belmokre, Adv. Chem. Eng. Sci. 1 (2011) 51–60
76. Thesis (BA), I. Schledowetz, Elektrochemische Untersuchung von Korrosionsprozessen an Zinklamellenbeschichtungen in Natriumchlorid-Lösung zur Entwicklung eines Korrosionsschnelltests, Niederrhein University of Applied Sciences, Krefeld, 2014
77. D. M. Brasher, A. H. Kingsbury, J. Appl. Chem. 1954, 62–72
78. L. Hartshorn, N. J. L. Megson, E. Rushton, J. Chem. Soc. Ind., London, 56 (1937) 266ff
79. J. Crank, The Mathematics of Diffusion, Clarendon Press, Oxford, 2. Ed., 1975
80. E. P. M. Van Westing, G.M. Ferrari, J. H. W. de Wit, Corros. Sci. 36, 6 (1994) 957
81. V.N. Nguyen, F. X. Perrin, J. L. Vernet, Corros. Sci. 47 (2005) 397–412
82. M. Dornbusch, S. Kirsch, C. Henzel, C, Deschamps, S. Overmeyer, K. Cox, M. Wiedow, M. Dargatz, U. Meisenburg, Prog. Org. Coat. 89 (2015) 332–343
83. J. E. G. Gonzalez, J.C. Mirza Rosca, J. Adhes. Sci. Technol. 13, 3 (1999) 379–391
84. J. H. Park, G. D. Lee, H. Ooshige, A. Nishikata, T. Tsuru, Corros. Sci. 45 (2003) 1881–1894
85. D. Beckert, G. Bockmair, Farbe& Lack 95 (1989) 318–323
86. J. D. Scantlebury, K. N. Ho, J. Oil Cot. Chem. Assoc. 62 (1979) 89–92
87. A. Amirudin, D. Thierry, Prog. Org. Coat. 26 (1995) 1–28
88. S. Pietsch, Plaste Kautschuk 32 (1985) 233–236
89. J. Vogelsang, W. Strunz, Mater. Corr. 52 (2001) 462–469
90. M. D. Danford, M. J. Mendrek, D. W. Walsh, NASA Technical Paper 3534, 1995
91. Q. Zhou, Y. Wang, G.P. Bierwagen, Corros. Sci. 55 (2012) 97–106.
92. Q. Zhou, Y. Wang, Prog. Org. Coat. 76 (2013) 1674–1682.
93. M. Dornbusch, R. Rasing, Christ, Epoxy Resins, Vincentz Network, Hanover, 2016
94. M. Dornbusch, M. Hickl, K. Wapner, L. Jandel, Millennium Steel 2008, 231–236
95. R. Posner, M. Santa, G. Grundmeier, J. Electrochem. Soc. 158, 3 (2011) C29–C35
96. Thesis (PhD), K. Wapner, Grenzflächenchemische und elektrochemische Untersuchungen zur Haftung und Enthaftung an modifizierten Klebstoff/Metall-Grenzflächen, University Bochum, 2006
97. Kelvin, Phil. Mag. and J. Sci. 46 (1898) 82
98. W. Zisman, Rev. Sci. Instrum., 3 (1932) 367

99. K. Wapner, B. Schoenberger, M. Stratmann, G. Grundmeier, J. Electrochem. Soc. 152, 3 (2005) E114–E122
100. R. Posner, G. Giza, R. Vlasak, G. Grundmeier, Electrochim. Acta 54 (2009) 4837–4843
101. M. Rohwerder, Scanning Kelvin Probe and Scanning Kelvin Probe Force Microscopy and their Application in Corrosion Science. In U. K. Mudali, & B. Raj (Eds.), Corrosion Science and Technology: Mechanism, Mitigation and Monitoring (pp. 468–499). London, UK: Taylor & Francis, 2008
102. M. Stratmann, H. Streckel, Corros. Sci. 30 (1990) 681
103. M. Stratmann, Die Korrosion von Metalloberflächen unter dünnen Elektrolytfilmen, VDI-Verlag GmbH, Düsseldorf, 1994
104. M. Cappadonia, K. Doblhofer, M. Jauch, Ber. Bunsenges. Phys. Chem. 92 (1988) 903
105. K. Doblhofer, R. D. Armstrong, Electrochim. Acta 33 (1988) 453
106. K. Doblhofer, Bull. Electrochem. 8 (1992) 96
107. A. Leng, H. Streckel, M. Stratmann, Corros. Sci. 41 (1999) 547–578
108. W. Fürbeth, M. Stratmann, Corros. Sci. 43, 2001, 243–254
109. A. Leng, H. Streckel, M. Stratmann, Corros. Sci. 41 (1999) 579–597
110. W. Fürbeth, M. Stratmann, Fresenius J. Anal. Chem. 353 (1995) 337–341
111. W. Fürbeth, M. Stratmann, Corros. Sci. 43 (2001) 207–227
112. M. Stratmann, R. Feser, Farbe&Lack 100 (1994) 93–99
113. P. P. Leblanc, G. S. Frankel, J. Electrochem. Soc. 151, 3 (2004) B105–B113
114. M. Dornbusch, Prog. Org. Coat. 61 (2008) 240–244
115. M. Dornbusch, The use of modern electrochemical methods in the development of corrosion protective coatings, Coatings Science International Conference 2007, Noordwijk, Netherlands, June 2007

3
Spectroscopic Methods

3.1 Spectroscopic Methods on the Metal Surface

3.1.1 X-Ray Photoelectron Spectroscopy

Basics

Only a brief introduction in the principles of the method and the necessary equipment is presented in the following paragraph. Detailed presentations of X-ray Photoelectron Spectroscopy (XPS) are given in ref. [1–3]. In literature the abbreviation ESCA (Electron Spectroscopy for Chemical Analysis) is often used instead of XPS [4].

XPS is based on the external photoelectric effect [5], where illumination of a sample with photons of defined energy larger than the ionization energy causes electrons to be emitted from the sample [2]. The absorption of x-ray photons with the energy E = hν causes the ionization of core level electrons with the binding energy E_B from the surface, whereas the resulting kinetic energy E_{kin} of the electron could be calculated by Eq. 3.1 [1]:

$$E_{kin} = h\nu - E_B - \Phi_{Sp} \wedge \Phi_{Sp}\text{: Spectrometer work function} \qquad \text{Eq. 3.1}$$

The spectrometer work function could be defined by calibration and therefore the binding energy could be investigated by measuring the kinetic energy of the emitted electrons and the binding energy is a characteristic value for each element. Because of the relative low energy E_{kin} of the electrons, only electrons close to the surface can escape from the sample. The inelastic free path d_{in} of electrons in the range of 100–2000 eV could be estimated by the following equation [1]:

$$d_{in} \approx 1.1\frac{E_{kin}^{0.5}}{\rho} \wedge \rho\text{: Density of the surface material} \qquad \text{Eq. 3.2}$$

The inelastic free path is in the range of 1–5 nm. To measure the kinetic energy accurately the experiment has been done in ultra-high vacuum with a pressure of $< 10^{-8}$ mbar [1].

Another effect caused the emission of additional electrons called Auger electrons [4]. When one electron from the core level E_1 is emitted by irradiation with x-rays, an electron from a higher level E_2 will fall in this deep free level. The difference of the energy levels causes x-ray fluorescence or as a competing process causes the emission of a further electron from a higher energy level E_3. The resulting kinetic energy of the so called Auger electron could be calculated by Eq. 3.3 [3]:

$$E_{1,2,3} = E_3 - E_2 - E_1 \quad \text{Eq. 3.3}$$

If the emitting electron was from the energy level L and the second from M and the third also from M, the energy of the Auger electron will be denoted as E_{LMM}. The energy of an Auger electron is also characteristic for every element, and often more sensitive to the chemical environment and therefore provides very useful additional information in the XPS spectrum. Both effects generating electrons form the surface are summarized in Fig. 3.1.

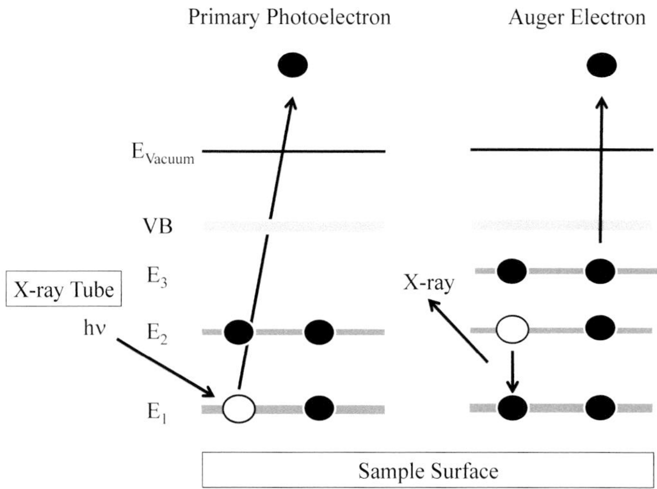

Fig. 3.1 Photoelectron and Auger electron emission.

The labelling of the primary photoelectrons is similar to the energy levels from where they emitted. Besides the s-levels all levels are doublets because of the spin-orbit splitting. The possible energetic levels and the resulting intensity rates are summarized in Table 3.1, whereas the total

Table 3.1 Parameters for the spin-orbit splitting.

Level	l	s	j	g	Intensity ratio
s	0	1/2	1/2	2	–
p	1	1/2	1/2	2	1:2
			3/2	4	
d	2	1/2	3/2	4	2:3
			5/2	6	
f	3	1/2	5/2	6	3:4
			7/2	8	

angular momentum j and the degree of degeneracy g are calculated with the following equations:

$j = l \pm s > 0 \wedge$ l: azimuthal quantum number s: spin quantum number Eq. 3.4

$$g = 2j + 1 \qquad \text{Eq. 3.5}$$

Therefore the integrated intensity of the $2p_{1/2}$ signal is half of that of $2p_{3/2}$. For the generation of satellite peaks and the distance of Auger signals see ref. [3].

Chemical Shift

The charge of an atom and the polarity of the chemical bonds to the environment cause slight changes of the energy of the emitted electron. This chemical shift is a powerful tool to analyze the surface. A positive charge increases the energy and a negative decreases it for some eV [3]. In Table 3.2 some data for oxygen in different environments are summarized. The challenge of the use of literature data is visible in the table and a detailed discussion of this challenge is given in Ex. 3.2. In Table 5.5 in the app. the XPS data of relevant substrates, conversion layers and organic compounds are summarized. The higher the electronegativity of the atoms in the environment the higher is the energy of the emitted electron. This

Table 3.2 Chemical shifts of the O1s signal in different chemical environments.

Compound	Energy [eV]
Water (adsorbed)	533.1–533.4
Oxide	530.0–530.7
Hydroxide	531.4–531.9

principle is useful for the interpretation of spectra especially from organic compounds because the chemical shifts vary from graphite with 284.5 eV to PTFE with 295.3 eV (see Table 5.5 in the app.).

Quantitative XPS

For the quantitative evaluation of XPS signals the appropriate background subtraction is necessary (see Fig. 3.2) and is meanwhile done by software, whereas the Shirley method [6] is most widely used [3]. Furthermore the band shape of an XPS signal behaves like a Gauss-Lorenz curve and could also be fitted by software. Therefore the quantitative separation of overlain signals by fitting procedures is possible, too. The resolution around 0.1 atomic% is relatively high in comparison with other methods like EDX (see Chap. 3.1.2).

In Fig. 3.2 the survey spectra of an alkaline cleaned electrolytic applied zinc layer on steel is shown (see Chap. 1.3.2). The spectrum shows the electron energy E_B (Eq. 3.1) in eV against the counts of the detector c/s. The expected zinc and oxygen signals are visible but in addition to a high amount of carbon. If a metal surface is handled on normal atmosphere, it will be contaminated by organic compounds from the air. On the one hand it is a disadvantage, if a thin organic layer has to be analyzed and

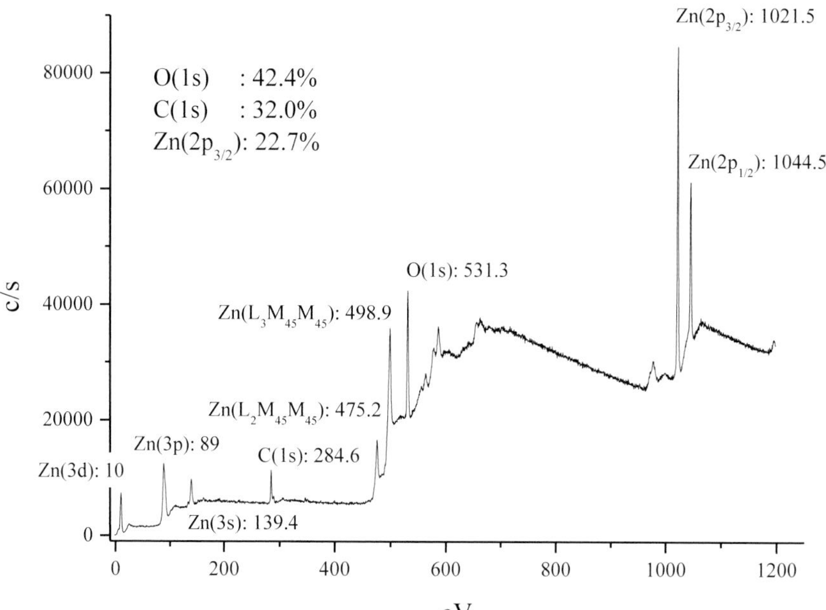

Fig. 3.2 XPS survey spectrum of alkaline cleaned electrolytic applied zinc on steel.

distinguished from the contamination, whereas it is often possible because of the different chemical shifts. On the other hand the always present contamination with a very constant energy but varying concentration could be used for calibration of the spectra.

Depth Profiling

To remove the contamination a cleaning process within the high vacuum, i.e., the removal of the first atom layers could be useful. Furthermore a continuous removal of material to investigate the thickness and possible chemical gradients in the layer is also interesting. This could be realized by a sputter profile, i.e., a beam of noble gas ions (often argon) continuously removes atoms from the surface with energies between 0.1 keV and several keV [3]. If a homogenous ion beam could be generated, a homogenous sputtering is possible. The XPS spectra measured after fixed intervals of sputtering time and also called sputter depth profile results. The thickness of the layer correlates with the sputter time. The challenge of the method is to define the absolute thickness of the layer. A calibration of the sputter rate could be done with common standards like Ta_2O_5 [3] or SiO_2 layers with defined thicknesses. The drawback of the calibration is that a similar sputter rate of the standard and the sample has to be assumed. A second independent method has to be used to prove the thickness illustrated in Ex. 3.9 (UV-VIS spectroscopy). Another disadvantage is the preferential sputtering of light elements like oxygen in oxide or hydroxide layers and therefore the possibility of chemical changes of the layer causing a misinterpretation of the spectra. All in all the depth profiling is a useful tool to get information about the chemical composition of a layer and rough information about the thickness (see Ex. 3.1).

Equipment

The principle setup for a XPS device is illustrated in Fig. 3.3. The photon energy of the x-ray anode has to be higher than the material work function to emit photoelectrons and the radiation has to be monochromatic. Besides the characteristic radiation of the anode (coated with Al, Mg, Zr, etc.) the bremsstrahlung is also emitted and suppressed by an aluminium window. The monochromator crystal reduces the full width at half maximum to 0.3 eV for a monochromatic aluminium anode [2]. The kinetic energy of the emitted electrons will be controlled by the lens to a fixed value, the pass energy, and only this type of electrons could pass the hemispherical analyzer with an inner negative and outer positive applied potential through the detector. This procedure increases the resolution up to 1meV [2]. The detector consists of a channeltron array. More details for the XPS device are illustrated in ref. [3].

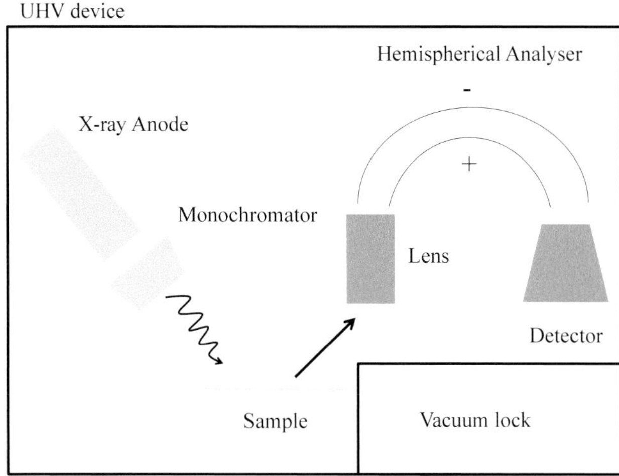

Fig. 3.3 Principle setup of a XPS device.

Example 3.1

Zinc coated steel is often combined with steel and aluminium especially in the automotive industry. An alkaline cleaning in a pH range of 12 removes oil from the metal forming process and prepares the surface for the pre-treatment process. Therefore car bodies and other mixed metal parts cleaned with alkaline solutions despite the dissolution of the zinc surface (see the Pourbaix diagram of zinc in Fig. 1.2). The chemical composition of the zinc surface after the alkaline cleaning could be analyzed with XPS [7].

The spectra in Fig. 3.4 have been referenced internally to the C(1s) signal of the always present carbon contamination. ZnO is available from the market (Merck) but $Zn(OH)_2$ has to be prepared according to ref. [8]. The spectra of the zinc surface in comparison with the data from the reference powder show that the native zinc surface consist of a mixture of ZnO and $Zn(OH)_2$ and the hydroxide should be enriched at the surface. After the alkaline cleaning process a pure $Zn(OH)_2$ layer results. The sputter-depth-profile of the surface shows that the thickness of the passive layer is also changed after the alkaline cleaning (Fig. 3.5). The sputter-depth-profile has been referenced to a standard SiO_2 sample with the assumption that the sputter rate of SiO_2 is the same as for zinc compounds. This is in principle wrong but gives at least an impression of the absolute thickness of the layer. The investigation of the thickness has been defined at the distance after 50% of the layer has been removed from the surface. The profile shows that the thickness of the passive layer increases from 3 to 9 nm. These changes of the chemical composition and

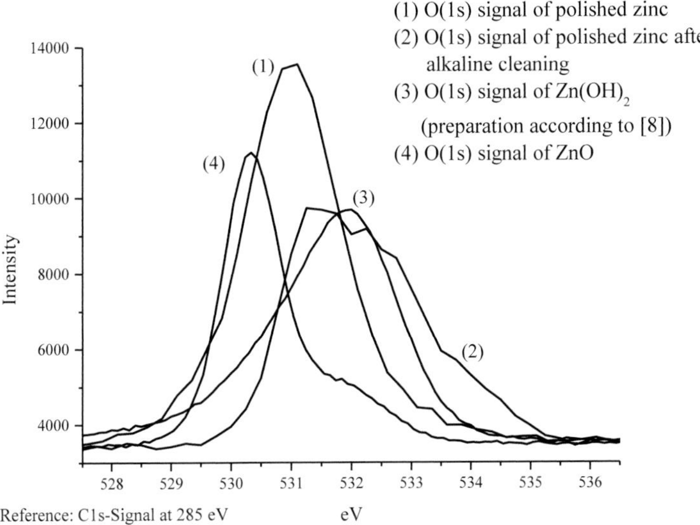

Fig. 3.4 XPS spectra of the O(1s) signal of zinc surfaces and reference materials according to [7].

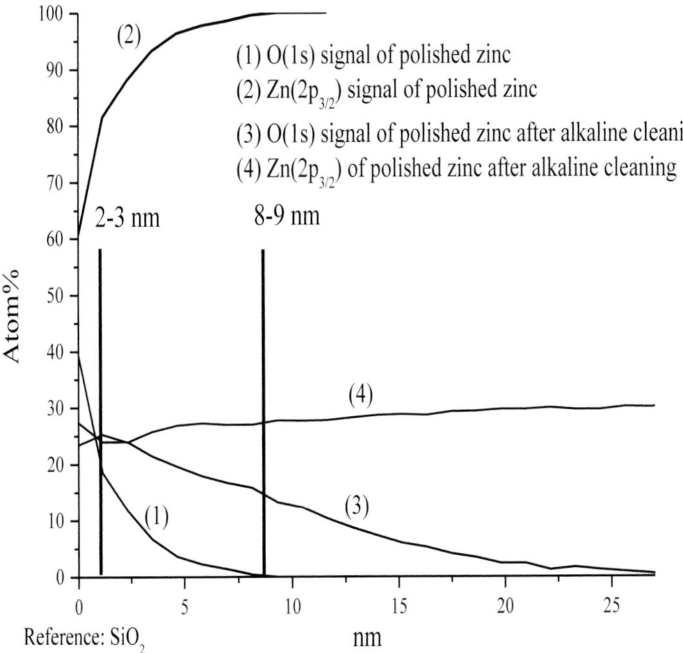

Fig. 3.5 XPS sputter-depth-profile of zinc surfaces before and after alkaline cleaning according to [7].

the thickness of the passive layer are important because the pre-treatment deposition rate will be influenced by the surface structure. The dissolution rate of the substrate based on acid corrosion on the conversion bath will be changed, if an oxide, a hydroxide or the metal is dissolute, and this causes a difference thickness and density of the resulting conversion layer. Therefore the final corrosion protective property of the coating system is influenced by the cleaning process. A detailed description of the cleaning process on the precipitation of a zirconium based conversion is illustrated in ref. [9]. The investigation of the chemical composition of the zinc surface could be also done by infrared spectroscopy (see Ex. 3.11).

Example 3.2

The structure of modern conversion layers based on the precipitation of zirconium compounds (see Reac. 1.17) is still under discussion. In Reac. 3.1 the most important reactions of the discussion are summarized.

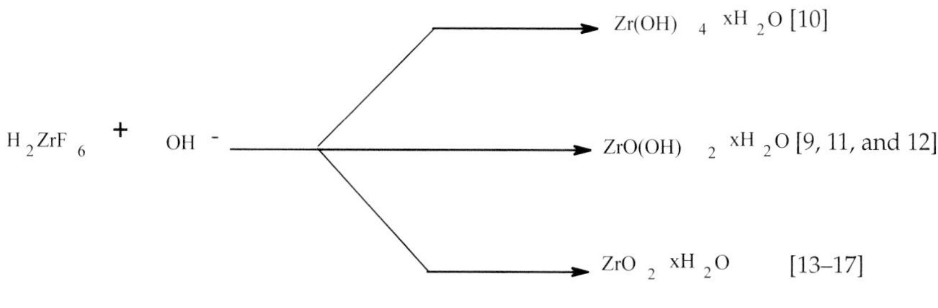

Reac. 3.1

The discussion includes statements from Cotton and Wilkinson [13] that $Zr(OH)_4$ cannot exist to clear spectroscopic data from $Zr(OH)_4$ [10]. The situation is additionally complicated by decomposition reactions at different temperatures summarized in the following reactions:

$$Zr(OH)_4 \cdot H_2O \xrightarrow{80^\circ C, 24h} "Zr(OH)_4" \text{ [14]} \qquad \text{Reac. 3.2}$$

$$"Zr(OH)_4" \xrightarrow{82^\circ C} ZrO_2 \cdot H_2O \text{ [14], } 85^\circ C \text{ [12]} \qquad \text{Reac. 3.3}$$

$$ZrO_2 \cdot H_2O \xrightarrow{158^\circ C} ZrO_2 \text{ [14]} \qquad \text{Reac. 3.4}$$

There is no doubt in literature that the decomposition reactions occur but the structure of the compounds, i.e., hydroxide or water containing oxide, is unclear. To solve the question XPS data have been collected from

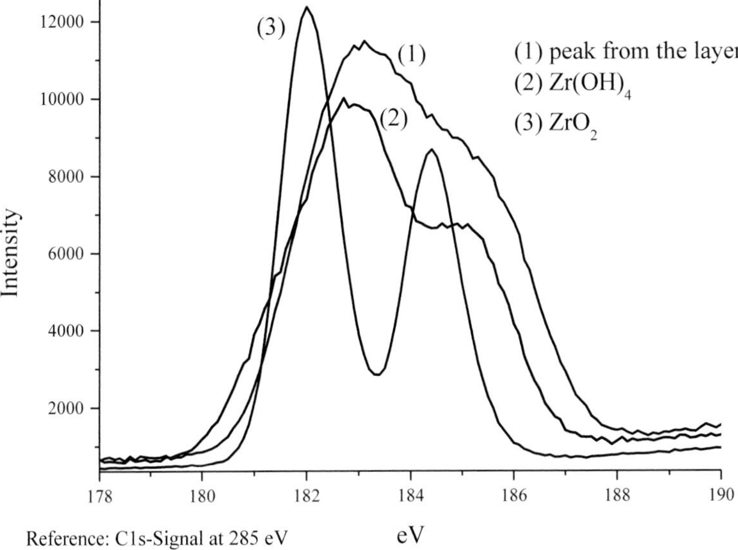

Fig. 3.6 XPS Zr3d-signals of a zirconium based conversion layer on zinc in comparison with standards (Aldrich).

zirconium based conversion layers on zinc or zinc coated steel, whereas a bath based on H_2ZrF_6 at a pH = 2.8–3.0 with a wetting agent has been used. In Fig. 3.6 the XPS spectra of conversion coatings on zinc in comparison to standard compounds are illustrated. The zinc surface dipped into the bath at 35°C for 3 minutes and dried at 80°C for 10 minutes. Disregarding the question what is $Zr(OH)_4$ (Aldrich) the data show that the conversion layer is not a oxide layer.

Furthermore the O1s-signals show (Fig. 3.7) that even after curing at 200°C the conversion layer is still more a hydroxide layer than an oxide layer. The move of the O1s-signal after drying at 200°C in comparison to the conversion layer dried at 80°C could be caused by the zinc hydroxide that also precipitated, because the $Zr(OH)_4$ signal fits exactly with the conversion layer signal dried at 200°C (Fig. 3.8). If a certain amount of the surface is removed by a sputter process, the composition of the zirconium compound does not change as shown in Fig. 3.8. Therefore the conversion layer consists of a hydroxide like zirconium compound that cannot be converted into ZrO_2 at 200°C.

The structure of "$Zr(OH)_4$" is still an open question and could be clarified with a second and third independent method illustrated in Exs. 3.12 and 3.9 with infrared and UV-VIS reflection spectroscopy, respectively.

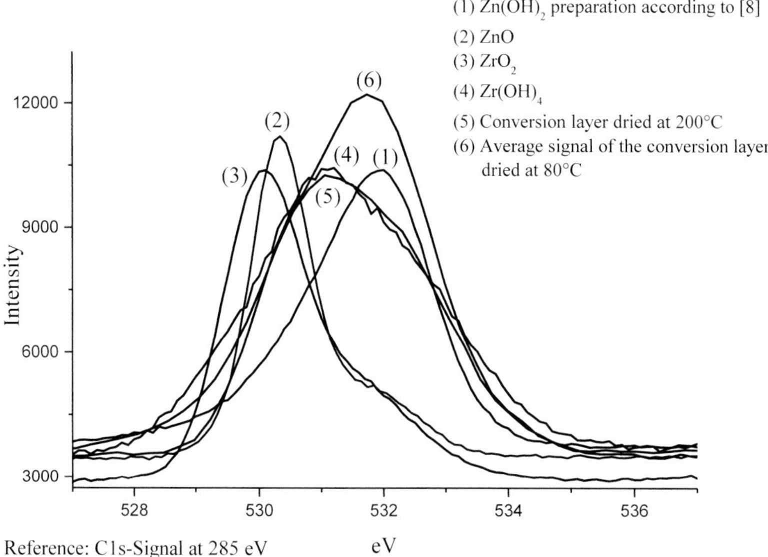

Fig. 3.7 XPS O1s-signals of a zirconium based conversion layer on zinc with drying at 200°C for 10 minutes in comparison with standards.

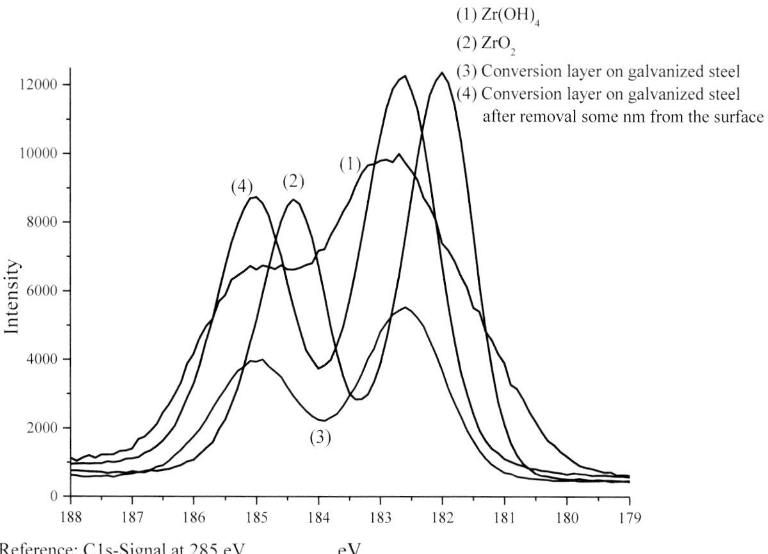

Fig. 3.8 XPS Zr3d-signals of a zirconium based conversion layer on galvanized steel with and without sputtering the surface.

The chemical composition is relevant because it defines the adhesion of the organic coating, the band gap energy (see Chap. 3.1.3) of the conversion layer, i.e., the barrier properties, and the solubility of the layer under corrosive conditions, i.e., acidic conditions near the local anode or alkaline conditions near the cathodic process.

Example 3.3

The conversion layer from Ex. 3.2 could be applied stepwise and the resulting thickness of the layer could be investigated by means of a sputter depth profile shown in Fig. 3.9. With the resulting time dependent layer thickness some kinetic data of the film formation are available. Because of the fact that for each step a new sheet has to be used with a certain roughness, i.e., certain reactivity, the noise of the depth profile data is in the range of the expected effect (increase of the conversion layer).

Therefore this data have been proved with a second independent method whereas a short drying is enough to perform the measurement and the stepwise application could be done on one sample realized with the UV-VIS reflection spectroscopy (Ex. 3.10) and infrared (IRRAS) spectroscopy (Ex. 3.12). This example shows the limitation of the method, because the high vacuum and the time consuming sample handling and

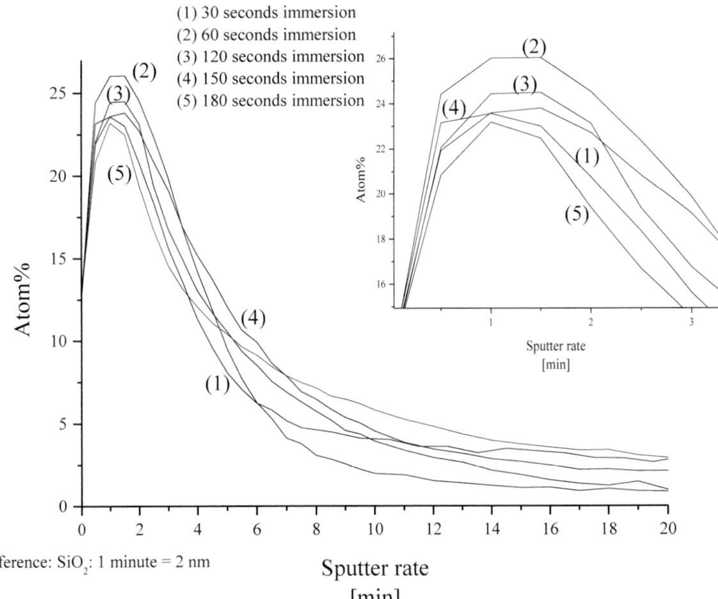

Fig. 3.9 XPS Zr3d-signal based sputter-depth-profiles of a zirconium based conversion layer on galvanized steel after different immersion times.

measuring time is a disadvantage in the analysis of water based coatings. On the other hand the thickness and the chemical composition of the conversion layer could be investigated and therefore the XPS could be used as reference for the UV-VIS and IR-spectroscopy and vice versa.

Example 3.4

The other limitation of the method is that small changes of the chemical composition on a rough technical surface that are hard to investigate. In the following example a scratched sample of an E coat (see Chap. 1.4.4) on zinc coated steel has been treated in a climate test according to VDA 621415 (see Table 4.6) and analyzed with XPS. In Fig. 3.10 the scratch of the sample after the climate test and after cleaning is shown with the analyzed areas. The E coat from the delaminated area has been removed from the sample to analyze the surface.

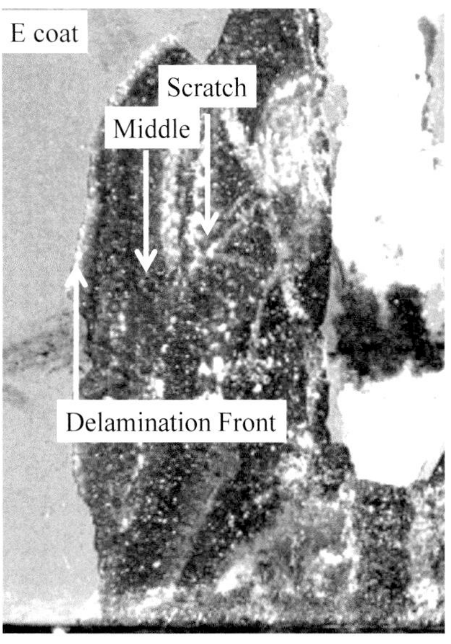

Fig. 3.10 Microscopic picture of the scratch on an E coated zinc surface after the alternating climate test.

An often detected corrosion product on zinc is simonkolleite ($Zn_5(OH)_8Cl_2$) and the corresponding element distribution measured with an XPS survey spectra:

Simonkolleite : Cl:Zn:O → 1:2.5:4

Delamination front : Cl:Zn:O → 1:2.5:3.5

Scratch : Cl:Zn:O → 1:1.8:2.9

fits passably with the theoretical distribution (Fig. 3.11). Small amounts of other corrosion products like oxides, hydroxides or, especially on zinc, carbonates are hard to detect. In high resolution element spectra of the O1s or $Zn2p_{3/2}$ signals a chemical and quantitative analysis of all compounds on the surface is possible but as already shown in Ex. 3.2 is not an easy process. A definitely investigation of the corrosion products is possible with the combination of the XPS with a second independent method like the Raman spectroscopy (see Ex. 3.15).

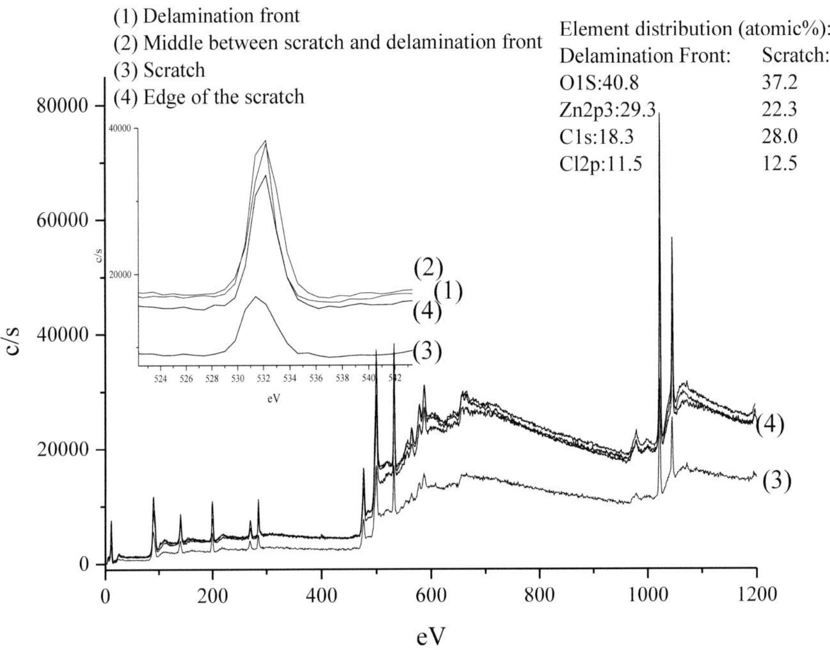

Fig. 3.11 XPS survey spectra of different areas on a scratched sample (E coat on zinc coated steel) after VDA 621415 climate test.

3.1.2 *EDX Spectroscopy in Combination with SEM*

Basics

Only a brief introduction in the principles of the method and the necessary equipment is presented in the following paragraph. Detailed presentations of Electron Dispersive X-ray spectroscopy (EDX) and the Scanning Electron Microscopy (SEM) are given in ref. [18–25]. From the corrosion analysis point of view the Transmission Electron Microscopy (TEM) is normally not possible because the thickness of the substrate and the coatings prohibit this method. Therefore the following chapter focuses on SEM.

The scanning electron microscope is a powerful tool to analyze metal and coated metal surfaces. A focussed electron beam in high vacuum scans the surface by means of deflection plates (Fig. 3.12) in both directions. The electron beam is generated by a hot cathode (Fig. 3.12 and compare Fig. 3.3) with an accelerating voltage E_0 from 1–50 kV and is focussed with magnetic lenses. The electrons generate backscattered (BSE) and Secondary Electrons (SE) in the sample used as an information source for the picture of the surface with a resolution below 10 nm especially with the use of the SE. The electrons emitted from the sample surface are subdivided regarding their energy in BSE (> 50 eV) and SE (< 50 eV) whereas the BSE generated from elastic and the SE from inelastic interactions with the atoms in the sample [19]. Because of the low energy the SE could leave the sample only from the first 0.5–2 nm (Fig. 3.13) and are very sensitive for the topography of the sample.

Furthermore the interaction of the electron beam with the surface produces element characteristic x-rays (compare Fig. 3.1) by emitting SE with an information depth of approximately 5 µm. The detection of the x-rays (EDX) allows the combination of microscopic information and chemical information from the surface illustrated in spectra or in mappings, i.e., colouring the topographic picture with the relative amount of a certain element [15, 19].

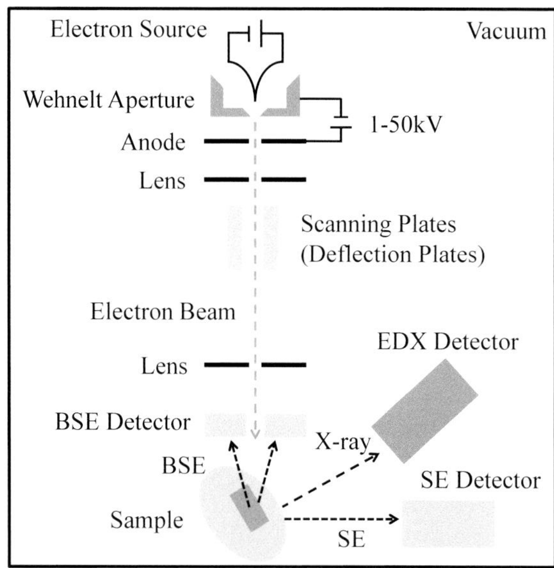

Fig. 3.12 Principle setup of a SEM/EDX device.

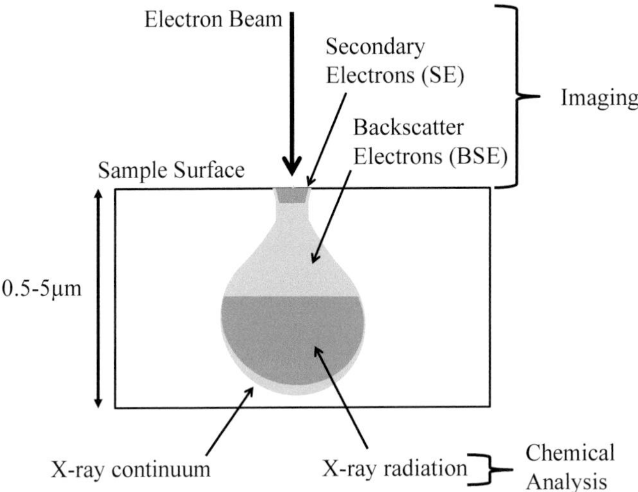

Fig. 3.13 Scheme of the emission volume for detectable signals in SEM/EDX according to [26].

The spatial resolution r of the EDX spectra depends on the density ρ, E_0 [keV] and the critical excitation voltage E_C and could be calculated according to the following equation [27]:

$$r[\mu m] = 0.231 \frac{\left(E_0^{2/3} - E_C^{3/2}\right)}{\rho} \quad \text{Eq. 3.6}$$

For example on a zinc coated steel substrate with a small amount of aluminium in the alloy the density of the surface could be estimated with the density of zinc (7.14 g/cm^3) and the spatial resolution for zinc and aluminium is in the range of 60 nm and 40 nm, respectively [28]. The quantitative analysis of the element distribution by means of EDX is possible but with a lower resolution as XPS with values about 1–2%. Another disadvantage is the fact that only the amount of an element could be investigated but no information about the chemical environment is available (chemical shift in the XPS). The lower resolution could be increased by means of the Wavelength Dispersive X-ray spectroscopy (WDX) with a 10 times higher resolution as the EDX but with much longer testing times.

Microscopic methods are important for corrosion research because of three reasons. First the picture of a surface allows investigating the density and the distribution of coatings especially of conversion layers (see Ex. 3.7). The second corrosion phenomenon is local effects (compare Figs. 1.3, 1.9

and 1.10), i.e., the anode and cathode area after corrosive treatment have a small dimension in the beginning of the corrosion process, and therefore a local analysis is necessary (see Ex. 3.6). Thirdly the metal substrate is not a homogenous surface regarding the reactivity. For instance the reactivity at the grain boundaries is higher than on the grain surface (see Ex. 3.5).

Example 3.5

In the following example the untreated surface of steel and zinc coated steel are illustrated. Visible on the steel surface is the inhomogeneous structure and the grain boundaries (Fig. 3.14). Hot Dip Galvanized steel (HDG steel) produced in a coil coating process is in principle a very smooth surface. A metal forming process needs a certain amount of oil on the surface to reduce the friction during the forming process. On a smooth surface oil is hard to deposit and therefore after the zinc coating a skin-pass process is followed generating a structure of the surface (Fig. 3.15). In the produced cavities the oil could be deposited to optimize the forming process. In principle industrially used metal substrates need a certain roughness because of metal forming process.

Edges, scratches and grain boundaries are more reactive than a smooth surface, i.e., these are the areas where corrosion processes begin. The initiation of a corrosion process on an organic coated surface from a defect starts at the grain boundaries and the fast propagation under the coating occurs along the grain boundaries followed by a spread into the grain surfaces illustrated in Fig. 3.16. This could be shown by means of local SKP experiments in combination with SEM/EDX [28–31].

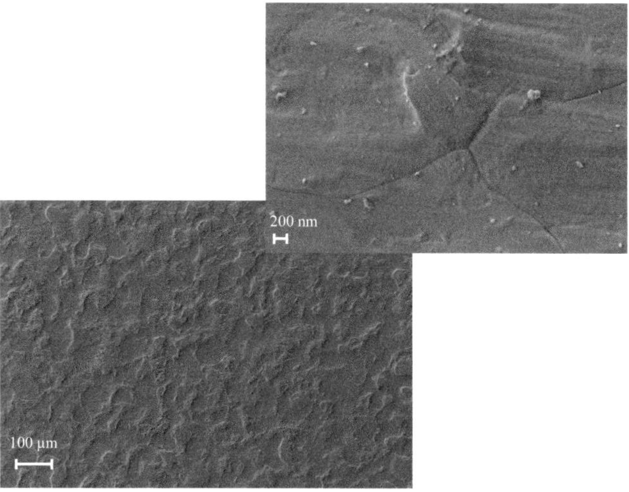

Fig. 3.14 SEM (SE) pictures of a common steel surface.

Fig. 3.15 SEM (SE) pictures of a common hot dip galvanized steel surfaces.

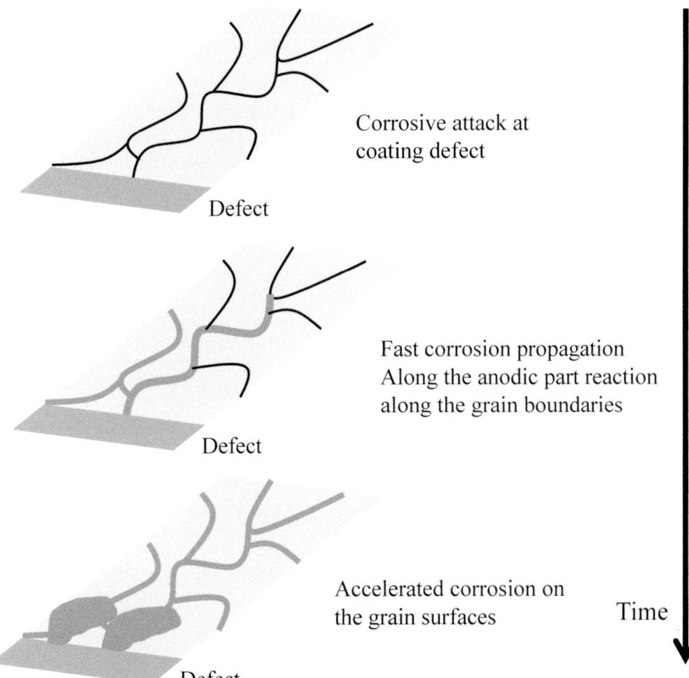

Fig. 3.16 Corrosion process on a metal surface with grain boundaries according to [28].

Furthermore at these areas the reaction for producing a conversion layer also begins. This could be used in principle to protect the whole surface by protecting only the most reactive areas of the surface [28–31].

Example 3.6

Corrosion is a local phenomenon illustrated on steel and phosphated steel in the Figs. 3.17 and 3.18, respectively. Because of the fact that the EDX measurement achieves information up to 5 µm the chemical composition of the upper surface is hard to analyze but possible with other spectroscopic methods like infrared or Raman spectroscopy (see Exs. 3.14 and 3.15). These methods have the disadvantage that they investigate only the chemical composition of the complete surface (microscopic methods for infrared and Raman spectroscopy are possible with a resolution in the µm scale) and therefore the combination of the microscopic and spectroscopic methods give the complete picture of the corrosion process. The surface sensitivity of EDX could be increased by decreasing the accelerated voltage but reduces the amount of detectable elements, too.

It is noteworthy that on steel the precipitation of corrosion products is radial around a local anode (Fig. 3.17) in contrast to the local enrichment of corrosion products on the phosphate surface on non covered areas (see Fig. 3.18).

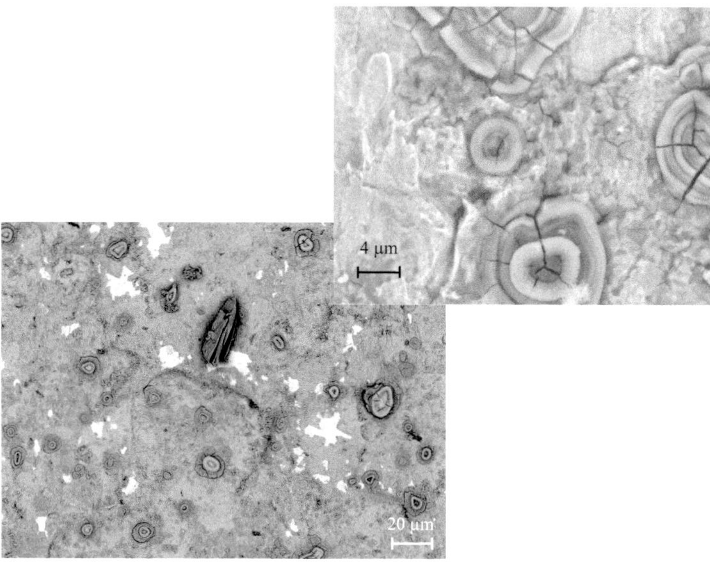

Fig. 3.17 SEM pictures of a common steel surface after two days immersion in 0.5 M NaCl solution.

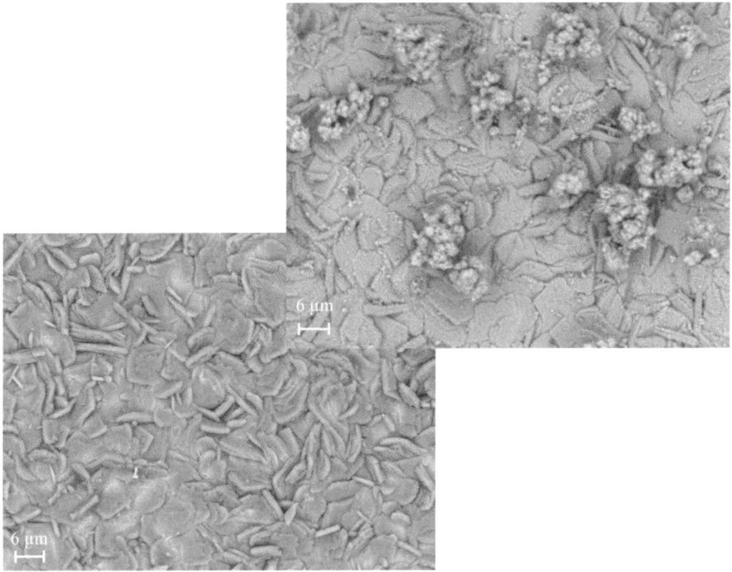

Fig. 3.18 SEM pictures of a tri-cation-phosphated (Hopeite) steel surface (below) and after two days immersion in 0.5 M NaCl solution (above).

Example 3.7

As already mentioned in Ex. 2.5 conversion layers based on phytic acid (Fig. 2.16) could be generated by immersing the metal surface into the acidic phytic acid solution. To optimize the precipitation of the conversion layer different pH values have been used and a surfactant and a corrosion inhibitor (molybdate) have been added to the solution [32].

The SEM pictures in Fig. 3.19 show a precipitation of a conversion layer with a lot of cracks and the amount of cracks reduced with decreasing pH.

The reducing of cracks is important because no phosphorus, i.e., phytic acid, in the cracks could be detected by EDX shown in Fig. 3.20. The amount of phosphorus on the surface increases with decreasing pH (Table 3.3) and therefore pH = 2 has been chosen for further optimization. To avoid the cracks a levelling agent has been added to the solution to achieve a smoother layer because a better contact of the immersion bath to the surface should improve a homogeneous reaction on the surface. The SEM of the surface (Fig. 3.21) proved it but there are still cracks in the layer.

Finally the addition of ammonium molybdate reduces the cracks to an acceptable minimum (Fig. 3.22) with a homogeneous distribution of molybdenum and phosphorus in the layer (Fig. 3.23).

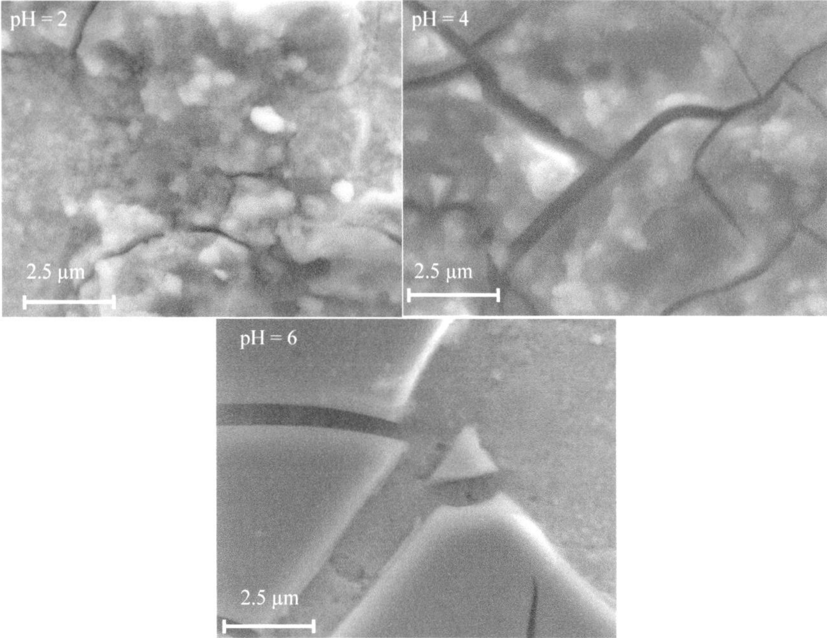

Fig. 3.19 SEM pictures of phytic acid based conversion layers precipitated with different pH values in the bath.

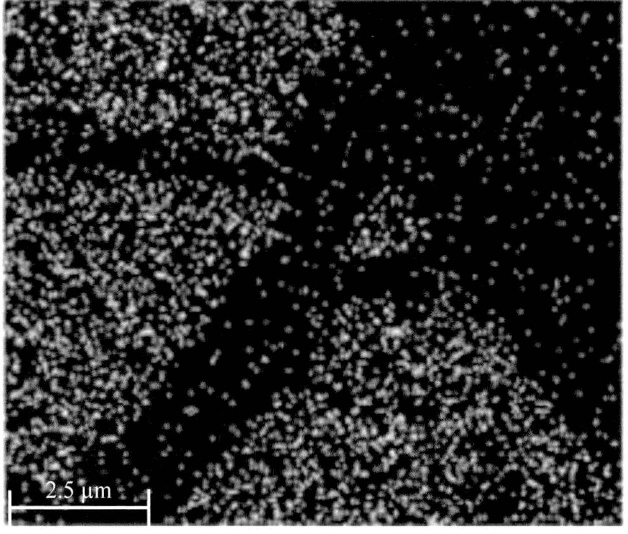

Fig. 3.20 SEM/EDX mapping of the PKα1 signal on the SEM picture of the conversion layer deposited at pH = 6 (Fig. 3.19) from [32]. With permission.

Table 3.3 Amount of phosphorus of the resulting conversion layers measured by means of EDX of different phytic acid solutions (for details of the immersion process see [32]).

Bath	Phosphorus [Mass%]	Molybdenum [Mass%]	Carbon [Mass%]
Phytic Acid, pH = 6	6.6 ± 0.3		12.1 ± 0.8
Phytic Acid, pH = 4	7.1 ± 0.2		10.0 ± 0.6
Phytic Acid, pH = 2	12.0 ± 0.3		17.3 ± 0.9
Phytic Acid + Levelling agent, pH = 2	5.7 ± 0.2		7.6 ± 0.5
Phytic Acid + Levelling agent + $(NH_4)_6Mo_7O_{24}$, pH = 2	8.4 ± 0.2	5.2 ± 0.4	9.3 ± 0.6

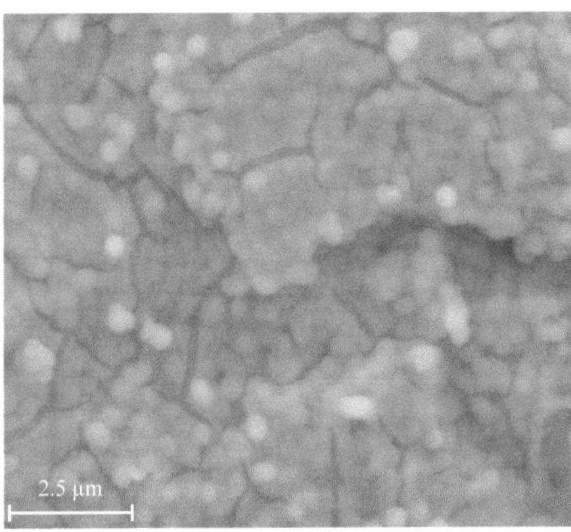

Fig. 3.21 SEM picture of the conversion layer precipitated with a surfactant in the immersion bath from [32]. With permission.

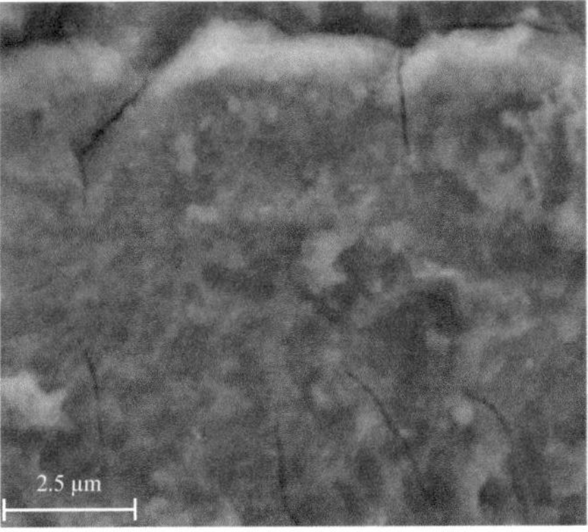

Fig. 3.22 SEM picture of the conversion layer precipitated with a surfactant and $(NH_4)_6Mo_7O_{24}$ in the immersion bath from [32]. With permission.

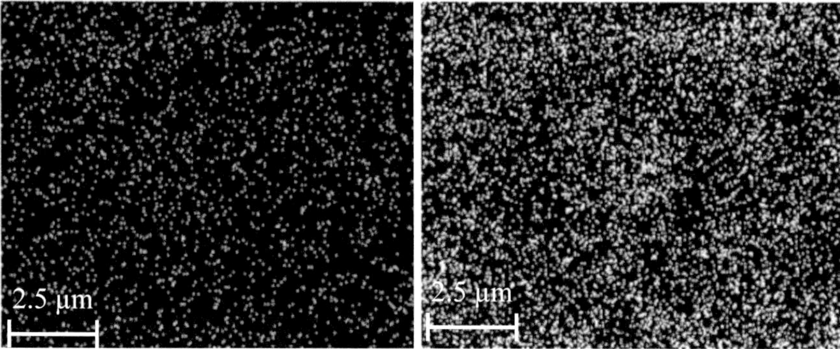

Fig. 3.23 SEM/EDX mapping of the PKα1 (right) and MoLα1 (left) signal of the picture of the conversion layer precipitated with a surfactant and $(NH_4)_6Mo_7O_{24}$ in the immersion bath (Fig. 3.22) from [32]. With permission.

A homogeneous layer without cracks is necessary for corrosion protective properties and evaluated with different organic coatings in the salt spray test (see Ex. 4.4). A dense layer increases the barrier properties, i.e., the insulating of the surface to avoid the galvanic coupling (see Fig. 1.11), and for a homogeneous adhesion of the organic coating a homogeneous conversion layer is also necessary.

3.1.3 UV-VIS Spectroscopy

Basics

Only a brief introduction in the principles of the method and the necessary equipment is presented in the following paragraph. Detailed presentations of UV-VIS spectroscopy in general are given in refs. [33–36]. The basics in reflection spectroscopy are presented in refs. [37, 38] with a detailed description of the Kubelka-Munk theory.

UV-VIS Spectroscopy in Transmission

Electromagnetic radiation is characterized by the wavelength λ and the frequency ν combined in the following equation:

$$\lambda \nu = c \wedge c = 2.99810^{10}\ cm\, s^{-1} \qquad \text{Eq. 3.7}$$

A photon with the frequency ν has the energy:

$$E = h\nu \ \wedge \ h = 6.6310^{-34}\ Js \qquad \text{Eq. 3.8}$$

The wavelength is not linear with the energy therefore the wavenumber $\bar{\nu}$ is often used to present spectral data in the UV-VIS range and is defined as:

$$\bar{\nu} = \frac{1}{\lambda} \quad \text{Eq. 3.9}$$

The energy of a photon in the ultra violet (UV, 10–400 nm) and the visible (VIS, 400–750 nm) range of the electromagnetic radiation causes the excitation of a valence electron of a molecule or an atom from the HOMO (Highest Occupied Molecular Orbital) to the LUMO (Lowest Unoccupied Molecular Orbital) and the excitation of the electron causes the absorption of light with the corresponding energy. The most important transitions are summarized in Fig. 3.24.

The polymers, additives and cross-linking agents in organic coatings show transitions from n, π, and σ Molecular Orbitals (MO) and are presented in detail in Chap. 3.2.2 (for details see refs. [35, 36]). In inorganic compounds, i.e., passive or conversion layers on the substrate or dissolute ions in the electrolyte, d-d-transitions of the ion or Ligand to Metal Charge Transfer (LMCT) and Metal to Ligand Charge Transfer (MLCT) transitions are relevant from the corrosion analysis point of view (for details see ref. [39]). A very useful transition for corrosion analysis is based on a charge

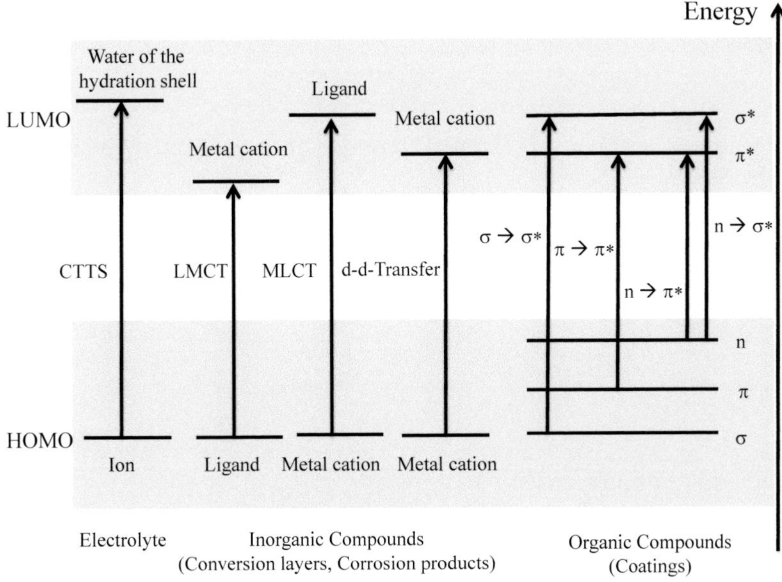

Fig. 3.24 Important electron transitions caused by UV-VIS excitation according to [35, 36, and 39].

transfer from a hydrated ion to the water molecules in the hydration shell called charge transfer to solvent band (CTTS) [40–46] and is presented in detail in Chap. 3.2.2.

If light is transmitted through a solution or a homogeneous sample, the intensity I_0 of the light is reduced by absorption of the sample I_{abs} to the transmission intensity I as follows:

$$I = I_0 - I_{abs} \qquad \text{Eq. 3.10}$$

Bouguer, Lambert and Beer developed with a different approach (Fig. 3.25) a correlation between the intensity change, the thickness d of the sample and the specific absorption α as follows:

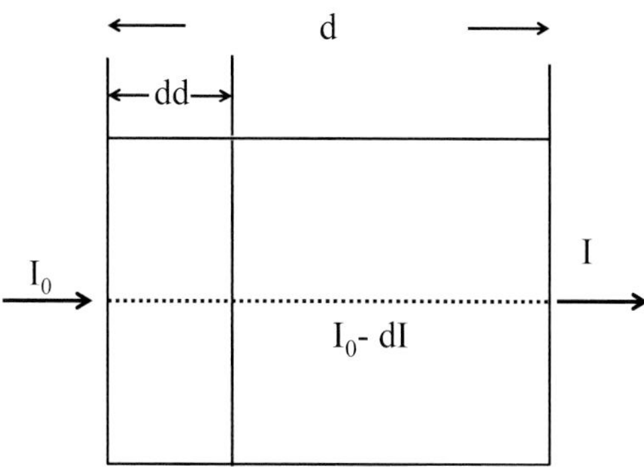

Fig. 3.25 Differential approach for the description of absorption processes of homogeneous samples.

As illustrated in Fig. 3.25 the differential change of the intensity could be described as:

$$dI = -\alpha\, I\, dd \;\Leftrightarrow\; \frac{dI}{I} = -\alpha\, dd \qquad \text{Eq. 3.11}$$

From this it follows:

$$\int_{I_0}^{I} \frac{dI}{I} = -\int_0^d \alpha\, dd \qquad \text{Eq. 3.12}$$

and the integration results:

$$I = I_0\, e^{-\alpha d} \qquad \text{Eq. 3.13}$$

In diluted solutions ($c < 10^{-2}$ mol/l) and with the assumption that only the compound with the concentration c absorbs, the specific absorption

α could be replaced by the concentration and the molar absorption coefficient ε as follows:

$$\alpha = 2.303\,\varepsilon c \qquad \text{Eq. 3.14}$$

Finally the Lambert-Beers law could be rearranged with the well-known equation:

$$\ln\frac{I_0}{I} = 2.303\,\varepsilon c d \Leftrightarrow A = \log\frac{I_0}{I} = \varepsilon c d \qquad \text{Eq. 3.15}$$

The absorbance A is correlated to the thickness and the concentration of the diluted compound. The absorption of a certain compound is described by the wavelength of maximum absorption λ_{max} and some data of organic and inorganic compounds are summarized in Table 5.2.3 in the app.

The absorption could be based on allowed or forbidden electronic transfers (an elaborately presentation is given in [35]) caused by the quantum mechanical selection rules (Laporte's rule etc.). Because of the fact that electronic states could interact with rotational and vibrational states (see Chap. 3.1.4) forbidden transition could be detected but with smaller ε values as the allowed transitions and therefore the molar extinction coefficient is a parameter to classify electronic transitions (see below).

The use of Lambert-Beers law is limited to dissolved compounds (Eq. 3.14) and the assumption that there is no interaction between the molecules or ions in the solution. If this is the fact, the absorption of a mixture of compounds is the sum of the absorptions. Furthermore Eq. 3.15 could only be used, if the sample shows no scattering because the model in Fig. 3.25 is only reasonable without it.

UV-VIS Spectroscopy in Reflectance

Corrosion is a phenomenon on solid state body and therefore the transitions cannot be caused by single MOs because the orbitals form bands, i.e., the overlapping of the MO from each atom in the crystal generate a band of certain energy (for details see refs. [4, 47]).

The so called band theory [47] allows describing two important properties of solids. First the absorption of UV-VIS light is caused by a transition of an electron occupied band to an unoccupied band similar to the MO in Fig. 3.24. Second the band theory allows describing the electrical conductivity of solids. The electrical conductivity is possible, if electrons excited to the conduction band CB (see Fig. 3.26), i.e., an unoccupied band above the Fermi energy E_F (highest energy of electrons), from an occupied band the so called Valence Band (VB). Three different situations could be

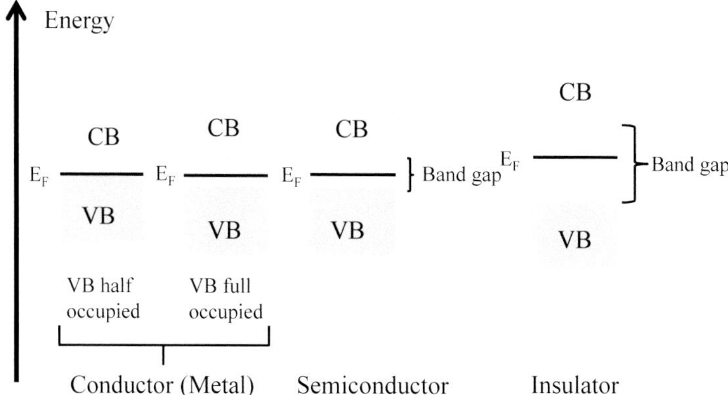

Fig. 3.26 Illustration of the conductivity based on the band theory.

the case. A conductor, e.g., a metal surface, shows an overlap of the CB with the VB or both are very close to each other. A semiconductor has a small band gap E_g between CB and the VB in the range of 0.7 to 1.5 eV [47] and finally an insulator has an E_g much higher than 2 eV (Fig. 3.26). Investigating the band gap of a surface could be achieved by means of measuring the UV-VIS absorption of the surface.

The band gap is an important value because the conductivity of the surface is necessary for corrosion processes because corrosive processes are not possible without a transfer of electrons through the surface. Insulating passive layers naturally the case on valve metals (compare Fig. 1.8) inhibits the corrosion process. The use of conversion or modified passive layers as corrosion protective layers is based on the insulating property of the resulting surface.

To investigate the E_g the UV-VIS spectra has to be converted according to the following equation named Tauc plot [3, 48–51]:

$$\alpha\, h\nu = A(h\nu - E_g)^n \qquad \text{Eq. 3.16}$$

α : Absorbance coefficient	h : $6.6310^{-34}\, Js$
ν : Frequency	A : Constant

The exponent n differentiates between different transitions as follows:

$n = \frac{1}{2}$ for allowed direct transitions

$n = \frac{3}{2}$ for forbidden direct transitions

$n = 3$ for forbidden indirect transitions

$n = 2$ for allowed indirect transitions, amorphous materials

A direct transition shows no energetic interaction of the photon with vibrations of the solid. In indirect transitions the photon interacts during the electron transition with phonons from the solid. In amorphous solids normally indirect transitions result and therefore n = 2 has to be used for the investigation of amorphous passive or conversion layers [3]. The situation is much more complex but cannot be explained here. A very elaborate illustration has been done by Di Quarto et al. [3]. The same author presented a very useful semi empirical concept for the calculation of band gaps based on the electronegativity (Pauling) of the atoms in the oxide or hydroxide layer presented in the following paragraph.

For amorphous oxide passive layers E_g could calculated as follows [52, 53]:

$$E_g^{cal} - \Delta E = A\left(\chi_C - \chi_O\right)^2 + B \wedge \chi_C = \frac{a\,\chi_A + b\,\chi_B}{a+b} \qquad \text{Eq. 3.17}$$

E_g^{cal} : Band gap (calculated)

ΔE : Difference between the band gap of the crystalline oxide and the amorphous layer in the range of 0.35 to 0.40 eV

A : Constant in the range of 1.41 (see refs. [3, 53])

χ_o : Electronegativity of oxygen

B : Constant in the range of 1.86 (see refs. [3, 53])

a, b : Stoichiometric factors of the cations in the oxide

χ_A, χ_B : Electronegativity of the cations in the oxide

For hydroxides the band gap depends of the metal cation in the layer. For sp-metals (see ref. [39]) the band gap could be calculated with the following equation [53, 54]:

$$E_g^{cal} = 1.21(\chi_M - \chi_{OH})^2 + 0.90 \qquad \text{Eq. 3.18}$$

χ_M : Electronegativity of the metal cation (sp-metal)

χ_{OH} : Electronegativity of the hydroxide anion = 2.85 (arithmetic mean of the values from oxygen and hydrogen)

For hydroxides from d-metals the following equation has to be use [53, 54]:

$$E_g^{cal} = 0.65(\chi_M - \chi_{OH})^2 + 1.38 \qquad \text{Eq. 3.19}$$

Finally with the assumption that there is a correlation between the band gap of oxides and the corresponding oxyhydroxides $MO_{(y-m)}OH_{(2m)}$ the following equation describes the correlation between the band gaps [53, 55]:

$$E_g^{hydrated} = \frac{E_g^{anhydrous}}{1 + k_m\, x_{OH}} \wedge x_{OH} = \frac{2m}{y+m} \qquad \text{Eq. 3.20}$$

The constant k_m has to be calculated for each system once.

The limitation of the equations is the need to define the parameters but it allows at least to prove measurements. Some E_g data from metal substrates (oxides, hydroxides, etc.) and conversion layer compounds are summarized in Table 5.9 in the app.

The measurement of the UV-VIS spectra of a rough surface cannot be analyzed with the Lambert-Beer's law because there is no transmission but reflection on the surface. Therefore the reflection of the light with the surface causes absorption and scattering (Fig. 3.27), i.e., diffuse reflection. This situation could be described with the Kubelka-Munk theory, if the following assumptions are fulfilled [37, 56]:

- Infinite thickness, i.e., the thickness of the sample has to be some millimetres
- Regular reflection has to be avoided → Diffuse reflectance spectroscopy
- The scattering of the sample has to be independent from the wavelength, i.e., particle size $d > \lambda$ (if $d < \lambda$ the scattering is wavelength dependent (Rayleigh scattering))

With these assumptions the measured reflectance spectrum could be transferred in an absorption spectrum [37]:

$$F(R_\infty) = \frac{(1 - R_\infty)^2}{2R_\infty} = \frac{K}{S} \wedge K = \varepsilon c \qquad \text{Eq. 3.21}$$

$F(R_\infty)$: Kubelka-Munk function

R_∞ : Reflectance, i.e., $R_\infty = \frac{R_{Sample}}{R_{Reference}}$

K : Absorption coefficient

S : Scattering coefficient

Unfortunately the assumptions for the Kubelka-Munk equation (Eq. 3.21) are not fulfilled by conversion or passive layers and on organic coatings because the layer thickness is too small (even the thickness of organic coatings normally do not exceed 500 μm) and particle sizes lower the wavelength (175–900 nm) are often also the case. The use of the Kubelka-Munk equation for organic coatings is illustrated in Ex. 3.20 and shows that it could be used although the assumptions of the Kubelka-Munk equation are not fulfilled. Nevertheless the diffuse reflectance spectroscopy could be used even for thin layers [57] and conversion layers [58] and the results (e.g., band gap according to Eq. 3.16) confirm the other spectroscopic methods. The diffuse reflectance spectroscopy is a useful tool to analyze pigments in organic coatings (compare [59]) and in some cases for resins in organic coatings. The experiment has to be done in an Ulbricht sphere (Fig. 3.27) to collect the diffuse reflectance completely. The disadvantage of the Ulbricht sphere is the fact that the material in the sphere absorbs itself (e.g., $BaSO_4$ < 300 nm and PTFE < 200 nm) at a certain wavelength and often above the absorption of the sample. The reference is in that case a white standard.

The alternative way is the regular reflectance with a small incident angle of around 5–7° and the interpretation of the spectrum with the Lambert-Beer-law (Fig. 3.28). The reference is in that case often an aluminium mirror. Of course the aluminium surface has an oxide layer absorbing at 190 nm but with a very low intensity because of the low thickness of 1–3 nm. The scattering of the sample has to be ignored in the experiment. Therefore for each type of sample the best compromise

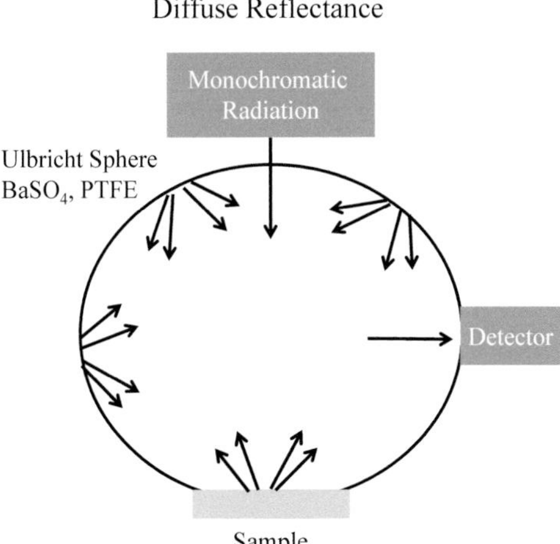

Fig. 3.27 Illustration of the experiment for diffuse reflectance spectroscopy in an Ulbricht sphere.

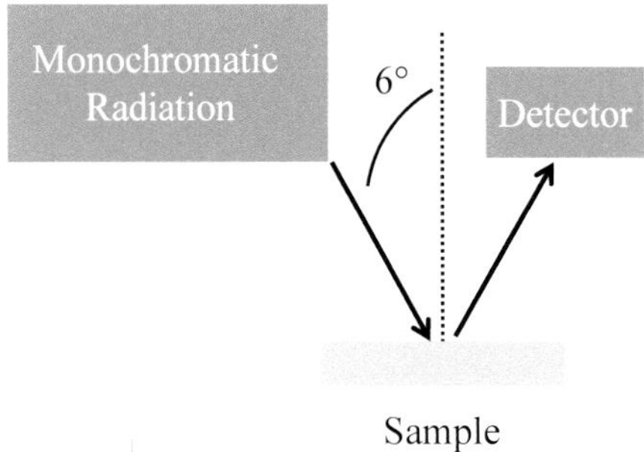

Fig. 3.28 Illustration of the experiment for regular reflectance spectroscopy on a reflection device with 6° incident angle.

has to be in finding a way to analyze the surface by means of UV-VIS spectroscopy.

The monochromatic radiation is generated with a grating monochromator and the light with wolfram lamps in the VIS-range and often with a Deuterium lamp for the UV-range therefore the lamps have to be switched on during the measurement. The emitted light is detected with electron multipliers. For solutions a two channel device is used with a simultaneous measurement of the reference (glass or quartz glass cuvette with solvent). In reflection the reference has to measured before the sample.

The reflectance spectroscopy has been used, e.g., to analyze corrosion products on copper and brass surfaces [60] whereas it is often used in combination with electrochemical methods because the UV-VIS-spectroscopy could be used in aqueous solution and on smooth electrodes the scattering could be ignored.

Example 3.8

The example shows the reflection spectrum of alkaline cleaned polished zinc (Fig. 3.29) with a $Zn(OH)_2$ layer (Fig. 3.5) with a thickness of 8–9 nm (Fig. 3.5). The spectrum was measured with a reflection device (7°) with an aluminium mirror as reference in a nitrogen flushed spectrometer because the absorption of oxygen in the atmosphere increases rapidly below 200 nm (see also Ex. 3.18) and prevents the measurement of an absorption with a λ_{max} = 205 nm. It is possible that underneath the broad absorption further absorption bands exist of ZnO (385–390 nm) or zinc oxy hydroxides (260 nm).

The investigation of the band gap with Eq. 3.16 and n = 2 is illustrated in Fig. 3.30. The calculated band gap of 1.82 eV is lower than literature

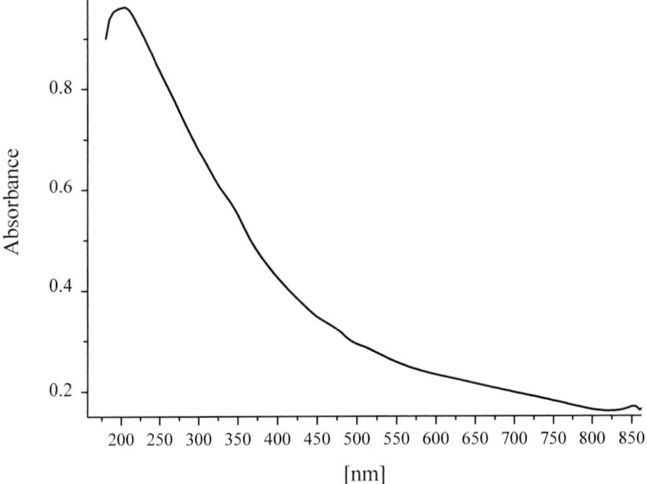

Fig. 3.29 UV-VIS reflection (6°) spectrum of polished alkaline cleaned zinc in nitrogen atmosphere. The reference was an aluminium mirror.

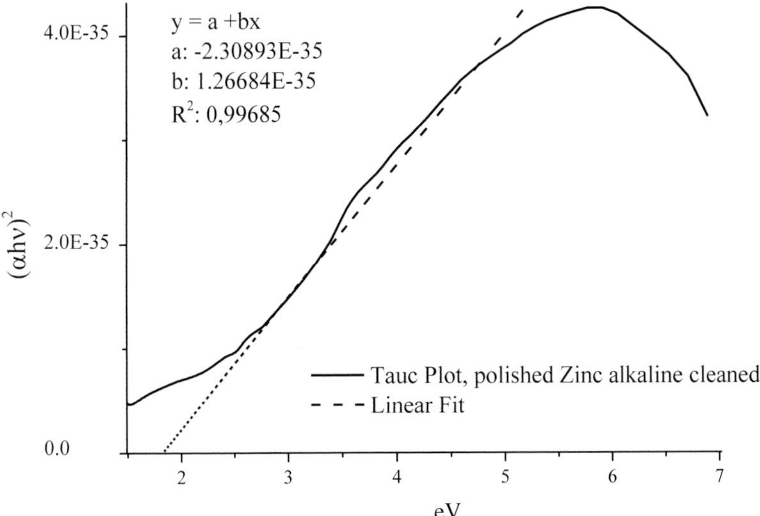

Fig. 3.30 Tauc plot of polished alkaline cleaned zinc based on the spectra of Fig. 3.29.

data but this is expected because on amorphous solids the band gap is often lower [3] and furthermore there is more than one way to make a linear fit on that curve in Fig. 3.30. Finally the surface is not insulated and corrosion processes could occur.

Example 3.9

The same measurement as in Ex. 3.8 has been conducted on a zirconium based conversion layer on a polished and alkaline cleaned zinc surface (Fig. 3.31). Besides the main absorption at 182–184 nm there is another absorption in the range of 250 nm correlating with the absorption edge of ZrO_2 at 230 nm or the absorption of zinc oxy hydroxides in the range of 260 nm (see Table 5.2.3 in the app.) with changing intensity in the two samples. A clearer picture is often available, if the spectra is drawn with wavenumbers because only in this graph energy correlates linear with the spectral data (Fig. 3.31). The spectra show that the conversion layer contains a mixture of zirconium oxy hydroxides as already mentioned [11, 61] and no zirconium oxide. On the other hand the band gap of the conversion layer according to Eq. 3.16 is in the range of literature data of ZrO_2 (see Table 5.2.3 in the app.) illustrated in Fig. 3.32 and too high for $Zr(OH)_4$ values from literature. The final analysis is possible with infrared spectroscopy (see Ex. 3.12). The band gap of the zirconium based conversion layers shows an insulating surface, a necessary property for a corrosion protective layer.

A comparison of the absorption in the range of 182–184 nm after different immersion times with the sputter depth profile from Fig. 3.9

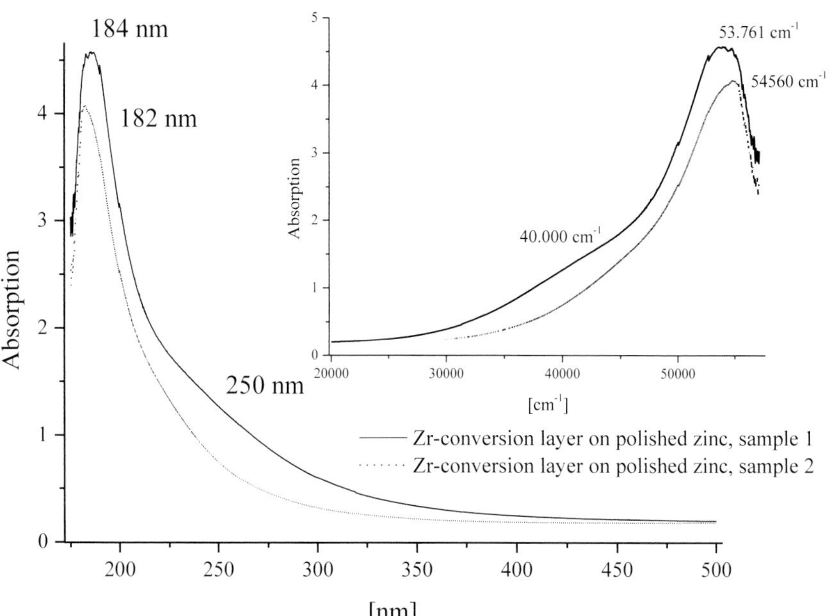

Fig. 3.31 UV-VIS reflection (6°) spectra of a zirconium based conversion layer on polished zinc in nitrogen atmosphere. The reference was an aluminium mirror.

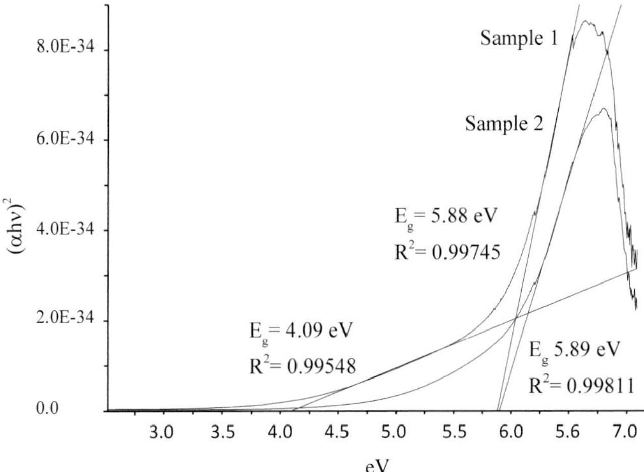

Fig. 3.32 Tauc plots of the zirconium based conversion layer on polished alkaline cleaned zinc based on the spectra of Fig. 3.31.

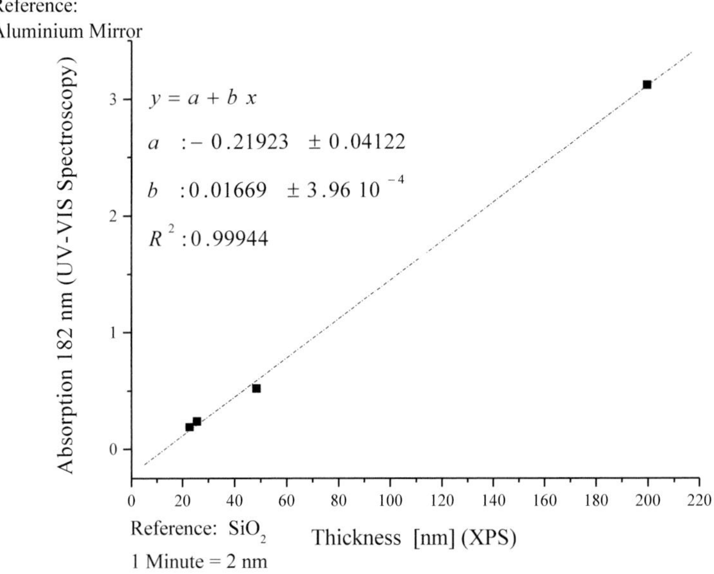

Fig. 3.33 Correlation of UV-VIS data from different conversion layers with the corresponding XPS sputter depth profiles (compare Fig. 3.9).

shows a good correlation of the data (Fig. 3.33). This correlation allows the investigation of the absorption coefficient ε of a layer, i.e., a calibration of the thickness with the XPS data to measure the layer thickness by means

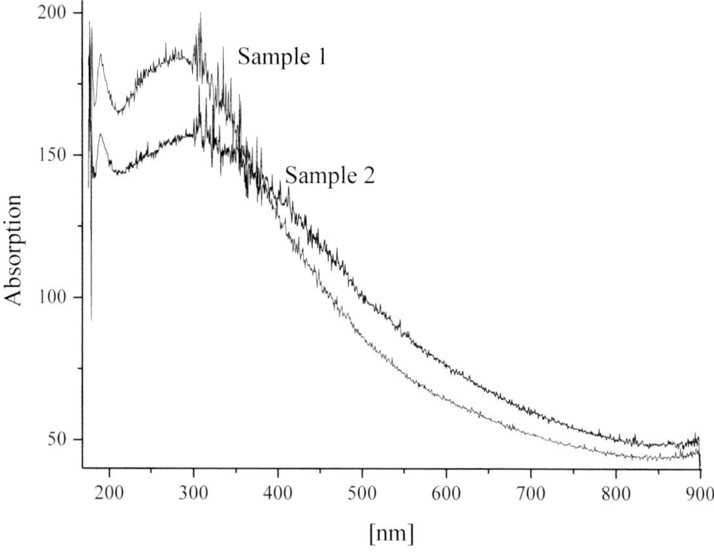

Fig. 3.34 UV-VIS reflection (6°) spectra of a zirconium based conversion layer on galvanized steel in nitrogen atmosphere. The reference was an aluminium mirror.

of UV-VIS spectroscopy. The advantage of the method is the low price and speed of the measurement at ambient conditions. Furthermore the XPS sputter depth profiles shows a higher amount of noise.

The limitation of the UV-VIS reflection spectroscopy is illustrated in Fig. 3.34. The measurement of an industrial—not polished—galvanized steel substrate is more or less impossible. The absorption band is still visible but the noise (caused by the roughness of the surface) dominated the spectra. Therefore a smooth surface is necessary or not for polishing of the metal substrate.

Example 3.10

Besides the investigation of the layer thickness, the chemical composition and the band gap of a layer the UV-VIS spectroscopy is a useful tool to measure kinetic data of passive or conversion layer deposition processes. In literature often the combination of electrochemical methods and the UV-VIS-spectroscopy have been used to study the oxide layer growth or the effect of corrosion inhibitors [62]. Only the UV-VIS spectroscopy and the Raman spectroscopy (see Chap. 3.1.5) are able to investigate *in situ* or quasi *in situ* measurements because water—necessary for corrosion analysis processes—absorb at 167 nm (see Table 5.9 in the app.) in the electronic and shows only weak absorption in the Raman spectrum. All other spectroscopic or microscopic methods need dry samples or the samples dried at high vacuum. As already mentioned the alkaline cleaning of a zinc surface generates a hydroxide layer (Fig. 3.4) with a thickness of 8–9 nm (Fig. 3.5). During a conversion process the hydroxide layer is dissolute in the acidic environment and causes the pH gradient on the surface and this is the driving force for the precipitation of a conversion layer (see Chap. 1.3.1). Therefore the substrate dissolution should be the rate-determining-step in the layer formation proved by UV-VIS spectroscopy in combination with infrared spectroscopy (see Ex. 3.12). In Fig. 3.35 the UV-VIS spectra at different times of immersion are illustrated. Based on the intensity at λ_{max} of the four fitted bands (based on the band analysis procedure in ref. [33]) shown in Fig. 3.35, the decrease of the layer could be analyzed with the resulting kinetic fits with a first order reaction. The average rate constant of all bands could be calculated to 0.21 l/mol*s [7]. The growth of the conversion layer has been investigated by means of infrared spectroscopy (see Ex. 3.12).

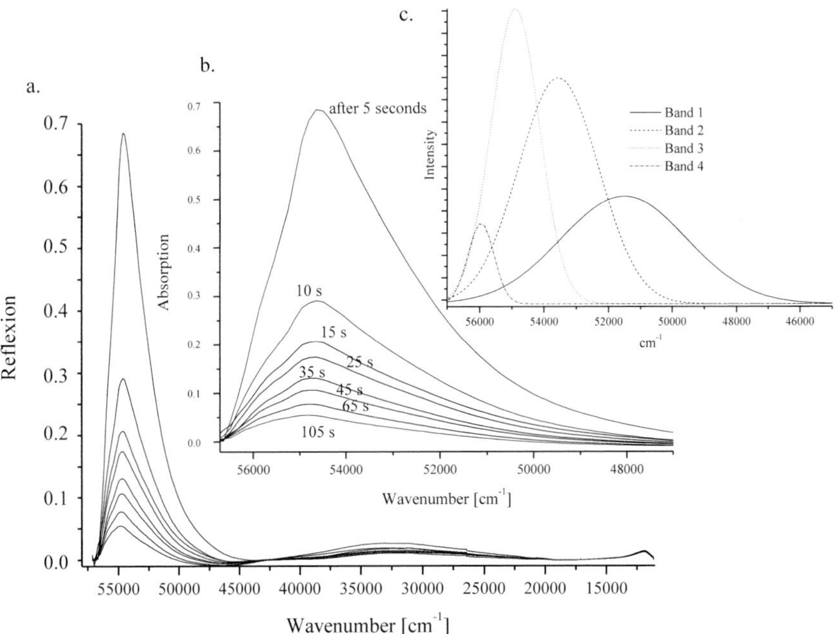

Fig. 3.35 UV-VIS reflection (6°) spectra of an alkaline cleaned polished zinc surface immersed in a zirconium based conversion bath (a). In (b) the detail of the absorption spectra with the different immersion times are illustrated and in (c) the band analysis according to [33] of the absorption spectra. The reference was an aluminium mirror.

3.1.4 Infrared Spectroscopy

Basics

Only a brief introduction in the principles of the method and the necessary equipment is presented in the following paragraph. Detailed presentations of infrared spectroscopy in general are given in refs. [3, 34, and 35]. The basics in reflection spectroscopy and an introduction of the use in corrosion research are presented in ref. [3].

The energy of a photon in the infrared (100–4000 cm^{-1}) range of the electromagnetic radiation causes the excitation of vibrational states of bindings in a molecule and the excitation of vibrational states causes the absorption of light of the corresponding wavelength or, more often used in the infrared spectroscopy, the corresponding wavenumber because it is proportional to energy. The most important vibrations of molecules are summarized in Fig. 3.36.

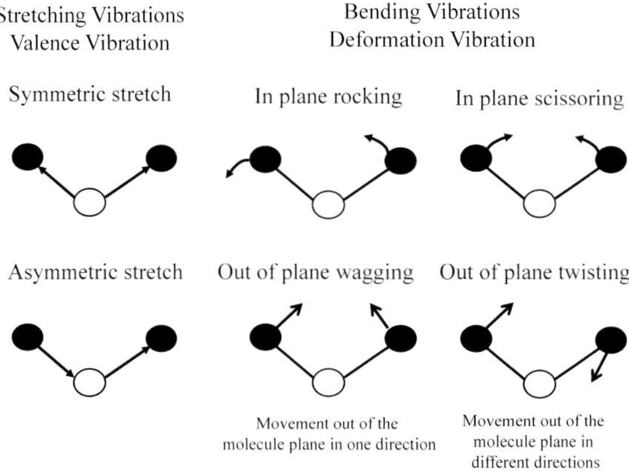

Fig. 3.36 Stretching and bending vibrations.

The transition of vibrational states is a quantized process with selection rules analogue to the UV-VIS spectroscopy. The basic quantum mechanics model for vibrational transitions for a diatomic molecule with the masses m_1 and m_2 for each atom is the harmonic oscillator describing the energy E_{vib} of the transition and the energy levels as follows [4]:

$$E_{vib} = \left(\nu + \frac{1}{2}\right) h\omega \;\wedge\; \omega = \left(\frac{k}{u}\right)^{\frac{1}{2}} \;\wedge\; u = \frac{m_1 m_2}{m_1 + m_2} \;\wedge\; \nu = 0,1,2,3... \qquad \text{Eq. 3.22}$$

v	:	Quantum number	u	:	Reduced mass
h	=	$6.6310^{-34}\,Js$	k	:	Force constant of the binding

The selection rules based on the harmonic oscillator are the following:

$$\Delta v = 1 \wedge \Delta\mu \neq 0 \qquad \text{Eq. 3.23}$$

The dipole moment μ of the molecule has to change during the transition for detection with infrared spectroscopy. This is normally the case for asymmetric vibrations. In Chap. 3.1.5 the Raman spectroscopy is introduced with the selection rule of a change of polarizability of the molecule and this is the case for symmetric vibrations. Therefore the combination of both methods gives the complete picture. The harmonic oscillator does not describe the reality because higher excitations, i.e., $\Delta v \geq 1$ (overtones), could be detected and the amount of energy in the system is limited to the dissociation energy D of the binding. A better description of the energy is possible with the so called Morse function [3]:

$$V(r)=D\left(1-e^{-a(r-r_{equ})}\right)^2 \wedge a=\left(\frac{1}{2}kD\right)^{\frac{1}{2}} \qquad \text{Eq. 3.24}$$

The difference $r-r_{equ}$ describes the deviation on the equilibrium distance of the atoms. The selection rules based on the so called anharmonic oscillator are the following:

$$\Delta v = 1, 2, 3 \wedge \Delta\mu \neq 0 \qquad \text{Eq. 3.25}$$

At ambient temperatures molecules are normally in the ground state, i.e., v = 0 and therefore the most important transition is the $1 \leftarrow 0$ transition in the infrared spectrum. During the excitation of vibrational states rotational states are also excited. In Fig. 3.37 the infrared spectrum of water vapour is illustrated and the rotational transitions are well separated. Condensed water and other solid or liquid material show broad signals because of two reasons. First the rotational states broaden the vibration absorption and overtones could be another reason for it.

Equipment

From the corrosion and corrosion protection point of view only reflection spectroscopy on the surface is useful therefore the following paragraph focusses on this method.

The state of the art method in infrared spectroscopy is the so called Fourier Transform (FT) or Fast Fourier Transform (FFT) spectroscopy. FT-IR spectrometers use Globar or Nernst glowers as sources for the infrared

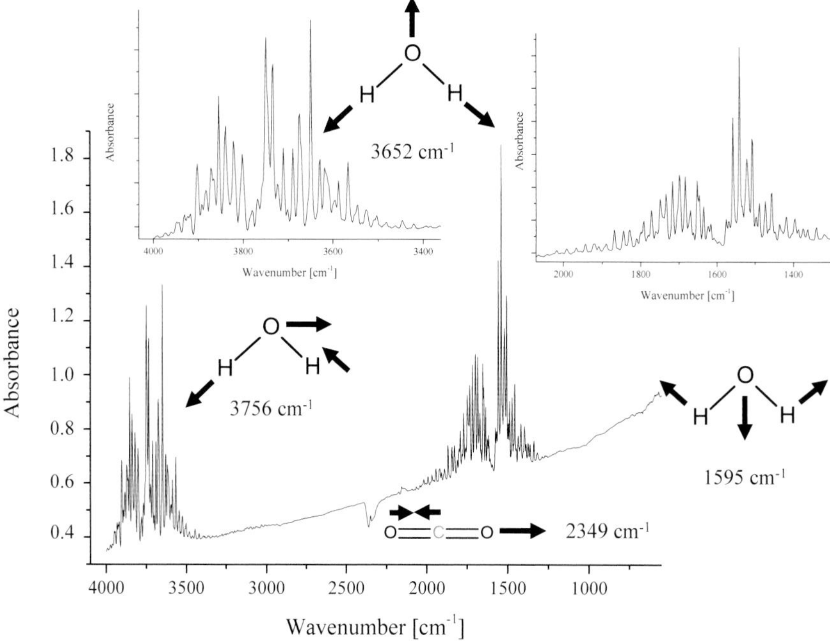

Fig. 3.37 Infrared spectrum of water vapour and carbon dioxide and the corresponding vibrations [4, 35]. The spectrum was measured with an ATR device with a bad contact of the sample to the crystal.

(mid-infrared) light. In the Michelson interferometer (Fig. 3.38) the polychromatic light is divided by the beam splitter in two beams and one is modulated by changing the distance of the movable mirror in the distance x to generate an interferogram, i.e., the resulting beam of the interference of both beams. The emitted beam consists of all frequencies in a time domain. After interaction with the sample the signal could be transferred with the Fourier transformation [63] in the frequency domain to achieve the infrared spectrum. The distance x could be measured very exactly with a laser (often 632 nm) to gain a resolution in the frequency domain of up to 0.01 cm^{-1}. Because of the fact that all frequencies are measured at the same time, the experiment is very fast and no monochromator reduces the light intensity.

The analysis of a metal surface could be performed with two methods. The IRRAS (infrared reflection absorption spectroscopy) and the ATR (Attenuated Total Reflection spectroscopy) are the most used methods for surface analysis in the field of corrosion and corrosion protection analysis. It should be noted that there are some more methods like diffuse reflectance infrared spectroscopy (DRIFT) and photoacoustic infrared spectroscopy (for details see ref. [63]) but not mentioned in this paragraph.

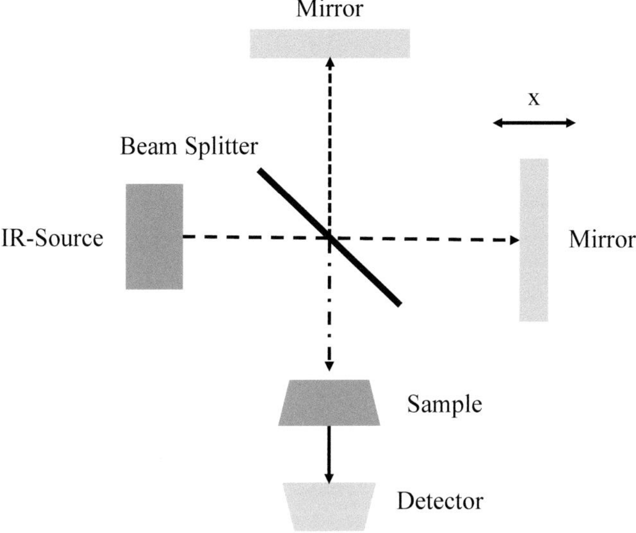

Fig. 3.38 Scheme of a Michelson interferometer.

IRRAS

The IRRAS or RAIRS (reflection-absorption infrared spectroscopy) is based on a regular reflection on the surface with an incident angle of 80° (Fig. 3.39). The high incident angle produces a large surface for interaction which increases the sensitivity especially for thin layers. Furthermore the transmission T:

$$T = \frac{I}{I_0} \wedge A = \log\frac{1}{T} \qquad \text{Eq. 3.26}$$

fulfils Lambert-Beers law (Eq. 3.15), if the layer thickness is above $\frac{1}{4}\lambda$ but even on thin layers the correlation gives reasonable results.

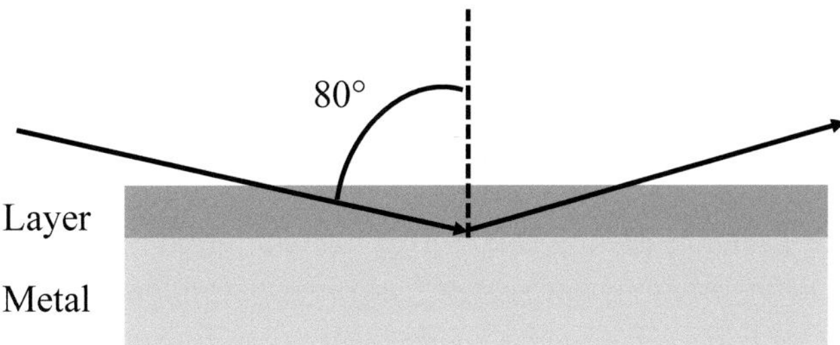

Fig. 3.39 Setup of the IRRAS method on metal surfaces.

Because of the fact that this method is similar to transmission the comparison with transmission spectra from databases is often possible. The disadvantage of the method is the roughness sensitivity but even on non-polished samples a measurement is possible (see Ex. 3.13). The PM-IRRAS (polarization modulation IRRAS) allows to investigate on very smooth surfaces the angle of molecules such as Self-Assembled Monolayers (SAM) to the metal surface (for further reading see refs. [64–67]). SAM layers are used as conversion layers in industry [68–71] because only monolayers of high oriented molecules are able to generate high barrier properties. The analytical challenge is the fact that the analysis of the layers could only be done on polished samples and not on the industrial rough metal. On polished samples *in situ* measurements of corrosion processes by means of IRRAS have been done [72]. The *in situ* measurements are possible, if the relevant signals are not affected by the intense water absorption.

In principle the infrared beam is reflected on the metal surface (Fig. 3.39) and therefore the complete layer is analyzed by this method in contrast to the ATR method.

ATR

The principle setup of the ATR method is illustrated in Fig. 3.40.

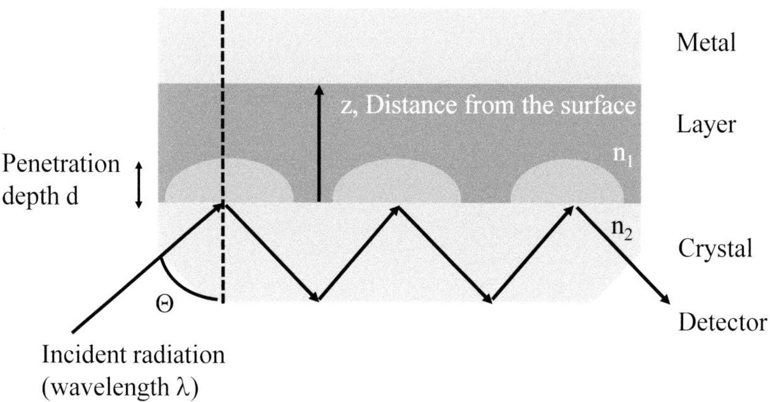

Fig. 3.40 Setup of the ATR method.

The internal reflection spectroscopy is based on the fact that, if two media of different refractive index n_1, n_2 an evanescent wave exists in the medium of the lower refractive index, if the light is introduced in the denser medium. The evanescent field E decays exponentially in the rarer medium according to the following equation [73]:

$$E = E_0 \, e^{-\frac{2\pi n_1}{\lambda}\sqrt{\sin^2\Theta - \left(\frac{n_2}{n_1}\right)^2}\cdot z}$$ Eq. 3.27

Therefore the penetration depth d depends on the refractive indexes and the wavelength as follows [63, 73–75]:

$$d = \frac{\lambda n_1}{2\pi}\frac{1}{\sqrt{\sin^2\Theta - \left(\frac{n_2}{n_1}\right)^2}}$$ Eq. 3.28

For example the penetration depth in a TiO_2 layer is around 2 µm [73] therefore the sample has to be in close contact to the crystal. The water spectrum of Fig. 3.37 was prepared with an organic coating sample at a distance of some µm to the crystal and only the spectrum from the atmosphere in between has been achieved. The crystal material needs a high refractive index and no or less absorption in the infrared region. In Table 3.4 some common materials are summarized. There is no perfect solution because every material has advantages and disadvantages.

Table 3.4 Materials used as ATR crystals (compare [63, 73, and 76]).

Crystal	Useful range [cm^{-1}]	n_2	Properties
KRS-5 (thallium iodide)	17,000–250	2.37	Soluble in bases; slightly soluble in water, insoluble in acids, highly toxic
ZnSe	20,000–500	2.4	Etched by diluted acids and bases; eroded by zinc complexing agents; insoluble in water
Ge	5,000–550	4.0	Resistant to diluted acids and bases
Diamond	45,000–2,500 and 1,667–33	2.4	Suitable to pH 1–14
Si	8,300–1,500 and 360–70	2.37	Insoluble in water, etched by strong acids and bases

The absorption of the diamond crystal in the range of 1667–2500 cm^{-1} could be a challenge especially for the investigation of organic compounds. This is not the case. In Fig. 5.1 in the app. spectra of the copolymer ABS are illustrated, measured with a germanium and a diamond crystal in an ATR device. The spectra show that the diamond ATR generates useful spectra even in the range of the diamond absorption with the advantage to measure in the wavenumber range below 500 cm^{-1}.

The polarization modulation in combination with the ATR method has also been described in literature [75]. For conversion or passive layers with a thickness in the nm up to low µm range the penetration depth is high enough to measure the complete layer but roughness of the surface in the µm range (standard industrial substrates) reduces the signal intensity.

For both methods a base line correction is always necessary. The spectrum of water in Fig. 3.37 illustrates the drift of the base line. There are a lot of methods to correct the base line but especially for IRRAS a handmade correction often gives the best results in comparison with a software correction especially, if the surface is changed, i.e., roughness and thickness of the layer is changed in a spectra series.

Example 3.11

In Ex. 3.1 the effect of the alkaline cleaning process has been analyzed by means of XPS. The result could be proved with IRRAS as second independent method. ZnO powder as reference has been measured as shown in Fig. 3.41. The signals fits with most of the published data (see Table 5.2.2 in

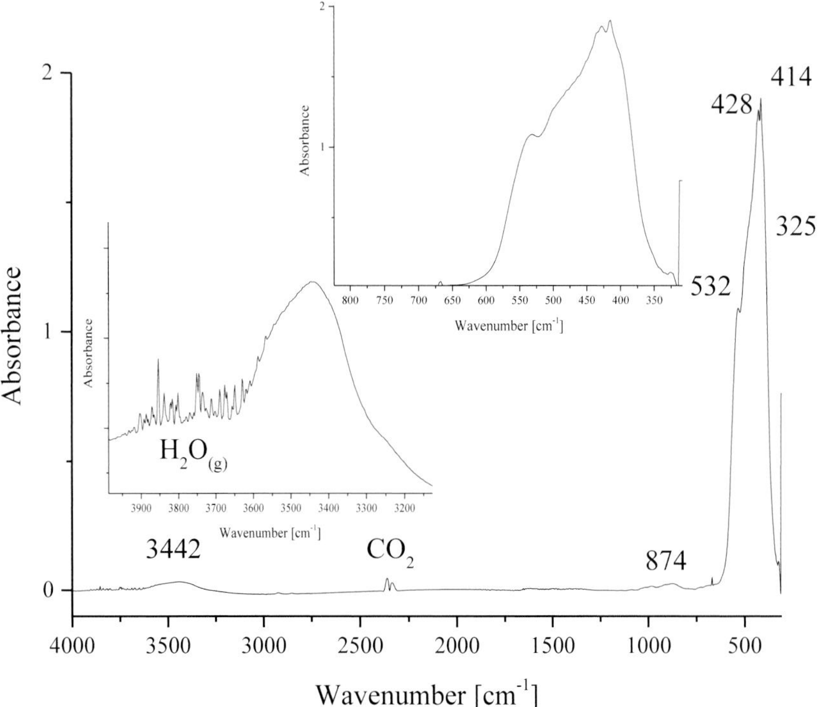

Fig. 3.41 IRRAS spectrum of ZnO pressed powder.

the app.). Water and carbon dioxide from the atmosphere are always present in IRRAS spectra with more or less intensity. One important result is the fact that even in ZnO OH-groups signals are present caused by interaction of the material with the atmosphere. The spectrum of the polished alkaline cleaned zinc surface (Fig. 3.42) shows the expected signals for OH-groups (3383 cm^{-1} and 3552 cm^{-1}) and Zn-O-vibrations in the lower wavenumber range and prove the XPS results. Furthermore the ZnO reference spectrum (Fig. 3.41) seems to be contaminated with the zinc hydroxide spectrum (Fig. 3.42) and exactly the same results achieved by means of XPS because the polished zinc surface—a zinc oxide layer—shows signals of both oxide and hydroxide groups. Therefore the zinc oxide layer contains hydroxyl groups and the alkaline cleaned surface consists of a zinc hydroxide layer. The amount of zinc oxide or oxyhydroxides visible in the UV-VIS spectra (Ex. 3.8) could be neither confirmed nor rebutted but the combination of all spectroscopic data (XPS, UV-VIS and IR) evidence a zinc hydroxide layer.

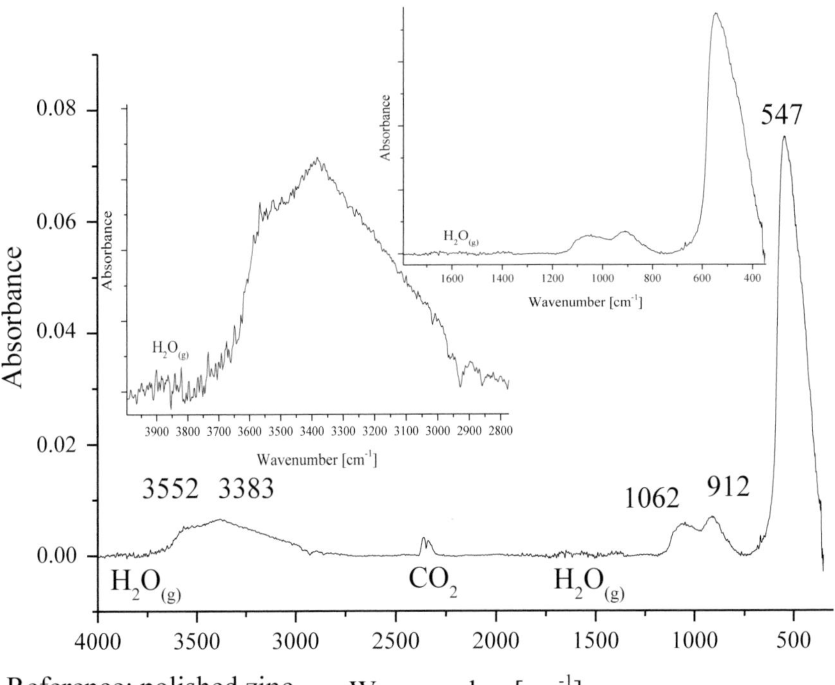

Fig. 3.42 IRRAS spectrum of an alkaline cleaned polished zinc surface.

Example 3.12

The zirconium based conversion layer analyzed by means of XPS (Ex. 3.2) and UV-VIS spectroscopy (Ex. 3.9) has also been analyzed by means of IRRAS, to clarify the structure of the conversion layer. To compare the infrared absorption of the conversion layer reference spectra (Fig. 3.43) have been performed. The data of the ZrO_2 sample fits with most of the published data (Table 5.7 in the app.) but the so called $Zr(OH)_4$ shows prominent water absorption at 1628 cm^{-1} and the high amount of water may cause the curious O-H-valence. Therefore the material is more a $Zr(OH)_2$ * x H_2O. Regarding the Zr3d-signal in the XPS (Figs. 3.6 and 3.8) this material fits with the conversion layer and the same is the fact for the O1s-signal (Fig. 3.7). It has to be noted that the XPS spectra measured in ultra-high vacuum and depending on the chemical environment of the water, could be evaporated and therefore it is not visible in the XPS spectra. The UV-VIS spectra (Fig. 3.31) show a mixture of hydroxide and oxide in the conversion layer.

The IRRAS spectra of the conversion layer (Fig. 3.44) show a mixture of oxide and hydroxide structures and a certain amount of water in the layer (OH-deformation at 1646 cm^{-1}) therefore the structure besides other compounds from the substrate is a $ZrO(OH)_2$ * x H_2O (compare Table 5.7 in the app.). Figure 3.44 shows the limitation of the method because the

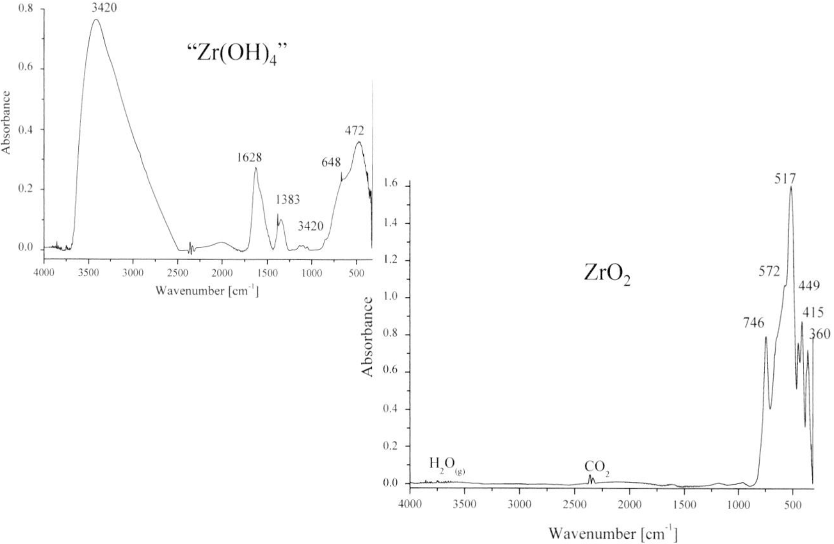

Fig. 3.43 IRRAS spectra of ZrO_2 (Aldrich) and $Zr(OH)_4$ (Aldrich) pressed powder.

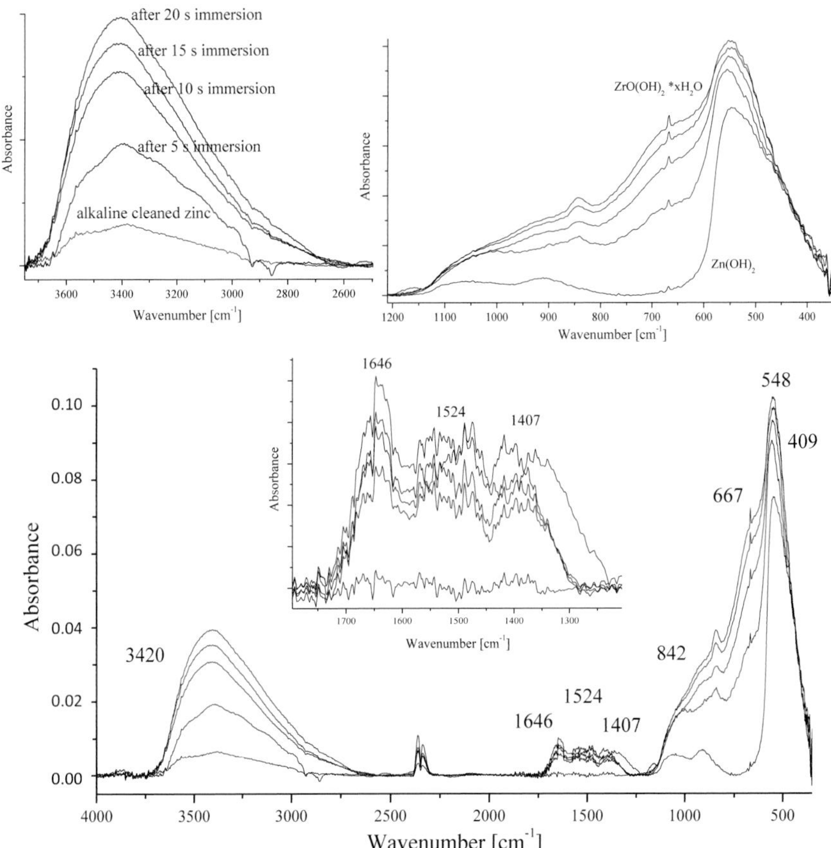

Fig. 3.44 IRRAS spectra of alkaline cleaned polished zinc surface with different times of immersion in a zirconium based conversion bath. The reference was the polished zinc surface [7].

spectrum of the alkaline cleaned zinc surface and the spectra at different times of the conversion layer formation are quite similar beside from the signals in the range 1300–1700 cm^{-1} because both are metal hydroxide structures and therefore the vibrations are in the same frequency range. Without a second method to analyze the element distribution the interpretation of the spectra are hardly possible.

The comparison of the kinetic data of the dissolution of the zinc hydroxide layer by means of UV-VIS spectroscopy and the zirconium based layer formation by means of IRRAS is summarized in Table 3.5. The data show that the rate of dissolution of the zinc hydroxide is similar

Table 3.5 Rate constants (first order reaction) of the dissolution of the zinc hydroxide layer in comparison with the film formation and the film formation without an alkaline cleaning process [7].

Pre-treatment	Band/ Wavenumber	k [l/mol*s]	Average k [l/mol*s]	Process
With alkaline cleaning	UV-VIS Band 1	0.2	0.21	Zinc hydroxide dissolution
	UV-VIS Band 2	0.27		
	UV-VIS Band 3	0.23		
	UV-VIS Band 4	0.15		
	IRRAS 675 cm^{-1}	0.21	0.2	Layer precipitation
	IRRAS 3428 cm^{-1}	0.22		
	IRRAS 848 cm^{-1}	0.17		
Without alkaline cleaning	IRRAS 667 cm^{-1}	0.12	0.13	Layer precipitation
	IRRAS 844 cm^{-1}	0.13		

to the precipitation and could be the rate-determining-step. Without an alkaline cleaning, i.e., the precipitation based on the dissolution of the thin oxide layer followed by acidic corrosion of the zinc substrate, the reaction rate is significantly lower therefore, as already mentioned, the cleaning process determines the film formation of the conversion layer.

Example 3.13

Phytic acid (Fig. 2.16) based conversion layers already mentioned in Exs. 2.5 and 3.19 with the characterization of the layer with CV and SEM/EDX, respectively. The barrier properties of the layer (Fig. 2.17) and the inhibition effect (Fig. 2.18) are investigated with electrochemical methods. The formation of a dense layer has been optimized by means of SEM/EDX but all methods do not describe the chemical composition. The IRRAS spectra of phytic acid (evaporation of the solution on an inert substrate) and the resulting conversion layers at different pH values are summarized in Fig. 3.45.

The spectra of the conversion layer have been performed on steel sheets without polishing. The surfaces have been sanded and the direction of the measurement has been selected parallel to the sanding direction [77]. As reference spectra the respective untreated surfaces have been used. This method allows the investigation of layers on technical substrates without polishing.

The spectra of phytic acid and conversion coatings present three peaks at 972, 1011, and 1047 cm^{-1} assigned to the -P-O-H bond. Only for

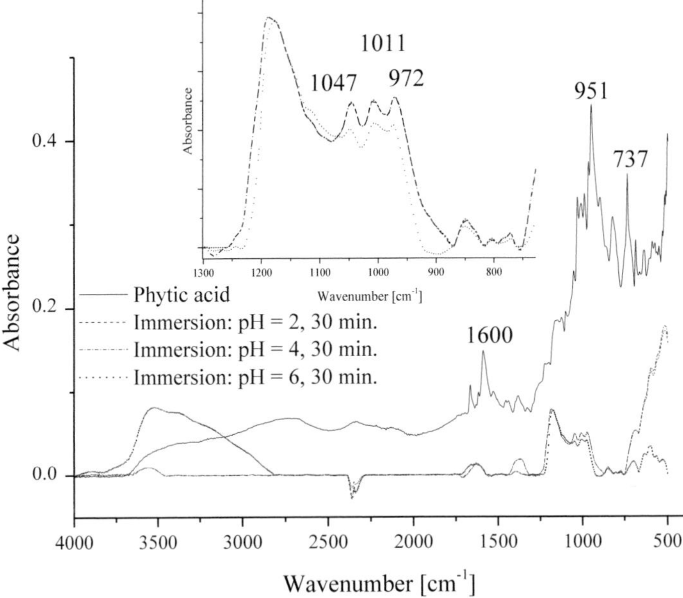

Fig. 3.45 IRRAS spectra of phytic acid and conversion layers of steel modified from [32].

the phytic acid spectrum three peaks at 951, 827, 737 cm^{-1} ($-PO_4^{3-}$) can be seen [78] (compare Table 5.7 in the app.). The peaks at 1375 (phytic acid) and 1371 (conversion coating) are assigned to the $-P = O$ bond [79]. The peaks around 1600 cm^{-1} are related to HPO_4^{2-} group [3]. For the conversion coating it is obvious that the phosphate hydrogen group is present while the phosphate group is absent. This indicates that the phosphate group of phytic acid can form complexes with metal ions such as Fe^{3+} resulting in the conversion coating on mild steel. The peaks in the spectra of the conversion coating below 700 cm^{-1} also support the thesis of the formation of metal-oxygen bonds [79]. It is reasonable that phytic acid was deposited on the metal surface. The samples of different pH-values were similar to each other. Only in the range of 3500 cm^{-1} the peaks show differences. At pH = 6 the peak for –OH is more intensive than at pH = 2 and 4. The reason for this is a lower deprotonation of phytic acid at pH 6 [32].

Example 3.14

Zinc surfaces have a low corrosion rate caused by the stable carbonate/hydroxide passive layers broadening the passive range of the metal (compare Chap. 1.1.1). On corroded zinc surfaces in CO_2 containing atmosphere and chloride containing electrolytes (sea water) often formed besides ZnO and $Zn(OH)_2$ complex compounds like $Zn_5(OH)_8Cl_2$ (simonkolleite), $ZnCO_3$ (smithsonite) and $Zn_5(OH)_6(CO_3)_2$ (hydrozincite) [80, 81]. The corresponding spectroscopic data of the corrosion products are summarized in Table 5.6 in the app. (for Raman data see also Ex. 3.15). Therefore carbonate based conversion layers on zinc and aluminium have been tried to precipitate on the surface to increase the corrosion protective performance of the surface. The zinc coated steel has been immersed into a carbon dioxide saturated solution at pH = 5 for 15 minutes (Fig. 3.46) and for the aluminium surface sodium fluoride has been added to the carbonate solution at pH = 3 for 15 minutes (Fig. 3.47). As already mentioned the samples have been sanded before the measurements to use the IRRAS methods to investigate the spectra.

In Fig. 3.46 hydrozincite as conversion layer could be investigated (compare Table 5.6 in the app.) but with a very low thickness even after 15 minutes immersion time. Therefore sanding should be a useful pre-treatment to use the IRRAS for the investigation of thin layers on technical substrates.

In Fig. 3.47 the aluminium surface is covered with a layer containing hydroxide structures visible with the corresponding bands at 3486 cm^{-1}, 1011 cm^{-1}, 768 cm^{-1} and 687 cm^{-1}. A carbonate structure like Dawsonite

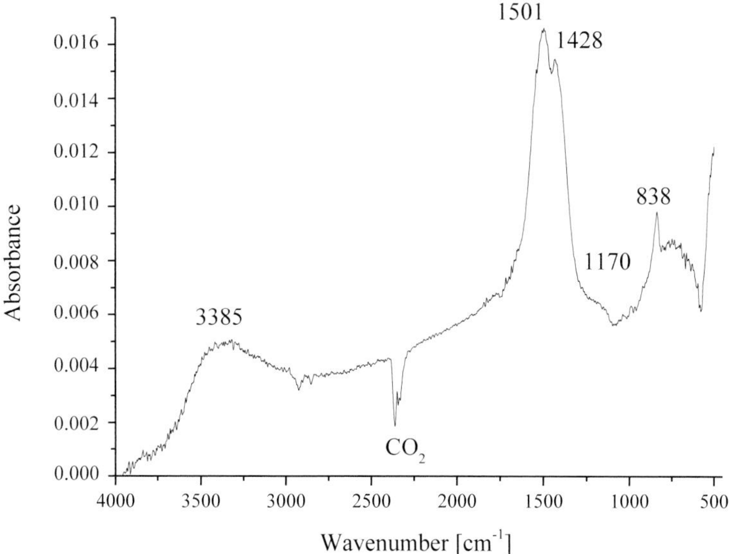

Fig. 3.46 IRRAS spectrum of a sanded zinc surface after treatment for 15 minutes in a carbonate bath at pH = 5.

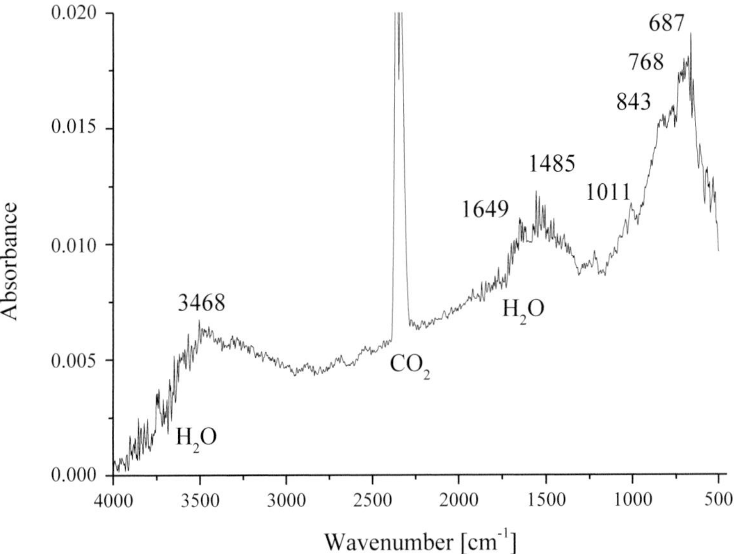

Fig. 3.47 IRRAS spectrum of a sanded aluminium surface after treatment for 15 minutes in a carbonate/NaF bath at pH = 3.

should be visible by the bands at 1649 cm^{-1}, 1485 cm^{-1} and 843 cm^{-1} (compare Table 5.6 in the app.). This spectrum shows the limitation of the method. If the layer is very thin and has no intensive signals, the noise caused by atmosphere dominates the spectrum.

3.1.5 Raman Spectroscopy

Basics

Only a brief introduction in the principles of the method and the necessary equipment is presented in the following paragraph. Detailed presentations of Raman spectroscopy in general are given in refs. [36, 82–84]. The basics in confocal Raman spectroscopy are presented in refs. [85, 86]. An overview about the surface enhanced Raman spectroscopy (SERS) is given in refs. [82, 86–89].

The name of the Raman Effect is based on the experiments done by C.V. Raman who first observed this phenomenon of liquid benzene in 1928 [90].

The Raman Effect is an inelastic light-scattering phenomenon, which changes the frequency of the incident light illuminating a sample [86] and the effect could be subdivided into three methods/effects summarized in Fig. 3.48.

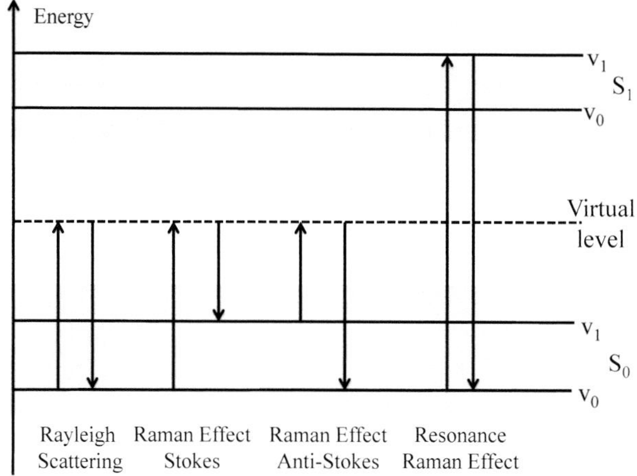

Fig. 3.48 Summary of the possible interaction in light-scattering.

Normal Raman Scattering

The electric field strength E of a light wave with the amplitude E_0 and the frequency v_0 at the time t is described as follows [86, 91]:

$$E = E_0 \cos 2\pi v_0 t \qquad \text{Eq. 3.29}$$

If the light incident on a molecule, a dipole moment μ is induced:

$$\mu = \alpha E \qquad \text{Eq. 3.30}$$

The proportionality constant α is called polarizability. The molecule itself is vibrating with the frequency v around the state of equilibrium q_0 of the normal coordinate as follows:

$$q = q_0 + \cos 2\pi v t \qquad \text{Eq. 3.31}$$

For small amplitudes of vibration, α is a linear function of q.

$$\alpha = \alpha_0 + \left(\frac{\delta\alpha}{\delta q}\right)_0 q + ... \qquad \text{Eq. 3.32}$$

Then it follows for the induced dipole moment:

$$\mu = \left[\alpha_0 + \left(\frac{\delta\alpha}{\delta q}\right)_0 q_0 + \cos 2\pi v t\right] E_0 \cos 2\pi v_0 t \qquad \text{Eq. 3.33}$$

With the trigonometric relation: $\cos x \cos y = \frac{1}{2}\left[\cos(x+y)+\cos(x-y)\right]$ it follows:

$$\mu = \alpha_0 E_0 \cos 2\pi \nu_0 t + \frac{1}{2}\left(\frac{\delta\alpha}{\delta q}\right)_0 q_0 E_0 \cos 2\pi(\nu_0 - \nu) + \frac{1}{2}\left(\frac{\delta\alpha}{\delta q}\right)_0 q_0 E_0 \cos 2\pi(\nu_0 + \nu)$$

Eq. 3.34

Rayleigh Scattering + Raman Stokes Scattering + Raman Anti-Stokes Scattering

The selection rule for the Raman Effect analogue to the infrared spectroscopy (Eq. 3.25) is the following:

$$\left(\frac{\delta\alpha}{\delta q}\right) \neq 0$$

Eq. 3.35

A vibration is Raman active, if the polarizability changes during the vibration. If the polarizability changes during the vibration, the Raman signal could be of higher (anti-Stokes) or lower frequency (stokes) as the most intensive Rayleigh scattering whereas the anti-stokes are less intensive because the molecules are at ambient conditions predominantly in the ground state (see Fig. 3.48). Therefore a Raman spectrum often presents only stokes shifts (Raman shift) and not anti-stokes. The Raman process can be described as a two-photon process with a timescale of 10^{-12} s [86]. The excitation is initiated by a first incident photon with frequency of ν_0 to a "virtual" state (see Fig. 3.48) between the ground and excited electronic state and simultaneous emission of the second photon as the decay from the virtual excited state back to the ground electronic state. In principle the Raman signals are very weak because of the fact that on 10^{10} incident photons one Raman photon could be detected [86] and therefore lasers used as light source for this spectroscopy and enhancement methods especially for thin layers on surfaces are used, if possible.

Resonance Raman Scattering

The Resonance Raman scattering (RR) could be realized, if the frequency of the incident light is in or very close to the range of an electronic absorption of the scattering molecule (see Fig. 3.48), whereby an enhancement of the Raman Effect with the factor of 10^6 is achievable. The reason for the enhancement is based on the longer resident time of the molecule in the electronic excited state compared to the virtual state increasing the probability of the Raman transition. The enhancement makes it possible to detect monolayer of molecules on surfaces. Another advantage is the simple spectrum because only the vibration modes coupled with the electronic transition are enhanced and therefore the number of signals in the spectrum is reduced. The disadvantage of the method is that most of the relevant molecules in corrosion research topics do not absorb in the range of the common laser wavelength (compare Table 3.6 with Table 5.9 in the app.).

Surface-enhanced Raman Scattering

To use the Surface-Enhanced Raman Scattering (SERS) in the field of surface analysis, the surface is covered with nanoparticles of noble metals such as silver, gold and copper. The enhancement factor is about 10^6 fold and could by combined in principle with the RR.

The exact mechanism is still under discussion in the literature but two theories dominate the discussion [83, 86]. The electromagnetic theory is based on the excitation of localized surface Plasmon's and the chemical theory proposes charge-transfer-complexes of chemical bonded molecules on the surface. Because of the fact that the nanoparticles have to be small (90 nm) in comparison to the incident wavelength and need a certain particle size, the SERS is not a standard method because of the complicated surface preparation. The use of aluminium nanoparticles in the UV range avoid the expensive noble metals particles [92] and should be a strategy to establish this method.

Equipment

To measure a Raman spectrum monochromatic light is necessary because every wavelength of a polychromatic light generates a spectrum. Furthermore a high intensity of the incident light is necessary to detect spectra because of the weak effect. Both requirements are fulfilled with a laser (single mode) as light source. The common lasers used in Raman spectroscopy are summarized in Table 3.6.

The right choice of the wavelength depends on the chemical composition of the sample because fluorescence of the sample has to be avoided. In principle the sensitivity increases with lower wavelength of the incident light because according to the quantum mechanical theory of light scattering the intensity $I_{1,2}$ of the Raman signal caused by two vibrational states ν_1 and ν_2 increases by the 4th power of frequency ν_0 as follows [86, 93]:

$$I_{1,2} = \frac{2^7 \pi^5}{3^2 c^4} I_{Laser} \left(\nu_0 - \nu_{1,2}\right)^4 \sum \left|(\alpha)_{1,2}\right|^2 \qquad \text{Eq. 3.36}$$

Table 3.6 Laser used in Raman spectroscopy.

Type	Medium	Wavelength [nm]
Ar-Laser, frequency doubled	Ar^+-Gas	244
Ar-Laser, frequency doubled	Ar^+-Gas	257
He-Cd-Laser	Cd-gaseous and He	325
Diode-Laser	e.g., InGaN-Laser Diode	473
Ar-Laser	Ar^+-Gas	488
Ar-Laser	Ar^+-Gas	514.5
Nd:YAG-Laser, frequency doubled	Nd in Y-Al-Garnet	532
Kr-Laser	Kr^+-Gas	568.2
He-Ne-Laser	Ne and He Gas	632.8
Kr-Laser	Kr^+-Gas	647.1
Diode-laser	e.g., AlGaAs-Laser Diode	758
Diode-laser	e.g., AlGaAs-Laser Diode	785
Nd:YAG-Laser	Nd in Y-Al-Garnet	1064

On the other hand the tendency to generate fluorescence increases with lower frequency and photochemical reactions on the surface caused by the incident laser light are more probable. Furthermore the heating of the sample by the incident light is another problem because the surface has to be analyzed and not modified by the Raman spectroscopy. Therefore often near infrared laser in the range of 750–790 nm is used to avoid these disadvantages. If the fluorescence of a sample dominates the Raman spectrum, the spectral analysis is hardly possible (see Fig. 3.53).

Particularly with regard to corrosion research topics the combination of the Raman spectroscopy with a confocal microscope allows the lateral resolution (see Ex. 3.17) to analyze the local corrosion phenomenon. The confocal spectroscopy allows a depth resolution to analyze the chemical composition in an organic coating and the change of it during transport or corrosion processes. The principle setup for the confocal Raman spectroscopy is illustrated in Fig. 3.49.

The line filter generates a monochromatic laser light and the spatial filter and the pinhole are necessary for the confocal resolution by eliminating light out of the focus of the objective. The Notch filter removes only the Rayleigh scattering, i.e., the wavelength of the laser. The adjustable mirror allows switching between spectroscopy and microscopic picture of the surface. Besides the classic analysis of the scattering with a monochromator and a detector (Photomultiplier, CCD), FT-spectroscopy analogue to the IR-spectroscopy with IR-lasers is already available.

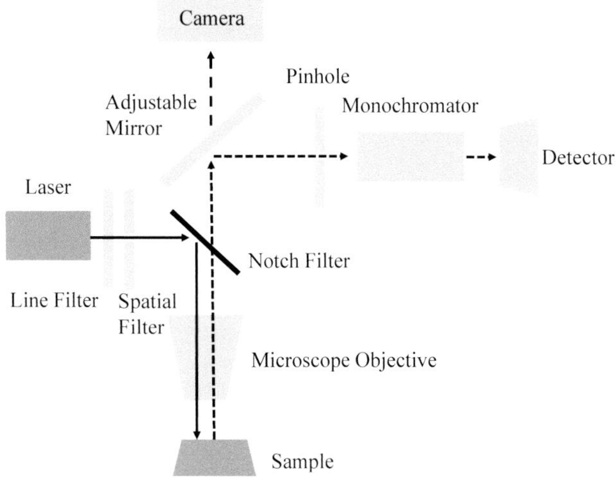

Fig. 3.49 Setup of a confocal Raman Microscope.

Example 3.15

A well-established use of the Raman spectroscopy is the analysis of corrosion products on metal surfaces [82, 94] and the combination with electrochemical methods to analyze passive layers [84, 86].

In Fig. 3.50 the possible corrosion products of zinc surfaces in chloride and sulphate containing electrolytes are summarized. Depending of the atmosphere, ions in the electrolyte and time different compounds could be detected. In the beginning of the corrosion process zinc hydroxide is precipitated on the surface followed by the generation of carbonates. In long term treatment (weeks to months) hydroxide-chlorides dominate the surface, if sulphate is absent.

In the following example the surface of the sample shown in Fig. 3.10 (Ex. 3.4) has been analyzed by means of Raman spectroscopy. The supposed simonkolleite $Zn_5(OH)_8Cl_2$ corrosion product on the surface analyzed by means of XPS could be proved with the Raman spectra (Fig. 3.52) because the signals fit with data from literature (compare Table 5.6). In addition the increase of the amount of corrosion products from the scratch to the middle of the defect is clearly visible (Fig. 3.51) with the Raman spectroscopy but hardly with XPS.

The precipitation of corrosion products occurs between the local electrodes (compare Figs. 1.3 and 1.9) therefore the increasing intensities are reasonable. Furthermore in Fig. 3.52 the signals 486, 546, 727 and 3456 cm^{-1} indicate β-$Zn(OH)_2$ or ZnO as corrosion products in the defect. The measurement of the surface is limited to the metal substrate because the E

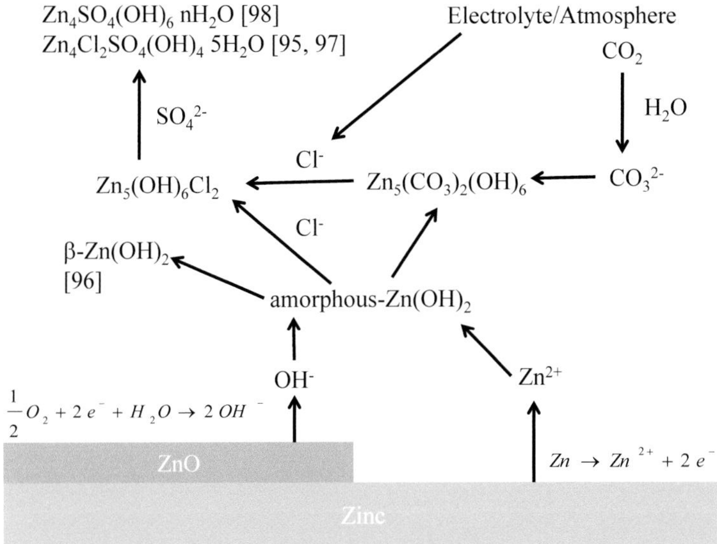

Fig. 3.50 Possible corrosion products on zinc in chloride/sulphate containing electrolyte (compare [95–98]).

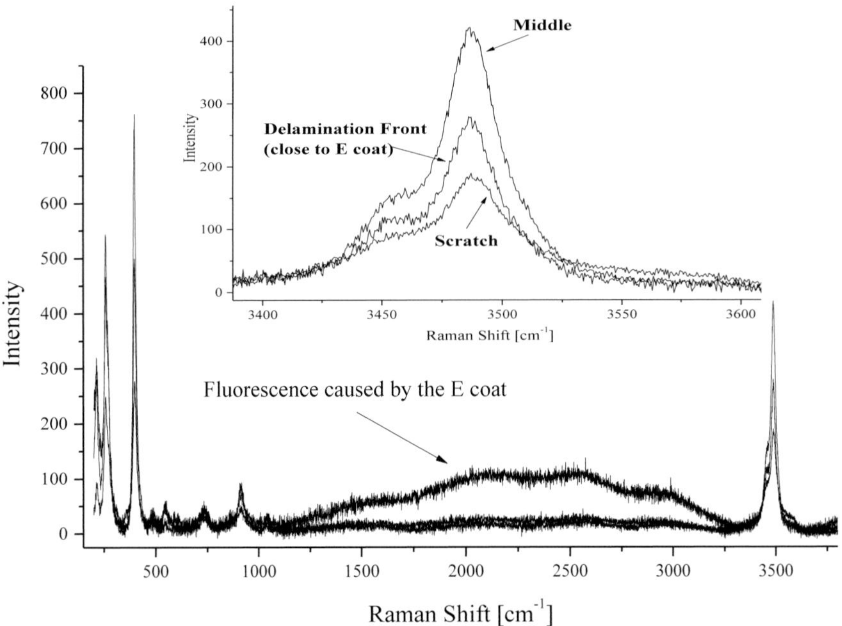

Fig. 3.51 Raman spectra of the sample illustrated in Fig. 3.10 (microscope picture) performed with a confocal microscope with an Argon-Laser (488 nm).

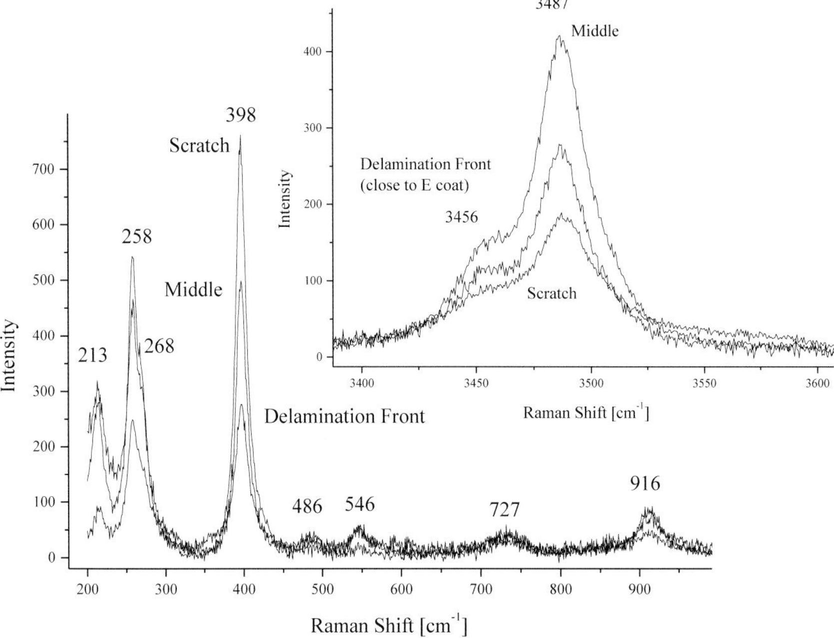

Fig. 3.52 Raman spectra of the sample illustrated in Fig. 3.10 (microscope picture) performed with a confocal microscope with an Argon-Laser (488 nm).

coat generates an intensive fluorescence with an Argon-Laser. At the edge of the coating the fluorescence increases so that the analysis of the spectra is hardly possible as illustrated in Fig. 3.53.

From a corrosion protective point of view investigating the mechanism of the delamination, anodic or cathodic (compare Fig. 1.11), is important but could not be performed in the example because no enrichment of chloride or zinc for an anodic area or hydroxide for the cathodic could be detected. If the shoulder at 268 cm^{-1} is caused by $Zn(OH)_2$, the increase in the direction to the delamination front is a hint for a cathodic delamination process.

Meanwhile infrared microscopes are available with a high spatial resolution of some µm. The infrared spectra show no fluorescence and therefore an analysis very close to an organic coating is possible to answer the question of the delamination mechanism. The SKP is the method of choice to analyze delamination processes (see Chap. 2.2.3) but only in atmospheric conditions producible in the measurement chamber. If the delamination mechanism is different in the salt spray test or in an alternating climate test in comparison to the conditions of the SKP, the SKP data has to be proved by means of spectroscopic methods on the samples after the treatment.

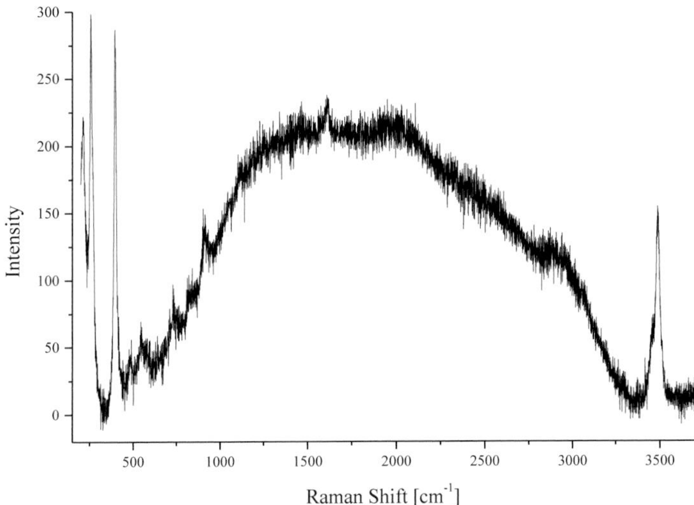

Fig. 3.53 Raman spectra of the sample illustrated in Fig. 3.10 (microscope picture) performed with a confocal microscope with an Argon-Laser (488 nm). The focus of the laser was partially on the E coat.

Example 3.16

The limitation of the Raman spectroscopy is the low sensitivity. The polished zinc surface with a passive layer in the range of 2–3 nm (Fig. 3.5) was easily analyzed by XPS (Fig. 3.4) but hardly with Raman spectroscopy. At least the ZnO surface analyzed by XPS could be proved with the Raman spectrum at different positions on the polished surface and therefore it could be use as a second independent method even on thin layers without enhancement methods illustrated in Fig. 3.54.

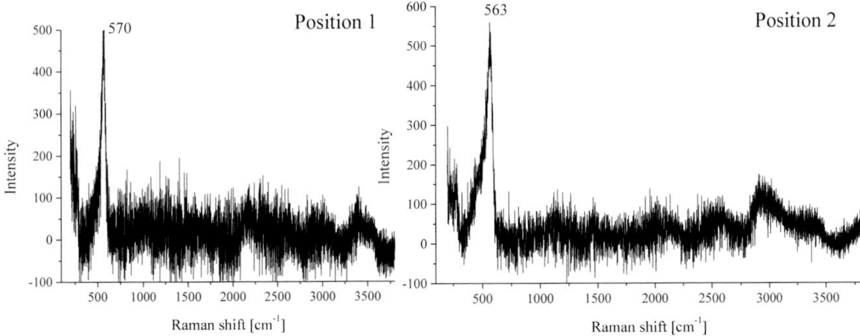

Fig. 3.54 Raman spectra of polished zinc performed with a confocal microscope with an Argon-Laser (488 nm).

3.2 Spectroscopic Methods on the Coating Surface

3.2.1 EDX Spectroscopy in Combination with SEM

Basics

The basics for EDX Spectroscopy and SEM are already explained in Chap. 3.1.2.

Focussed Ion Beam (FIB)

Only a brief introduction in the principles of the method and the necessary equipment is presented in the following paragraph. Detailed presentations of FIB-SEM-EDX in general are given in refs. [99–101].

The Focussed Ion Beam (FIB) device (Fig. 3.55) is similar to that of a SEM (Fig. 3.12). The major difference is the use of a gallium ion (Ga^{+}) instead of an electron beam. The ion beam is generated from a liquid gallium source in an electric field (7kV) [101]. The FIB could strip materials with a spatial resolution in the nm-range but also with depth of some µm, if necessary. From the experimental point of view it is the alternative to the XPS sputter depth profile in the µm-range (see Chap. 3.1.1). In principle three different experiments are possible with the combination of the FIB with SEM/EDX. The milling of the surface could be done simultaneously with the SEM/EDX and a 3-dimensional picture and elemental distribution results [102]. Second the FIB is used to prepare a defined defect in the surface as shown in Fig. 3.56 and analyzed by SEM/EDX or third the FIB is used to generate a very thin sample of the surface to analyze it with TEM [99].

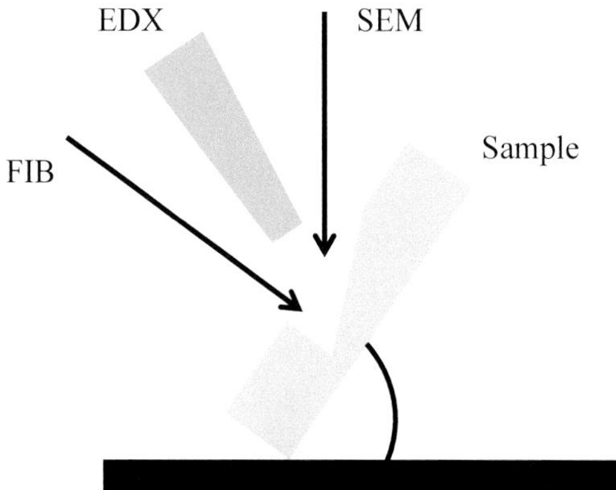

Fig. 3.55 Setup of the FIB-SEM-EDX [99, 102].

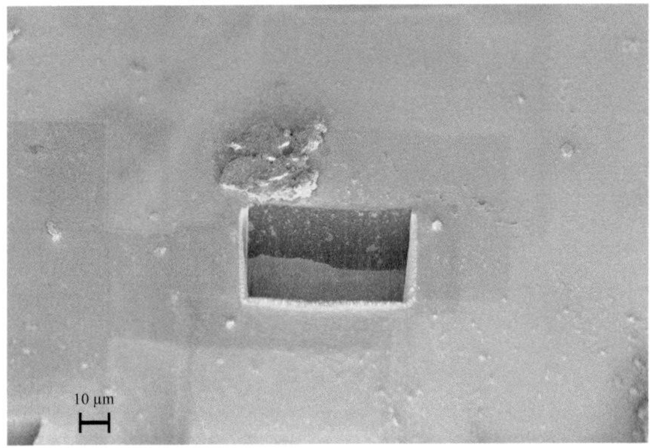

Fig. 3.56 FIB cut of an organic coating.

Example 3.17

Cataphoretic coatings (E coat) applied with a DC current between 200–400 V by decomposing water at the electrodes and the resulting increase of the pH values near the cathode as follows (for details see [103, 104]):

$$2H_2O + 4e^- \rightarrow 2H_2 + 4OH^- \qquad \text{Reac. 3.5}$$

The resin (see Fig. 1.16) is de-protonated near the surface and precipitates because of the missing electrostatic stabilization (for details see ref. [105]) of the dispersion as follows:

$$Phenoxy - NH_3^+ + OH^- \rightarrow Phenoxy - NH_2 + H_2O \qquad \text{Reac. 3.6}$$

A detailed presentation of the cataphoretic deposition of E coat with electrochemical methods is presented in ref. [106].

Because of the amphoteric character of aluminium and zinc these substrates dissolute during the cataphoretic application as follows (for thermodynamic data see Table 5.2):

$$Zn + 4OH^- \rightarrow Zn(OH)_4^{2-} + 2e^- \qquad \text{Reac. 3.7}$$

$$Al + 4OH^- \rightarrow Al(OH)_4^- + 3e^- \qquad \text{Reac. 3.8}$$

Steel substrates are passive in alkaline conditions and this causes the high stability of steel in concrete.

If E coats are analyzed with the combination of FIB and SEM/EDX, the amount of metal ions in the organic coating could be analyzed

(Figs. 3.57 and 3.58). The high amount of metal ions (Fig. 3.58) measured with EDX in the prepared defect in the coating is caused by the missing pre-treatment (zinc phosphatation). The pre-treatment reduces the free metal surface up to 95% and therefore the alkaline dissolution of the substrate is reduced to the same amount. The precipitation of the organic coating also reduces the free substrate surface, causing the gradual decrease of the ion content in the organic coating. The content of iron in the E coat on the steel

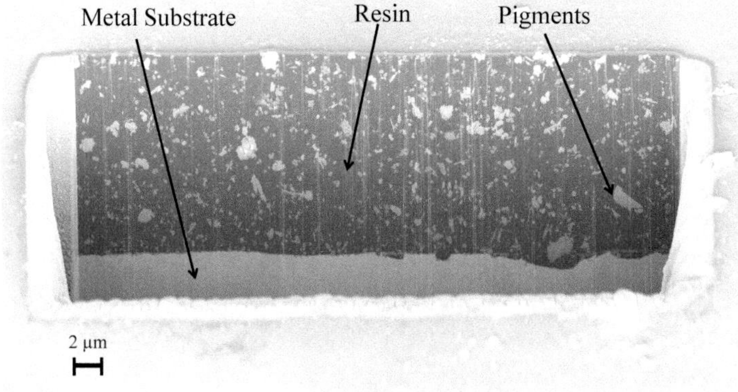

Fig. 3.57 FIB cut of an E coat for the EDX depth profile in Fig. 3.58.

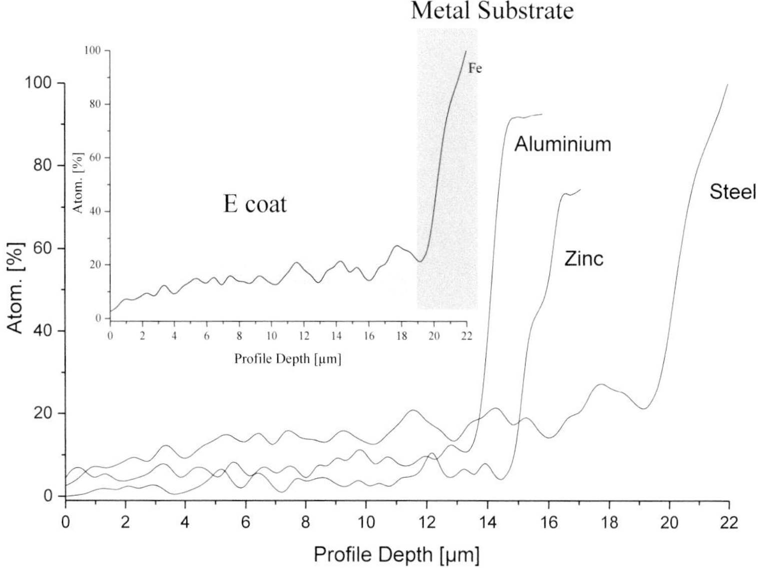

Fig. 3.58 EDX depth profile of E coats on different metal substrates (modified from [107]).

surface is surprising at first look because steel should be passive under these conditions but it has been detected in clear and pigmented E coats with a higher content in comparison to zinc and aluminium [107].

In the battery industry alkaline corrosion of steel is a well-known effect in alkaline batteries. At high pH values (40% KOH electrolyte) and elevated temperatures the battery case corroded and the electrolyte escape from the battery. Serebrennikova shows [108] that above 60°C under alkaline conditions steel dissolute as follows:

$$Fe + 4H_2O \rightarrow FeO_4^{2-} + 8H^+ + 6e^- \wedge \mathrm{E}_0 = +1.2V \qquad \text{Reac. 3.9}$$

$$Fe^{3+} + 4H_2O \rightarrow FeO_4^{2-} + 8H^+ + 3e^- \wedge \mathrm{E}_0 = +1.9V \qquad \text{Reac. 3.10}$$

The ferrate ions are not stable and decompose as follows:

$$FeO_4^{2-} \rightarrow Fe^{3+} + O_2 \qquad \text{Reac. 3.11}$$

$$2FeO_4^{2-} + 10H^+ \rightarrow 2Fe^{3+} + 1.5O_2 + 5H_2O \qquad \text{Reac. 3.12}$$

Therefore during the application of the cataphoretic coating the temperature near the substrate has to be above 60°C. This is reasonable because of the high current and potential during the application [107]. Normally high ion content in organic coatings has to be prevented because the barrier properties, i.e., the amount and rate of water/electrolyte uptake increases and the corrosion protective properties decrease. Why in an E coat the high amount of ions (may be bounded at the resin) even without a pre-treatment do not reduce the outstanding barrier properties and corrosion protection even without a pre-treatment [107] is still an open question.

3.2.2 UV-VIS Spectroscopy

Basics

The basics for UV-VIS spectroscopy are already explained in Chap. 3.1.3.

The UV-VIS-spectroscopy in corrosion research is often used to describe the photochemical degradation of organic coatings [109, 110]. Besides the photochemical degradation the UV-VIS spectroscopy is a useful tool to investigate transport and electrochemical processes of organic and inorganic coatings presented in the following paragraph.

CTTS

As already mentioned in Chap. 3.1.3, ions in aqueous solution show a CTTS [111] absorption (Fig. 3.24) in the far-UV range. Spectra in the far-UV range available in nitrogen atmosphere because oxygen absorbs below 220 nm [112] and below 188 nm UV-VIS-spectroscopy is not possible anymore.

The spectra are dominated by the anions with a slight effect of the cations illustrated in Fig. 3.59. Up to concentrations of 1–3*10^{-3} M the assumptions of Lampert-Beers-law are fulfilled (Eq. 3.15) and the concentration of the solution could be defined (Table 3.7).

Hydrochloric acid in Fig. 3.60 shows at low concentrations a negative band at 203 nm (49304 cm^{-1}) possibly caused by CO_3^{2-} or HCO_3^- in the reference solution. In pure water carbon dioxide diffuses from the atmosphere in the solution and condensed with water to carbonic acid (see Eqs. 2.1–2.3). The resulting carbonate or hydrogen carbonate generates a CTTS band (54555 cm^{-1}) and carbonate or hydrogen carbonate itself absorbs nearly 203 nm (49261 cm^{-1}) and 255 nm (39262 cm^{-1}). The spectra in Fig. 3.62 investigated water treated in atmosphere against fresh ultra-pure water (compare [113]). Therefore the water has to be as pure as possible (see Chap. 2.1.1) or the signal could be used to determine the carbonate concentration in the solution.

Table 3.7 Molar extinction coefficients and λ_{max} of the CTTS transition of different chlorides in comparison with literature. In [114] and [115] the measurements have been done at ambient conditions.

Compound	ε [l/mol*cm] at λ_{max}	Linearity [mol/l]* 10^{-3}	λ_{max} [cm^{-1}] (nm)	ε [l/mol*cm] at 193 nm from [114]	ε [l/mol*cm] at 210 nm from [115]
LiCl	1281 ± 55	1	54171 (185)		
KCl	1291 ± 108	1	54112 (185)	673	0.154
NaCl	1431 ± 22	1	54171 (185)	527	0.163
CsCl	1267 ± 97	1	54171 (185)		
$AlCl_3$	3000 ± 164	0.5	53763 (186)		
$ZnCl_2$	3118 ± 100	0.5	53879 (186)		
HCl	533 ± 21	2	54289 (184)		
KOH	2940 ± 167	0.5	53590 (187)		
NaOH	2846 ± 92	1	53476 (187)		
HCO_3^-, CO_3^{2-}			54555 (183)		3.33 ($NaHCO_3$)

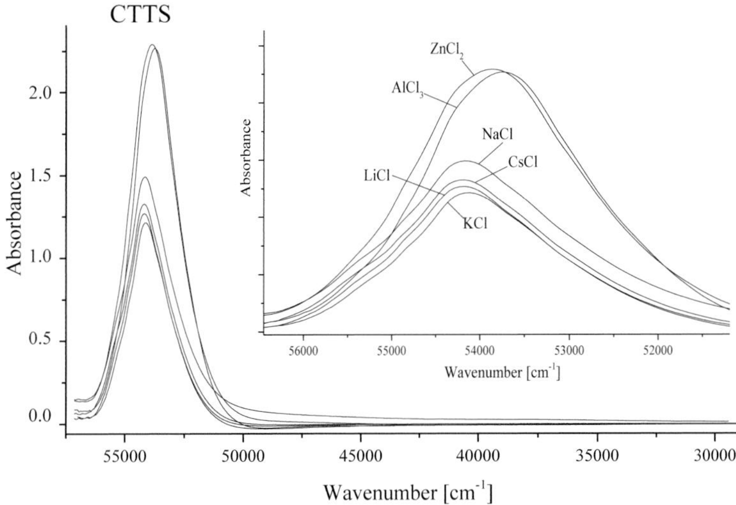

Fig. 3.59 UV-VIS spectra in nitrogen atmosphere of different chlorides with a concentration of 10^{-3} M in water.

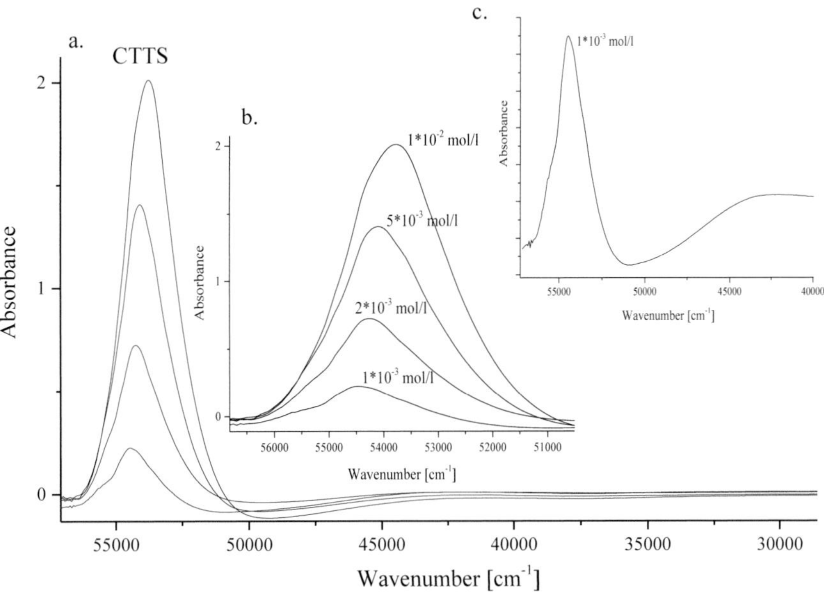

Fig. 3.60 UV-VIS spectra in nitrogen atmosphere of hydrochloric acid with different concentrations in water. a: survey spectra, b: CTTS band of different concentrations, c: possibly carbonate/hydrogen carbonate absorption in the reference.

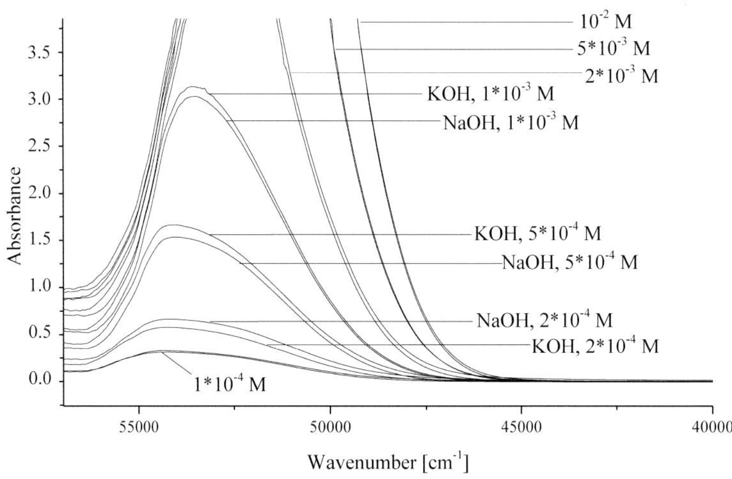

Fig. 3.61 UV-VIS spectra in nitrogen atmosphere of sodium and potassium hydroxide with different concentrations in water.

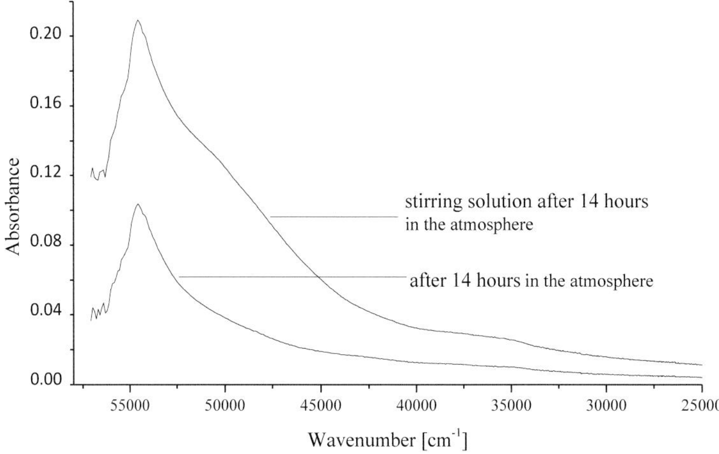

Fig. 3.62 UV-VIS spectra in nitrogen atmosphere of water with/without stirring in normal atmosphere against fresh ultra-pure water as reference.

Sodium and potassium hydroxide solutions show an absorption shift with increasing concentration, i.e., with increasing pH (Fig. 3.61) and the same effect is visible in hydrochloric acid (Fig. 3.60) and sodium chloride (Fig. 3.63) therefore it should be a general effect of the CTTS band. The molar extinction coefficient of the CTTS of hydroxide is more than twice of the chloride absorption (Table 3.7).

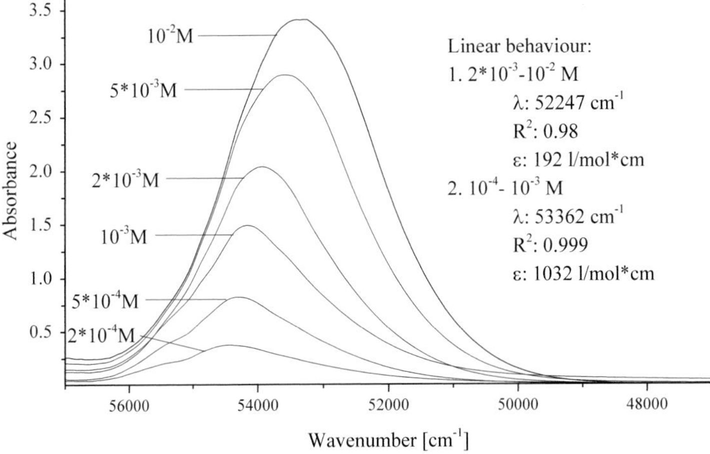

Fig. 3.63 UV-VIS spectra in nitrogen atmosphere of NaCl in water.

In corrosion research a standard electrolyte to simulate corrosive conditions is a 3% NaCl solution, i.e., 0.5 M NaCl. In Fig. 3.63 UV-VIS spectra of different concentrations of NaCl solutions are illustrated and show that an investigation of the CTTS transition at 54171 cm^{-1} is not possible in a 3% NaCl electrolyte but at higher wavelengths. At 189 nm the concentration of NaCl could be investigated with a lower resolution but with the advantage to measure in normal atmosphere (see Ex. 3.18). In literature [114, 115] wavelengths at 193 or 210 nm has been used to investigate electrolyte concentrations. There is at any concentration a linear behaviour at a certain wavelength illustrated in Fig. 3.63.

Besides the advantages of the CTTS bands the intense signal especially in saturated electrolytes could overlap or disrupt over absorption bands illustrated in Ex. 3.19.

Example 3.18

The transport of water or electrolyte through an organic coating is besides the oxygen transport the most important property regarding the corrosion protection of a coating. Therefore a lot of methods have been used to investigate the transport phenomenon. It has to be differentiated between the transport of water vapour and liquid water. In Table 3.8 some water vapour permeability's of organic coatings or composites are summarized.

The addition of inorganic pigments or fibres reduces the permeability of water vapour and also the increase of network density of the polymeric network. The disadvantage of the investigation of the water vapour permeability is the need of a free film for the experiment [117].

Table 3.8 Water vapour permeability's of different pigmented and non-pigmented organic networks.

Coating	Inorganic pigment, amount in weight [%]	Temp. [°C]	Water vapour permeability [10^{-12} g/m*s*Pa]	Lit.
Paraffin wax		25	0.2	[116]
Microcrystalline wax		25	0.3	[116]
Vinylester resin (Novolak basis)	Flake glass, 17	70	0.40	[117]
Vinylester resin (Novolak basis)	Flake glass, 15	70	1.33	[117]
Epoxy resin (BPA basis), Amine cross-linked	Glass fibre, 43	70	0.88	[117]
Vinylester resin (BPA basis)	Glass fibre, 27	70	1.96	[117]
PE-HD		70	0.46	[117]
PP-H		70	1.25	[117]
PVDF		70	2.50	[117]
Vinylester resin (Novolak basis)	Mica, 15	70	1.38	[117]
Alkyd paint		22	2.5	[118]
Vinylester resin (Novolak basis)		70	3.13	[117]
Unsaturated Polyester	Glass fibre, 46	70	3.33	[117]
Shellac		30	4.2 to 10.3	[116]
Shellac + Carnauba wax		30	1.8	[116]
Phenolic resin	Glass fibre, 34	70	4.44	[117]
Linseed oil paint		22	5.8 to 10.4	[118]
Polyurethane, highly cross-linked	Quartz, 18	50	9.79	[117]
Polyurethane, low cross-linked		40	120.84	[117]

The liquid water uptake could easily be defined by weight after immersion in water or electrolyte. To reduce the total weight especially for low amounts free films without the metal substrate have been used but there is no correlation between the behaviour of a free film and an organic coating on a metal substrate [119–122]. Therefore the EIS has been established as a standard method in the last two decades to investigate the water uptake of organic coatings on a metal substrate. The water uptake of an organic coating could easily be defined by means of EIS but to differentiate, if water or electrolyte diffuses inside the coating, is hard to analyze.

Some authors assume a pure water uptake [123–126] to describe the EIS data and others assume a stepwise uptake from water followed by a slow electrolyte uptake [127, 128]. On a free film a slow electrolyte uptake could be detected by means of radio tracer methods [121, 129] and on a coated ATR crystal the ion transport are as fast as the water diffusion [130] because both are detected at the ATR crystal at the same time. If a transport process of an electrolyte occurs, charges moves through the coating and therefore Warburg impedance should be visible and could be used to describe the electrolyte uptake illustrated in [131] (see also Chap. 2.2.2). The importance to measure the electrolyte uptake is stressed by the fact that the "water" uptake increases with decreasing ion strength (see Table 2.7) of the solution [132–134]. Finally the direct measurement of the ion transport in an organic coating on a metal substrate is not possible with electrochemical methods.

One approach is the investigation of the electrolyte uptake by means of UV-VIS spectroscopy of the electrolyte above the coating [132]. The principle is illustrated in Fig. 3.64. If only water absorbs from the organic coating, the chloride concentration of the above solution increases and also the corresponding CTTS band of chloride. With the assumption from a stepwise process according to [127, 128] the electrolyte uptake takes place after longer treatment followed by a decrease of the CTTS band. If the electrolyte is transported into the organic coating, no change of the chloride concentration could be measured. To decrease the electrolyte volume/coating surface ratio, i.e., increase the sensitivity of the method, the coating applied on a mesh wire to increase the surface area of the coating (for details see [132]).

The results for an acrylic dispersion based clear coat is illustrated in Fig. 3.65. The treatment of a 3% NaCl solution shows an increase of the UV-VIS absorption followed by a constant behaviour similar to the corresponding EIS data. In contrast with the 5% NaCl solution after a short increase of the chloride concentration the UV-VIS data shows a fast electrolyte uptake followed by a constant behaviour. The corresponding EIS data illustrate that the assumptions for the Brasher-Kingsbury equation (Eq. 2.30) are not fulfilled but besides the noise of the data a constant behaviour after the fast increase is reasonable. For comparison the treatment with a 1% NaCl shows ideal behaviour according to the Brasher-Kingsbury assumptions and a higher water uptake already mentioned in literature [132–134]. For long term measurements as illustrated in Fig. 3.66 the treatment with the 3% NaCl shows a slow electrolyte uptake as assumed in literature [127, 128]. Therefore the treatment of organic coatings with lower concentrations for short periods (< 100 hours) the assumption of a water uptake is useful. For longer treatments and for higher electrolyte concentrations an electrolyte uptake is the case at least in dispersion based coatings.

The phase (see Fig. 2.45) in the first 175 hours is stable on level 1, i.e., no Warburg impedance could be detected. Because of the fact that corrosion protection has to a be a long term effect the electrolyte uptake is

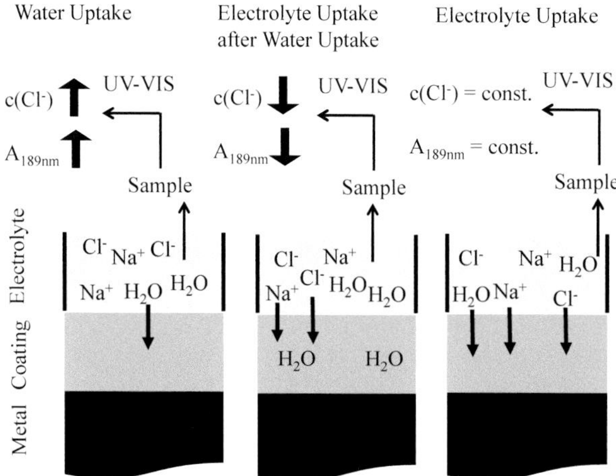

Fig. 3.64 Principle of the investigation of the electrolyte uptake of organic coatings [135].

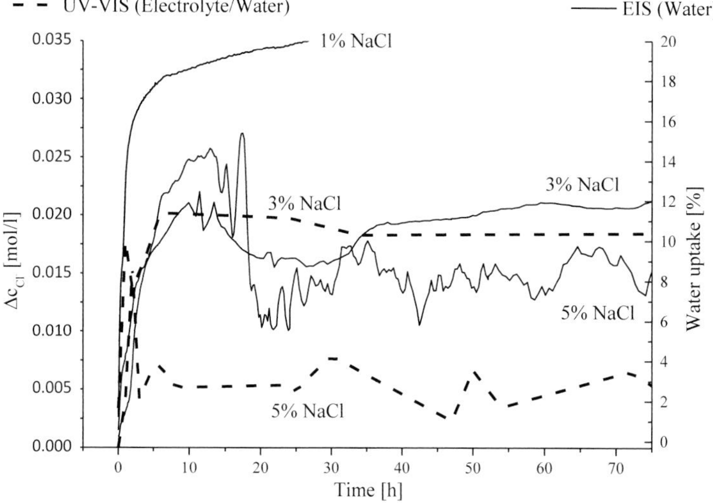

Fig. 3.65 Correlation of the chloride concentration change investigated with UV-VIS data at 189 nm from the above electrolyte on a dispersion based clear coat with the water uptake based on Eq. 2.30 with EIS data for 75 hours (modified from [132]).

the case sooner or later depending on the network density and chemical composition of the polymeric network and the amount of inorganic pigments in the coating (see Table 3.8). The limitation of the method is illustrated in ref. [132]. For coatings with a very low water/electrolyte uptake such as E coats the resolution of the UV-VIS measurement at

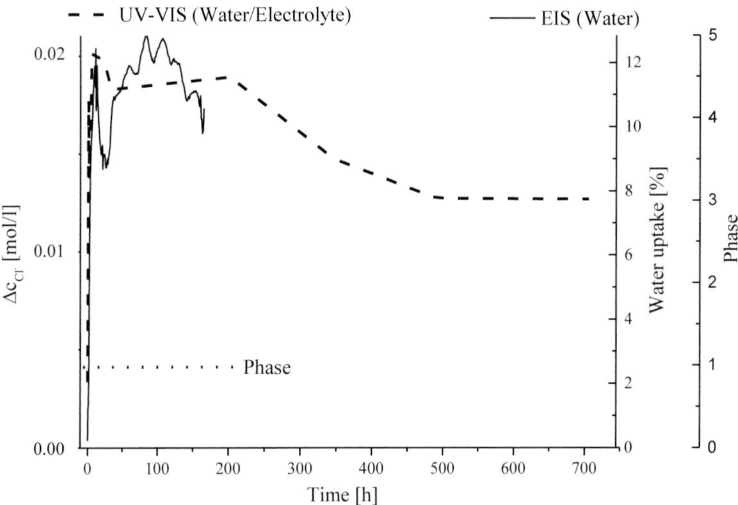

Fig. 3.66 Correlation of the chloride concentration change investigated with UV-VIS data at 189 nm from the above electrolyte (3% NaCl) on a dispersion based clear coat with the water uptake based on Eq. 2.30 with EIS data and the Phase according to Fig. 2.45 for 750 hours (modified from [132]).

189 nm of the electrolyte on the coating is inadequate for the investigation of the electrolyte uptake.

Example 3.19

In Exs. 2.3 and 2.6 the Russian mud corrosion and some methods for the investigation of the corrosion mechanism has already been illustrated. In the mechanism from Kiefer et al. [136] chromate as first corrosion product from the chromium surface has been assumed. To investigate the corrosion products from the chromium surface UV-VIS measurements from the deaerated electrolyte has been done to clarify the mechanism [137, 138]. The measurements have been performed in a nearly saturated $CaCl_2$ solution (see Ex. 2.3) generating an intense CTTS signal illustrated in Fig. 3.67. Nevertheless the corrosion product of the chromium surface under these conditions could be identified as $[Cr(H_2O)_6]^{3+}$ [139–141] at 23641 cm^{-1} (423 nm) and 16978 cm^{-1} (589 nm) and not the assumed Cr(VI) compounds at 277–280 nm and 317–370 nm [141]. In the case of chromium corrosion products the CTTS signal does not disturb the measurements even in highly concentrated electrolytes. In Table 5.9 in the app. the data for ions and corrosions products of important metal surfaces are summarized.

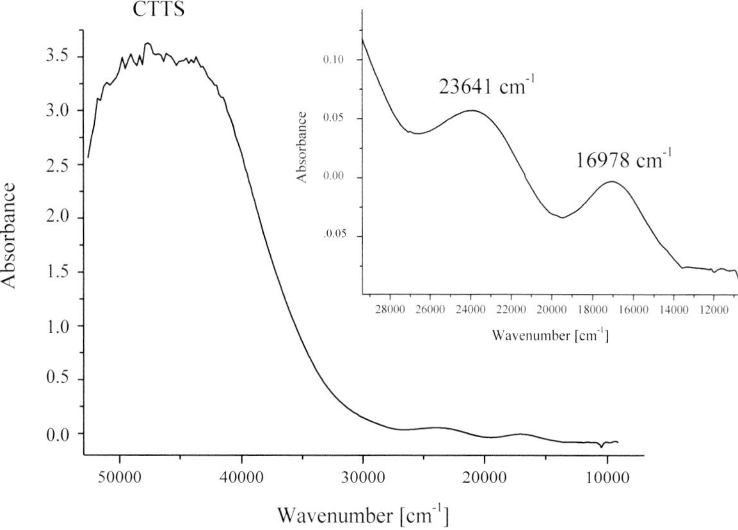

Fig. 3.67 UV-VIS spectrum in a deaerated 5.3 M $CaCl_2$ of corrosion products from a chromium surface treated in the solution [137]. With permission.

Example 3.20

Besides the cataphoretic coating another application method is used for corrosion protection based on an autophoretic process on steel substrates [103]. The application is initiated by a complex combination of different processes [142, 143] summarized in Fig. 3.68. The acidic solution generates an acidic corrosion process on the surface, i.e., dissolution of iron(II) cations. Finally the Fe^{2+} dissolutes from the surface into the bath caused the precipitation of the coating. The cations destabilize the dispersion of the resin and pigment resin particles and cause the precipitation of the organic coating [142, 143]. The dispersion precipitates on the surface forming a coating layer between 10–30 µm in the dry film because the first layer is porous and the diffusion controlled [144] iron dissolution could occur under the coating to proceed the autophoretic deposition [143]. The advantage of the process in comparison to a cataphoretic application is the absence of a potential to apply the coating and therefore there is no limitation caused by faradaic cage effects on the metal part especially for pipes. The unavoidable incorporation of ions into the coating caused by the acid corrosion process is not a disadvantage because it is in principle similar to the E coat process (see Ex. 3.17).

In the following example autophoretic coatings on steel with different pigment concentrations have been prepared and analyzed by means on

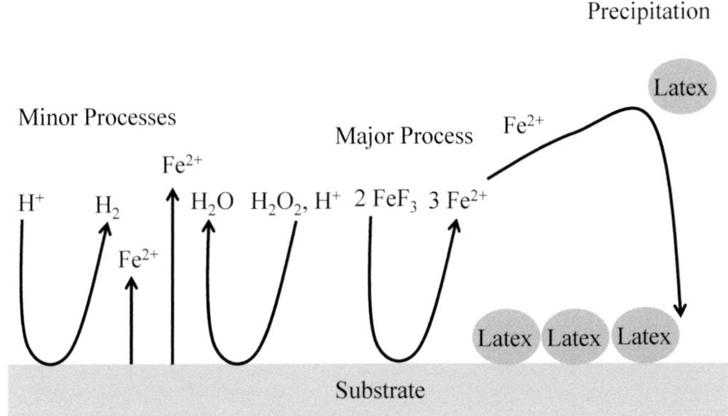

Fig. 3.68 Scheme of the initiation and the precipitation of autophoretic organic coatings.

UV-VIS-spectroscopy with the use of the Kubelka-Munk-Theory (Chap. 3.1.3 and Eq. 3.21).

To use the Kubelka-Munk-Theory some requirements of the sample have to be fulfilled [145]:

- Infinite thickness of the sample, i.e., some millimetres
- Low absorption of the sample
- Similar refractive index of the particles and the matrix

All requirements are not fulfilled for organic coatings because the thickness is far below a millimetre, the organic matrix absorbs at least in the UV-range and the refractive index of the resin and the pigments are different. If the Kubelka-Munk-Theory is applicable for organic coatings, the reflection spectrum intensity correlates with the pigment concentration as follows [146]:

$$f(R_{diffuse}) = const. \cdot c \wedge c = \text{Pigment concentration} \qquad \text{Eq. 3.37}$$

To test the usability of the Kubelka-Munk-Theory for organic coatings, autophoretic coatings with a different amount of pigments (mainly carbon black) have been prepared with a thickness of 15 μm on a steel substrate. The resulting spectra have been measured in an Ulbricht sphere (Fig. 3.27) illustrated in Fig. 3.69 and have been calculated with the Kubelka-Munk equation (Eq. 3.21) against a white standard. With the assumption that the matrix is represented by the clear coating, the pigment coating could be corrected by division with the clear coating spectrum according to the resin concentration in the formulation. The resulting spectra are shown in Fig. 3.70. These spectra show the diffusion reflection and absorption of the pigments in the coating. At the maximum at 354 nm the intensity

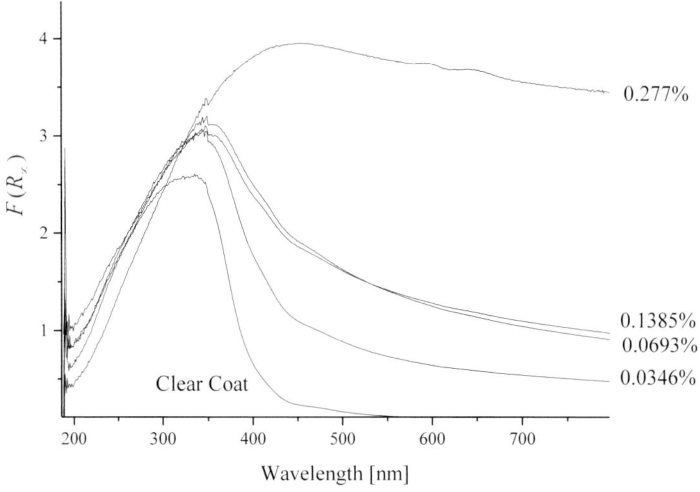

Fig. 3.69 UV-VIS spectra of autophoretic coatings (thickness 15 μm) on steel with different pigment concentrations measured in an Ulbricht sphere with a white standard.

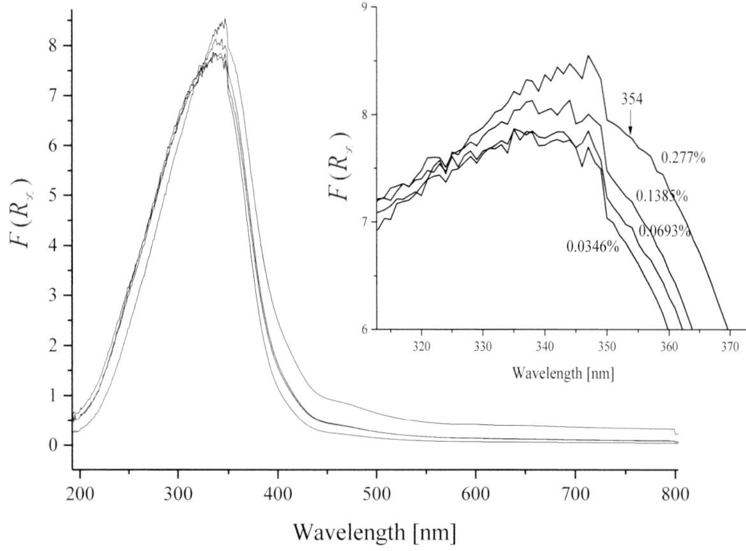

Fig. 3.70 Normalized UV-VIS spectra of autophoretic coatings (thickness 15 μm) on steel with different pigment concentrations measured in an Ulbricht sphere with a white standard.

of the spectra change with the pigment concentration. In Fig. 3.71 the correlation of the maxima to the pigment concentration is illustrated and a linear behaviour could be detected and confirmed Eq. 3.37. Therefore

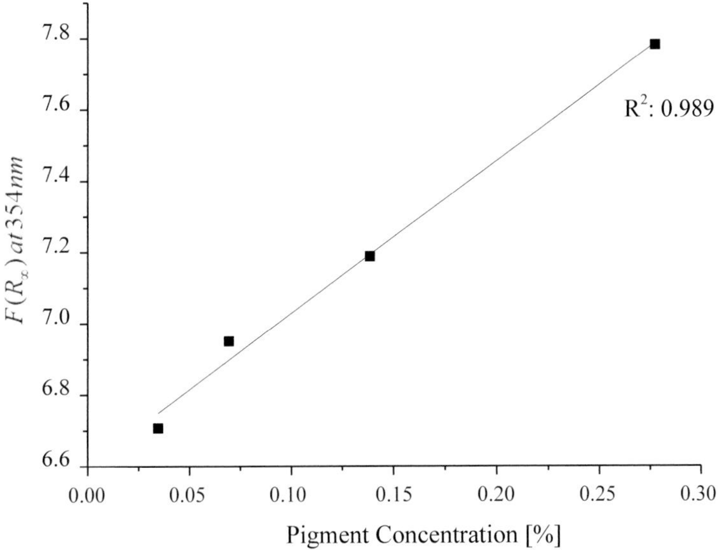

Fig. 3.71 Correlation of the intensity of the diffuse reflection at 354 nm (Fig. 3.70) with the pigment concentration according to Eq. 3.37.

the Kubelka-Munk-Theory could be used at least for autophoretic organic coatings.

There are easier methods to define the pigment concentration in an organic coating, but changes or degradation of the coating, e.g., after UV or corrosive treatment could be analyzed and sedimentation or enrichment of pigments in the coating could also be investigated. The advantage of the measurement in the Ulbricht sphere and the use of the Kubelka-Munk-Theory is that realistic coatings, i.e., applied on the metal substrate, could be used in contrast to free films, which could be analyzed by UV-VIS spectroscopy in transmission [147]. The measurement with the reflection device (Fig. 3.28) shows only a low signal on pigment coatings and therefore an increase of noise. The disadvantage of the Ulbricht sphere is the limitation of the wavelength in the UV range caused by the absorption of the sphere. For the investigation of data in the deep UV range the reflection device is the only possibility.

3.2.3 *Infrared Spectroscopy*

Basics

The basics for Infrared spectroscopy are already explained in Chap. 3.1.4.

The most important method to analyze organic coatings with infrared spectroscopy is the ATR method (see Chap. 3.1.4) because it is a fast and easy method to generate IR spectra from the surface. The penetration depth of the evanescent wave in the ATR spectroscopy depends on the wavenumber (Eq. 3.28). For example with an angle of 45° at a ZnSe ATR crystal the penetration depth for a polymer layer is 6.3 μm at 900 cm^{-1}, 3.5 μm at 1600 cm^{-1} and 1.9 μm at 3000 cm^{-1} [63]. This gets attention in the interpretation of the spectra especially for the enrichment of compounds on the coating surface or the analysis of gradients in the coating. To avoid this disadvantage IRRAS (see Chap. 3.1.4) has to be used or microtome slices have to be prepared to analyze gradients in organic coatings illustrated in ref. [148]. To analyze the polymer matrix the ATR is the method of choice.

Example 3.21

In the following example E coat samples on steel have been applied and cured for 20 minutes at different temperatures. The curing reaction (Fig. 3.72) is based on the elimination of the blocking agent on the NCO-cross-linking agent in the coating followed by the formation of a urethane group between the BPA based phenoxy resins (Fig. 1.16) and the cross-linking-agent. This elimination-addition mechanism illustrated in Fig. 3.72 is the most popular explanation of the reaction [149].

Therefore the amount of OH-groups, visible at 3302 cm^{-1}, in the coating should decrease with increasing curing temperature illustrated in Fig. 3.75 corresponding to the Arrhenius theory (Eq. 1.29). A disadvantage of the ATR method is the sensitivity to the roughness of the surface because the crystal needs a close contact to the sample. In Figs. 3.73 and 3.74 the intensities of more or less all signals changes with the curing temperature.

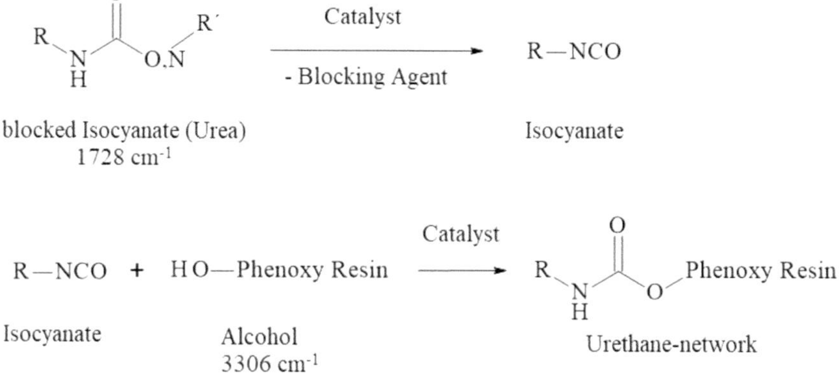

Fig. 3.72 De-blocking and cross-linking-reaction of isocyanates (compare [149]). The correspondent infrared signals illustrated in Figs. 3.74 and 3.75.

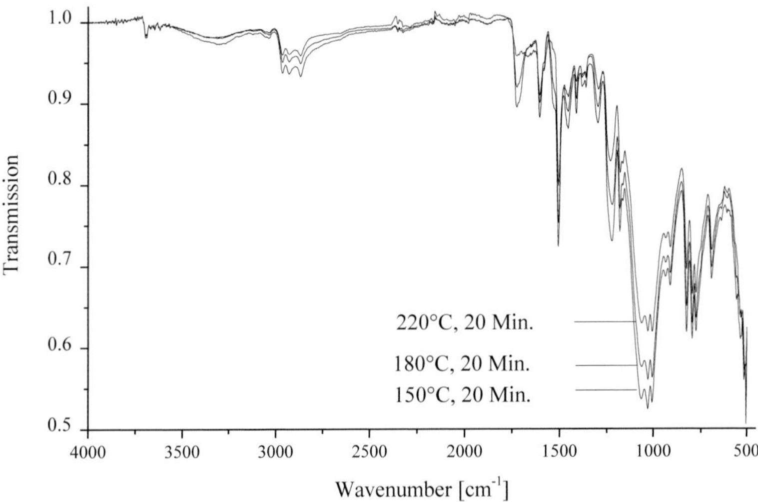

Fig. 3.73 ATR-spectra of E coats cured with different temperatures on steel measured with a germanium ATR-crystal.

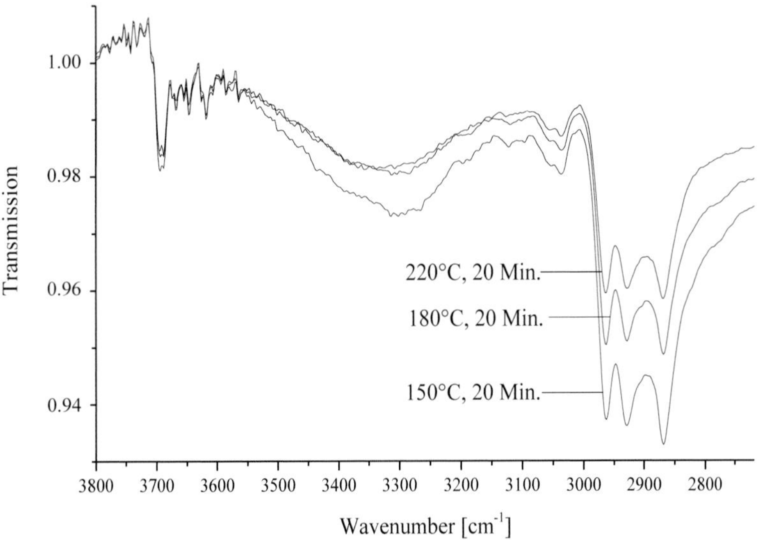

Fig. 3.74 Detail of the ATR-spectra in Fig. 3.73.

The C-H valence band at 2965 cm^{-1} (compare Table 5.8) is not influenced by the cross-linking reaction and could be used as an internal standard. The signals are close together and therefore the frequency dependence

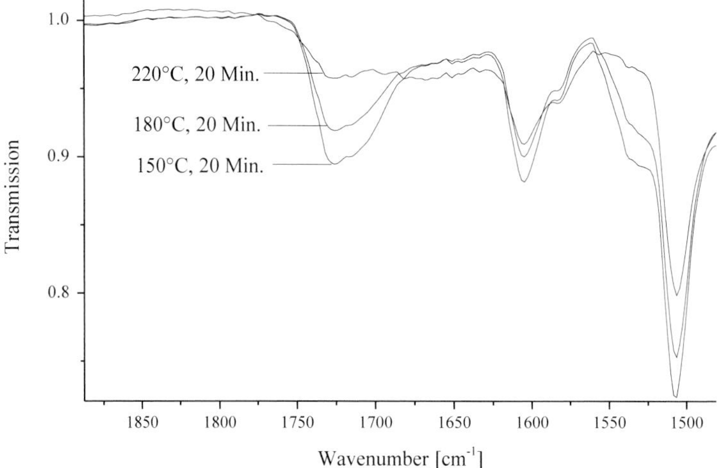

Fig. 3.75 Details of the ATR-spectra in Fig. 3.73.

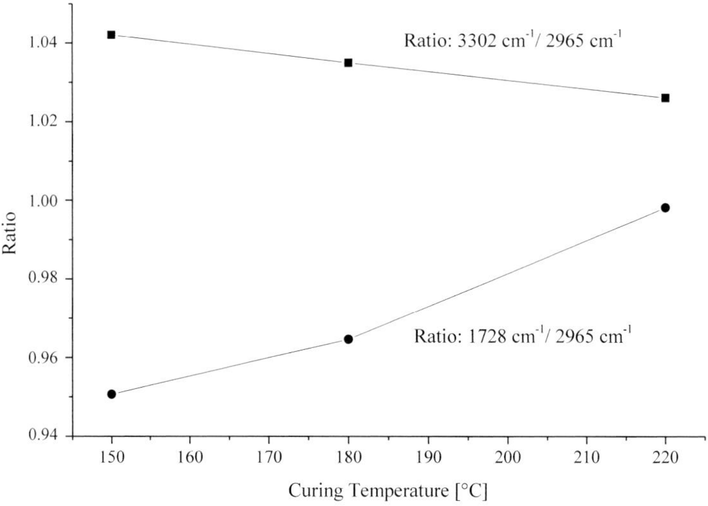

Fig. 3.76 Ratios of the infrared signal of the E coats cured with different temperatures from Figs. 3.74 and 3.75.

of the penetration depth is not relevant. In Fig. 3.76 the relation of the signal at 3302 cm^{-1} of the OH-group and the C-H-signal at 2965 cm^{-1} is illustrated and proves the assumption that the OH-group content decreases with increasing curing temperature. This behaviour could be proved by the correlation of the assumed urethane signal of the blocking agent at 1728 cm^{-1} (see Table 5.8) which shows the contrary behaviour in Fig. 3.76. Therefore the cross-linking density of the organic coating could be investigated by means of the ATR method.

As already mentioned the cross-linking density determines the barrier properties of the organic network and finally the corrosion protective properties of the organic coating.

The need for an internal standard is illustrated in Fig. 3.77. The E coat samples have been treated in a salt spray test (see Chap. 4.2.2) for 19 days and the surface has been analyzed by means of ATR afterwards. The spectra assume an increase of the OH-group, i.e., an increase of water in the coating. The relation of the OH valence band to the C-H valence band shows (Table 3.9) that there is no significant difference between the coatings and if, the increase of water is the same at a curing temperature of 150°C and 220°C. The result could be proved with the absence of the OH deformation signal of water in the range between 1625 and 1650 cm^{-1} in Fig. 3.77 (see Table 5.8). Therefore the change of the intensities is based on a change of the roughness of the coating surface illu^trated in ref. [107].

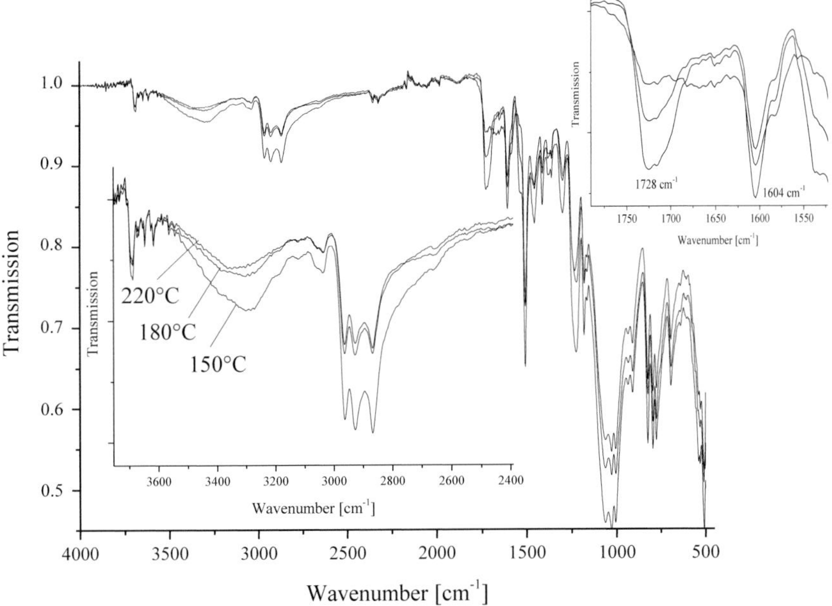

Fig. 3.77 ATR-spectra of E coats after treatment for 19 days in a salt spray test with different curing temperatures on steel measured with a germanium ATR-crystal.

Table 3.9 Change of the OH-signal at 3302 cm^{-1} after 19 days treatment in the salt spray test.

Curing temperature [°C]	Change of the signal at 3302 cm^{-1} according to the untreated coating with the signal at 2965 cm^{-1} as internal standard in [%]
150	101.5
180	100.3
220	101.4

Example 3.22

Wire enamels often based on polyamideimide (PAI) resins (see Fig. 1.16) solved in N-Methylpyrrolidone (NMP) or N-Ethylpyrrolidone (NEP). The advantages of PAI based coatings are the outstanding electric insulating properties and the high barrier properties against chemicals and corrosive environments combined with a very high thermal stability [150, 151]. The disadvantage is low solubility and the need of NMP or NEP as solvents. Both solvents have or there is the risk for a reproductive toxicity and should be replaced by non-toxic solvents. One approach is described in refs. [152, 153] with the use of N-Formylmorpholine (NFM) to replace NMP or NEP. The synthesis [152] generates a polymer solution similar to the pyrrolidone based solvents. To investigate the barrier properties ATR measurements of the cured coating have been done based on the structure-property-relation described in ref. [154]. During the polymerization of the resin and in the curing process amide and imide-groups take place illustrated in Fig. 3.78. This can be disregarded, if the amount of amide groups in the 1- and 2-position of the aromatic ring increases or the chain length based on reactions in the 5-position cause the higher amide group contained in the cured film, the ratio between the amide and the imide groups defines the thermic, mechanical and barrier properties of PAI based coatings [154].

The ATR spectra of cured coatings based on PAI synthesized in NFM and NEP are illustrated in Fig. 3.79. The resulting amide/imide ratios summarized in Table 3.10 show that the resins generated in NFM could produce coatings with higher thermal, mechanic and barrier properties caused by a different polymer structure achieved in the synthesis and/or according to [155] by different activation energies of the imide reaction in the curing process caused by the solvent. Because of the fact that the ratio between the amide and imide signal has been analyzed no internal standard is necessary.

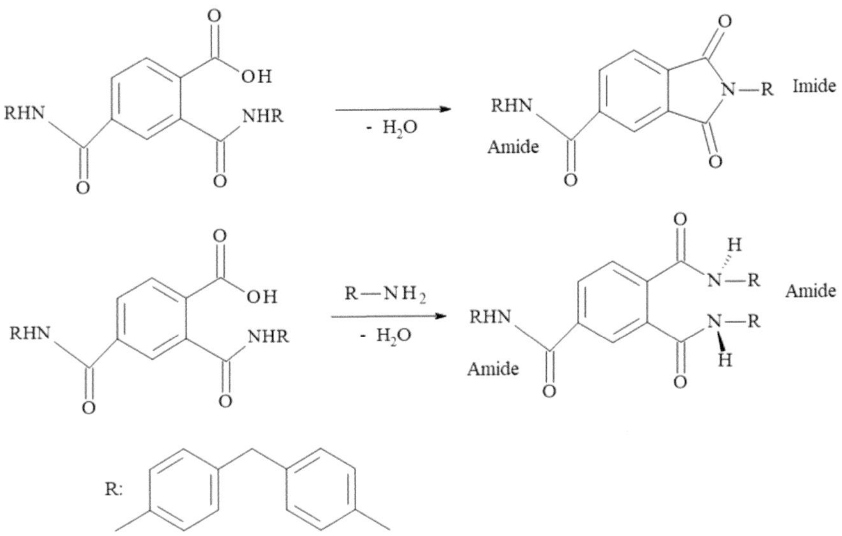

RHN NHR OH O O O → $- H_2O$ RHN Amide O O O N—R Imide

RHN NHR OH O O O $R—NH_2$ → $- H_2O$ RHN Amide O O O H N—R N—R H Amide

R:

Fig. 3.78 Polymer- and cross-linking-reaction of polyamideimides.

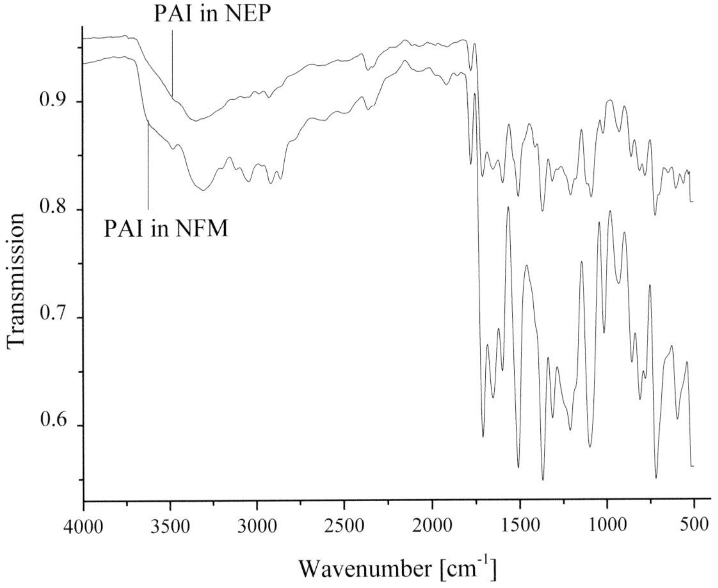

Fig. 3.79 ATR-spectra measured with a germanium ATR-crystal of cured PAI films on steel synthesized in different solvents.

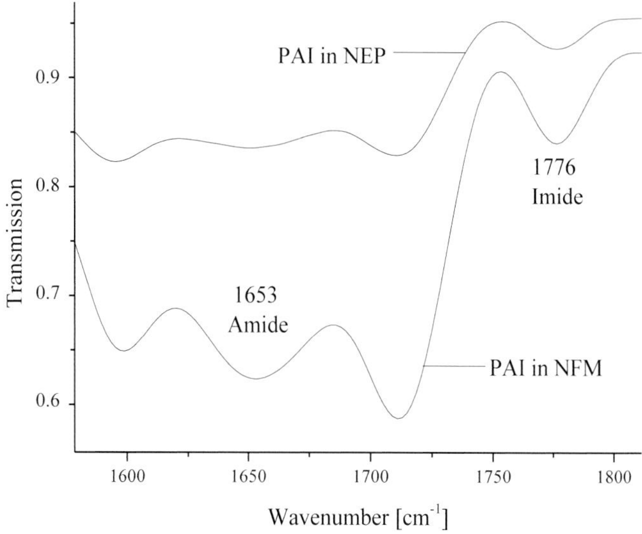

Fig. 3.80 Detail of the ATR-spectra from Fig. 3.79.

Table 3.10 Ratio of the amide signal at 1653 cm^{-1} and the imide signal at 1776 cm^{-1} of cured PAI films in different solvents [152].

Solvent	Amide/Imide ratio of the ATR signals
NEP	1.11
NFM	1.35

3.2.4 Raman Spectroscopy

Basics

The basics for Raman spectroscopy are already explained in Chap. 3.1.5.

The confocal Raman spectroscopy is a useful tool to analyze organic coatings especially coating systems with more than one layer [156–158]. The curing process could be analyzed [158] and also the chemical changes of the polymer such as crystallinity in the film [158]. From the corrosion analysis point of view the analysis of organic coated samples during corrosive treatment is a field of current research.

Electrochemical methods are useful to get information about corrosion processes at the metal surface and allow analyzing the corrosion mechanism. Spectroscopic methods are useful to get information about the chemical changes during the corrosive treatment on the surface and in an organic coating. The combination of both, i.e., the SKP and the confocal Raman spectroscopy was realized first in ref. [159] by applying Na_2SO_4 crystals as sensors at the interface of an organic coated steel surface. With this setup the diffusion of electrolyte at the interface and the electrochemical changes could be analyzed simultaneously. Without a sensor but with an unmodified interface it is hardly possible as illustrated in the next example.

Example 3.23

The advantage of the Raman spectroscopy is the impassiveness of the method against humidity, i.e., the possibility to measure in electrolyte/ water swollen organic coatings. The second advantage is the possibility to use a confocal optic to achieve spectra inside or from the corrosion research point of view more important from the metal/coating interface. The disadvantage is the low sensitivity of the method and therefore a correlation with a second independent method is necessary to avoid misinterpretation of the spectra illustrated for an E coat in Fig. 3.81. The Raman signal (laser 785 nm) from the bulk could be correlated to the corresponding infrared signals from the surface.

With the checked Raman signals the measurement of the metal/ coating interface of a sample with a defect similar to Fig. 2.63 could be performed with the use of a confocal Raman microscope (see Fig. 3.49). The electrolyte diffuses from the defect into the interface and changes the chemical situation visible with small shifts of the Raman signals shown in Fig. 3.82. The intensity and the signal/noise ratio of the signals are low, the shift is small and therefore the interpretation of the data is hardly possible. There is no second independent method because all mentioned spectroscopic methods cannot be used in a humid atmosphere or with a confocal optic. Furthermore electrochemical methods such as SKP give no chemical structure information.

With the use of a laser with a different emission wavelength (633 nm) a second method is available—not independent but at least a proof of the experiment. In Fig. 3.83 the Raman spectra measured by means of a 633 nm laser on the same samples shows similar signals (compare [161]). In Table 3.11 the signals of all spectra are summarized. The shift of the signals at the wet interface according to the bulk spectra could be proved with the second Raman experiment. Furthermore there could be an effect caused by different pH values of the electrolyte. The shoulder at 810 cm^{-1} in the interface spectra shows an increase of OH-groups at the interface.

With the results the need for enhanced methods illustrated in Chap. 3.1.5 is reasonable but for instance the SERS method changes the surface and therefore the interface but it could be one option [162]. Only the

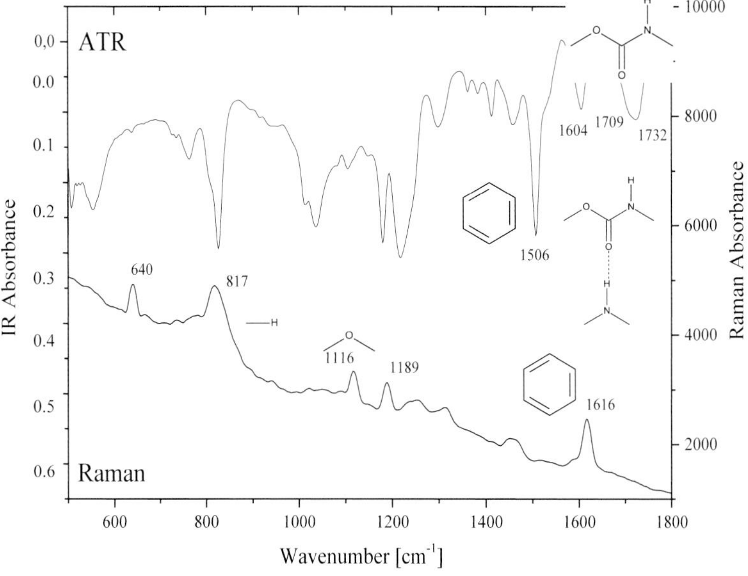

Fig. 3.81 Comparison of the ATR spectrum and the Raman spectrum (785 nm) of a clear E coat on an aluminium substrate (compare [160]).

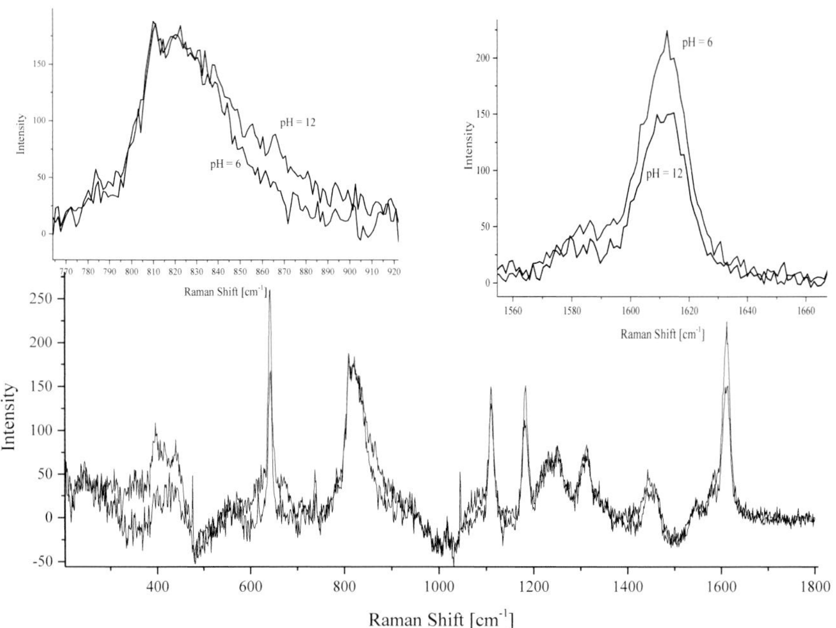

Fig. 3.82 Raman spectra (785 nm) focussed at the E coat aluminium interface near the defect with different pH values of the 1 M Na_2SO_4 electrolyte (compare [160]).

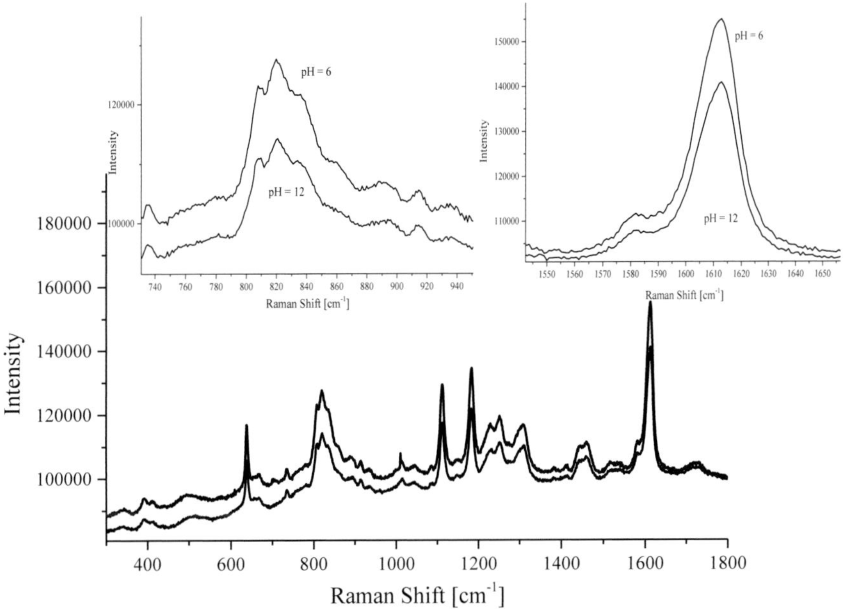

Fig. 3.83 Raman spectra (633 nm) focussed at the E coat aluminium interface near the defect with different pH values of the 1 M Na_2SO_4 electrolyte (compare [160]).

Table 3.11 Raman signals of the bulk and the humid interface in normal atmosphere with different pH values of the 1 M Na_2SO_4 electrolyte (compare Table 5.8).

Signal	Bulk 785 nm [cm^{-1}]	pH = 6 785 nm [cm^{-1}]	pH = 12 785 nm [cm^{-1}]	pH = 6 633 nm [cm^{-1}]	pH = 12 633 nm [cm^{-1}]
Arom. ring (C=C)	1615.8	1612.5	1611.9	1612.9	1612.9
C-H wagging/C-O-C stretching arom. Ring	1188.7	1183.8	1181.5	1183.5	1182.6
C-O-C stretching	1116.1	1100.6	1110.6	1112.1	1112.1
C-H wagging	817.4	822.5	820.2	819.9	820.9
C-C-OH stretching	/	810.0	811.1	810.9	809.9

resonance Raman spectroscopy do not change the interface but even the Ar-Laser (Table 3.6) with an emission at 244 nm could hardly excite the functional groups at the resin (see Table 5.9) but with a risk of photochemical reactions in the organic coating caused by the laser treatment.

Example 3.24

The advantage of the high sensitivity of the infrared spectroscopy comprised the risk of artefacts caused by the atmosphere (see Fig. 3.37) or impurities on the surface. The Raman spectroscopy allows to prove the infrared spectra as a second independent method as illustrated in Fig. 3.84 for hexakis(methoxymethyl) melamine (HMMM) because water and carbon dioxide do not disturb the spectrum and with data from literature from both methods (see Table 5.8 in the app.) the analysis of the structure or at least the main functional groups of the compounds could be investigated and could be proved by each other.

The ideal structure of HMMM (R = Methyl in Fig. 1.17) is realized in technical products for more than 95% but a certain amount of free OH- or NH-groups are visible in both spectra in the range between 3000–3500 cm^{-1}. Zhang et al. [156] used the confocal Raman spectroscopy to analyze HMMM cluster in clear and pigmented coil coating. The confocal scanning microscope allows a 3D analysis of the organic coating, i.e., the analysis of the complete coating in comparision to the ATR or IRRAS. The challenge of the method is to find the relevant signal as basis for the 3D mapping.

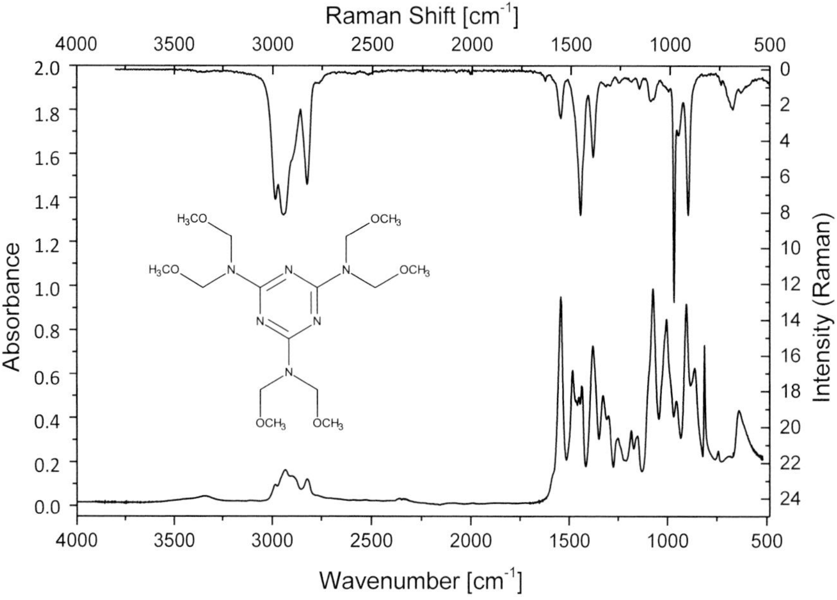

Fig. 3.84 Raman and ATR spectrum of pure HMMM.

3.3 Literature

1. B. J. Wood, Mater. Forum 35, 1 (2011) 12–19
2. Methods in Physical Chemistry, First Edition, Edited by R. Schäfer, P. C. Schmidt, Wiley-VCH, 2012, A. Klein, T. Mayer, A. Thissen, W. Jaegermann, Photoelectron Spectroscopy in Materials Science and Physical Chemistry: Analysis of Composition, Chemical Bonding, and Electronic Structure of Surfaces and Interfaces
3. P. Marcus, F. Mansfeld, Analytical Methods in Corrosion Science and Engineering, CRC Press, 2006
4. P. W. Atkins, Physical Chemistry, Oxford University Press, 1978
5. A. Einstein, Ann. Phys. 17 (1905) 132
6. D. A. Shirley, Phys. Rev. 135 (1972) 4709
7. M. Dornbusch, G. Grundmeier, Formation of thin inorganic amorphous films from aqueous solution, Poster Session Bunsentagung, Kiel, 2003
8. R. A. Reichle et al., Can. J. Chem. 53 (1975) 3841
9. P. Taher, K. Lill, J. H. W. de Wit, M. C. Mol, H. Terryn, J. Phys. Chem. C 116 (2012) 8426–843
10. C. Huang, Z. Tang, Z. Zhang, J. Am. Ceram. Soc., 84 [7] 1637–38 (2001)
11. C. Stromberg, P. Thissen, I. Klueppel, N. Fink, G. Grundmeier, Electrochim. Acta 52 (2006) 804–815
12. A. Clearfield, G. P. D. Serrette, A. H. Khazi-Syed, Catalysis Today 20 (1994) 295–312
13. F. A. Cotton, G. Wilkinson, Advanced Inorganic Chemistry, Wiley-Interscience, New York, 1962
14. A. A. M. Ali, M.I. Zaki, Thermochim. Acta 336 (1999) 17–25
15. Thesis (PhD), T. Lostak, Elektrochemische und grenzflächenchemische Untersuchungen an Zr-basierenden Konversionsschichten auf verzinktem Bandstahl, University Duisburg-Essen, 2013
16. H. E. Mohammadloo, A. A. Sarabi, R. M. Hosserini, M. Sarayloo, H. Sameie, R. Salimi, Prog. Org. Coat. 77 (2014) 322–330
17. P. Puomi, H. M. Fagerholm, J. B. Rosenholm, R. Sipilä, Surf. Coat. Technol. 115 (1999) 79–86
18. S. K. Sharma, X-Ray Spectroscopy, InTech open, 2011
19. M. Mulisch, U. Welsch, Romeis Mikroskopische Technik 18. Aufl., Spektrum Akademischer Verlag, Heidelberg, 2010
20. V. D. Scott, G. Love, S. J. R. Reed, Quantitative Electron-Probe Microanalysis, 2nd Ed., Ellis Horwood, New York, 1995
21. S. J. R. Reed, Electron Probe Microanalysis, 2nd Ed., Cambridge University Press, Cambridge, 1993
22. N. Taylor, Energy Dispersive Spectroscopy 2nd Ed. in Microscopy EKB Series Ed. J. Heath, John Wiley & Sons Ltd., Southern Gate, Chichester, West Sussex, 2015
23. K. F. J. Heinrich, D. E. Newbury, R. L. Myklebust, Energy Dispersive X-Ray Spectrometry, NBS Special Publication 604, 1981
24. L. J. Allen, A. J. D´Alfonso, B. Freitag, D. O. Klenov, Mater. Res. Soc. 37, 2012, 47–52
25. A. C. Thompson, D. Vaughan, X-Ray Data Booklet 2nd Ed., Lawrence Berkely National Laboratory, University California, Berkeley, 2001
26. W. Zhou, Z. L. Wang, Scanning Microscopy for Nanotechnology—Techniques and Applications, Springer Science + Business Media, LLC, New York, 2006
27. M. Smoluchowski, Phys. Chem. 92 (1917) 129
28. Thesis (PhD), S. Toews, Corrosion Protection by Selective Addressing of Polymer Dispersions to Electrochemical Active Sites, University Paderborn, 2010
29. W. Bremser, M. Dornbusch, H. Hintze-Brüning, S. Toews, Verfahren zur autophoretischen Beschichtung, Beschichtungsmittel und Mehrschichtlackierung, DE102010019245.7

30. S. Toews, W. Bremser, H. Hintze-Brüning, M. Dornbusch, New concepts for corrosion protection, EUROCORR 2010 – The European Corrosion Congress, Moscow, 2010
31. S. Toews, W. Bremser, H. Hintze-Brüning, S. Sinnwell, M. Dornbusch, R. Bautista-Mester, W. Kreis, Smart Functionalized Polymer Dispersions for effective mapping of heterogeneous metal surfaces: New Concepts for corrosion protection, Coatings Science International Conference Noordwijk, 2010
32. M. Dornbusch, T. Biehler, M. Conrad, A. Greiwe, D. Momper, L. Schmidt, M. Wiedow, JUnQ, 6, 2, 1-7, 2016
33. H.-H. Perkampus, UV-VIS-Spektroskopie und ihre Anwendungen, Springer-Verlag, Berlin, 1986
34. R. G. Mortimer, Physical Chemistry, The Benjamin/Cummings Publishing Company, Redwood City, 1993
35. M. Hesse, H. Meier, B. Zeeh, Spektroskopische Methoden in der organischen Chemie, Georg Thieme Verlag, Stuttgart, 2012
36. J. B. Lambert, S. Gronert, H.S. Shurvell, D.A. Lightner, Spektroskopie, 2. Aufl., Pearson, München, 2012
37. G. Kortüm, Reflectance Spectroscopy/Reflexionsspektroskopie Springer, Berlin 1969
38. W. W. Wendlandt, H.G. Hecht, Reflectance Spectroscopy, Interscience, New York, 1966
39. Shriver & Atkins, Inorganic Chemistry, 5th Ed. Oxford University Press, 2010
40. H. Benderly, M. Halmann, J. Phys. Chem. 71, 4 (1967) 1053–1160
41. R. E. Verrall, W.A. Senior, J. Chem. Phys. 50 (1969) 2746–2750
42. F. Williams, S. P. Varma, S. Hillenius, J. Chem. Phys. 64 (1976) 1549–1554
43. D. Majumdar, J. Kim, K.S. Kim, J. Chem. Phys. 112 (2000) 101–105
44. V. Di Noto, M. Mecozzi, Appl. Spectrosc. 51, 9 (1997) 1294–1302
45. G. T. Dair, R. A. Ashman, R. H. Eikelboom, F. Reinholz, P. P. van Saarloos, Arch. Ophthalmol. 119 (2001) 533–537
46. B.-H. Chai, J.-M. Zheng, Q. Zhao, G. H. Pollack, J. Phys. Chem. A 112 (2008) 2242–2247
47. J. S. Winn, Physical Chemistry, Harper Collins College Publishers, New York, 1995
48. N. M. Saadatabadi, M. R. Nateghi, M. B. Zarandi, Polymer Science Ser. A, 57, 4 (2015) 480–488
49. R. Lopez, R. Gomez, J. Sol-Gel Sci. Technol. 61 (2012) 1–7
50. E. Marquez, J. M. Gonzalez-Leal, A. M. Bernal-Oliva, R. Prieto-Alcon, J. C. Navarro-Delgado, M. Vleek, Surf. Coat. Technol. 122 (1999) 60–66
51. J. Tauc, Amorphous and Liquid Semiconductors, Plenum Press, London, 1974
52. M. Santamaria, D. Huerta, S. Piazza, C. Sunseri, F. Di Quarto, J. Electrochem. Soc. 147 (2000) 1366
53. F. Di Quarto, F. La Mantia, M. Santamaria, Eds. S.-I. Pyun, J.-W. Lee, Modern Aspects of Electrochemistry No. 46, Progress in Corrosion Science and Engineering I, Springer, Heidelberg, 2009
54. F. Di Quarto, M. C. Romano, M. Santamaria, S. Piazza, C. Sunseri, Russ. J. Electrochem. 36 (2000) 1203
55. F. Di Quarto, C. Sunseri, S. Piazza, M.C. Romano, J. Phys. Chem. B 101 (1997) 2519
56. J. P. Bonnelle, et al. Surface, Properties and Catalysis of Non-Metals, F.S. Stone, UV-Visible Diffuse Reflectance Spectroscopy applied to bulk and surface properties of Oxides related solids 237-272, D. Reidel Publishing Company, 1983
57. D. Boulainine, A. Kabir, I. Bouanane, B. Boudjema, G. Schmerber, J. Electron. Mater. 45, 8 (2016) 4357–4363
58. F. Petit, H. Debontride, M. Lenglet, G. Juhel, D. Verchere, Surface Modification Technologies VIII, 1995, Edited by T.S. Sudarshan, M. Jeandin, The Institute of Materials, 199–203
59. E. Pere, H. Cardy, O. Cairon, M. Simon, S. Lacombe, Vib. Spectrosc. 25 (2001) 163–175
60. M. B. Valcare, S. R. De Sanchez, M. Vazquez, J. Mater. Sci. 41 (2006) 1999–2007
61. A. Roustila, J. Chene, C. Severac, J. Alloys Compd. 356-357 (2003) 330–335

62. B. Beden, Mater. Sci. Forum, Vols. 192–194, pp 277–290, Trans Tech Publications, Switzerland, 1995
63. B. Stuart, Infrared Spectroscopy: Fundamentals and Applications, Wiley & Sons, Chichester, 2004
64. G. Steiner, V. Sablinskas, M. Kitsche, R. Salzer, Anal. Chem. 78, 8 (2006) 2487–93
65. W. P. Ulrich, H. Vogel, Biophys. J. 76, 3 (1999) 1639–1647
66. R. Mendelsohn, J.W. Brauner, A. Gericke, Annu. Rev. Phys. Chem. 46 (1995) 305–34
67. I. Cornut, B. Desbat, J.M. Turlet, J. Dufourcq, J. Biophys. 70, 1 (1996) 305–312
68. R. D. Andouri, H.-J. Kim, R.-H. Yoon, J. Zhang, Method of protecting metals from corrosion using thiol compounds, EP000001568800A1
69. Molecular assemblies as protective barriers and adhesion promotion interlayer, US000005487792A
70. H.-J. P. Adler, C. Bram, R. Feser, E. Jaehne, C. Jung, I. Maege, J. Rudolph, L. Sebralla , M. Stratmann , Method for treating metallic surfaces, WO001998029580A1
71. D. Crotty, J. Girard, T. Nahlawi, Inhibiting aluminium corrosion with mercapto-substituted silanes, WO002002072283A1
72. Q. Dai, A. Freedman, G. N. Robinson, Surface Modification Technologies IX, T. S. Sudarshan, W. Reitz, J. J. Stiglich, Metals & Materials Society, 1996
73. G. Lefevre, Adv. Colloid Interface Sci. 107 (2004) 109–123
74. H. Ishida, J. Adhesive and Sealant Council, Inc. 1996, Spring convention, Vol. 1, 521–533
75. Handbook of thin film materials, edited by H. S. Nalwa, Volume 2: Characterisation and Spectroscopy of Thin Films, Academic Press, 2002, F. Urs, ATR Spectroscopy of thin films, 191–229
76. Perkin Elmer, FT-IR-spectroscopy, 007024_01, 2004
77. Thesis (BA), S. Burchert, Analyse von modifizierten Zinkoberflächen mit der IRRAS-Methode, Niederrhein University of Applied Sciences Krefeld, 2015
78. H. F. Gao, H. Q. Tan, J. Li, Y. Q. Wang, J. Q. Xun; Surf. Coat. Technol. 212 (2012) 32–36
79. K. A. Saburov, Kh. M. Kamilov; Chem. Nat. Compd. 25 (1989) 695–698
80. H. Tanaka, T. Kitazawa, N. Hatanaka, T. Ishikawa, T. Nakayama, Ind. Eng. Chem. Res. 51 (2012) 248–254
81. T. Biestek, Atmospheric corrosion testing of electrodeposited zinc and cadmium coatings, Atmospheric Corrosion, W.H. Ailor, Ed. Wiley, New York, 1982, 631–643
82. T. Devine, Proceedings of Corrosion/97, Research Topical Symposia, New Orleans, March 1997, 131–162
83. Kirk-Othmer encyclopedia of chemical technology, E.W. Smith, Raman scattering, Vol. 21. Wiley, Hoboken, 2006, 321--330
84. NATO ASI Series, Electrochemical and Optical Techniques for the Study and Monitoring of Metallic Corrosion. Edited by M.G.S. Ferreira, C.A. Melendres, C.A. Melendres, Laser Raman Spectroscopy, Principles and Applications to corrosion Studies, Kluwer Academic Publishers, Netherlands, 1991, 355–388
85. N. J. Everall, JCT Coatings Tech (2005) 38–44
86. Encyclopedia of Electrochemistry, Edited by A.J. Bard and M. Stratmann, 2007 Wiley-VCH Verlag GmbH & Co. KGaA, Weinheim, Vol. 3, Instrumentation and Electroanalytical Chemistry, Chap. 2.6 Z.-Q. Tian, B. Ren, Raman Spectroscopy of Electrode Surfaces, 572–659
87. M. Fleischmann, P. J. Hendra, A. J. McQuillan, Chem. Phys. Lett. 26 (1974) 167
88. R. K. Chang, T. E. Furtak, eds., Surface Enhanced Raman Scattering, Plenum Press, New York, 1982
89. A. Otto, Light Scattering in Solids IV, Topics in Applied Physics, 54, M. Cardona, G. Guntherodt, eds. p. 289, Springer-Verlag, Heidelberg, Germany, 1984
90. C. V. Raman, K. S. Krishnan, Nature 121 (1928) 169
91. Thesis (PhD), T. Vosgröne, Untersuchungen zum oberflächenverstärkten Raman-Effekt auf Einzelmolekülebene, University Siegen, 2004

92. T. Dörfer, M. Schmitt, J. Popp, J. Raman. Spectrosc. 38 (11): 1379–1382
93. J. Behringer, J. Brandmuller, Z. Elektrochem. 60 (1956) 643–673
94. K. P. J. Williams, I. P. Hayward, Proceedings of the International Conference on Advances Surfaces Treatment: Research & Applications (ASTRA), November 2003, Hyderabad, Indian, 3-6, 435-441
95. Y. Y. Chen, S. C. Chung, H. C. Shih, Corros. Sci. 48 (2006) 3547–3564
96. W. Feitknecht, Werkst. Korros. 1(1965) 15–26
97. I. O. Wallinder, C. Leygraf, Reaction sequences in atmospheric corrosion of zinc, in: W. W. Kirk, H. H. Lawson (Eds.), Atmospheric Corrosion, ASTM STP 1239, ASTM Philadelphia, 1995, p. 215
98. Encyclopaedia of Electrochemistry, Edited by A.J. Bard and M. Stratmann, 2007 Wiley-VCH Verlag GmbH & Co. KGaA, Weinheim, Vol. 4, Corrosion and Oxide Films, Chap. 3.1, C. Leygraf, 191–215
99. V. G. M. Sivel, J. van den Brand, W. R. Wang, H. Mohdadi, F. D. Tichelaar, P. F. A. Alkemade, H. W. Zandbergen, J. Microsc. 214, 3 (2004) 237–245
100. C. A. Volkert, A. M. Minor, MRS BULLETIN 32 (2007) 389–399
101. S. Reyntjens, R. Puers, J. Micromech. Microeng. 11 (2001) 287–300
102. M. Kato, T. Ito, Y. Aoyama, K. Sawa, T. Kaneko, N. Kawase, H. Jinnai, J. Polym. Sci., Part B: Polym. Phys. 45 (2007) 677–683
103. M. Dornbusch, R. Rasing, U. Christ, Epoxy Resins, Vincentz Network, Hanover, 2016
104. G. E. F. Brewer, Chap. 26 Electrodeposition of Polymers, 2006 by Taylor & Francis Group, LLC
105. T. Brock, M. Groteklaes, P. Mischke, European Coatings Handbook, Vincentz Network, Hanover, 2010
106. F. Beck, H. Guder, J. Appl. Electrochem. 15 (1985) 825–836
107. M. Reichinger, W. Bremser, M. Dornbusch, Electrochim. Acta 231 (2017) 135–152
108. I. Serebrennikova, I. Paramasivam, P. Roy, W. Wei, S. Virtanenb, P. Schmuki, Electrochim. Acta 54 (2009) 5216–5222
109. Y. Wang, Z. Sun, J. Tian, H. Wang, H. Wang, Y. Ji, MATERIALS SCIENCE (MEDŽIAGOTYRA). 22, 2 (2016) 290–294
110. ACS Symposium Series, 805, Service Life Prediction, Washington, 2002, K. Adamsons, Chemical Depth Profiling of Automotive Coating Systems Using IR; UV-VIS, and HPLC Methods, 185–211
111. D. Majumdar, J. Kim, K. S. Kim, J. Chem. Phys. 112 (2000) 101–105
112. R. E. Huffman, Can. J. Chem. 47 (1969) 1823–1834
113. A. Ghadami, J. Ghadam, M. Idrees, Iran. J. Chem. Chem. Eng. 32, 3 (2013) 27–35
114. G. T. Dair, R.A. Ashman, R. H. Eikelboom, F. Reinholz, P. P. van Saarloos, Arch. Ophthalmol. 119 (2001) 533–537
115. V. Di Noto, M. Mecozzi, Appl. Spectrosc. 51, 9 (1997) 1294–1302
116. M. E. Embuscado, K. C. Huber, Edible Films and Coatings for Food Applications, Springer, Dordrecht, Heidelberg, New York, 2009
117. B. Gibbesch, D. Schedlitzki, Siershahn, Kautschuk gummi kunstst. 49, 6 (1996) 452–547
118. A. Ruus, P. Peetsalu, E. Tohvri, T. Lepasaar, K. Kirtsi, H. Muoni, J. Resev, E. Tungel, T. Kabanen, Agronomy Research Biosystem Engineering Special Issue 1, 197–201, 2011, 195–201
119. A. S. Castela, A. M. Simões, Prog. Org. Coat. 46 (2003) 130–134
120. M. Sun, F. Liu, H. Shi, E. Han, Metall. Sin. (Engl. Lett.) 22, 1(2009) 27–34
121. Encyclopaedia of Electrochemistry, Edited by A. J. Bard and M. Stratmann, 2007 Wiley-VCH Verlag GmbH & Co. KGaA, Weinheim, Vol. 4, Corrosion and Oxide Films, Chap. 5.4 G. Grundmeier, A. Simoes, Corrosion Protection by Organic Coatings, 500–566
122. S. Shreepathi, S. M. Naik, M. R. Vattipalli, J. Coat. Technol. Res. 9, 4 (2012) 411–422
123. M. Stratmann, R. Feser, Farbe&Lack 100 (1994) 93–99

124. G. Bierwagen, D. Tallmann, J. Li, L. He, C. Jeffcoate, Prog. Org. Coat. 46 (2003) 148–157
125. A. S. Castela, A. M. Simões, Prog. Org. Coat. 46 (2003) 55–61
126. R. E. Lobnig, V. Bonitz, K. Goll, W. Villalba, R. Schmidt, P. Zanger, J. Vogelsang, I. Winkels, Prog. Org. Coat. 60 (2007) 77–89
127. J. E. G. Gonzalez, J. C. Mirza Rosca, J. Adhesion Sci. Technol. 13, 3 (1999) 379–391
128. G. Lendvay-Gyorik, T. Pajkossy, B. Lengyel, Prog. Org. Coat. 59 (2007) 95–99
129. C. A. Kumins, J. Polym. Sci. Part C 10 (1965) 1–9
130. L. Philippe, C. Sammon, S. B. Lyon, J. Yarwood, Prog. Org. Coat. 49 (2004) 302–314
131. S. Skale, V. Dolecek, M. Slemnik, Corros. Sci. 49 (2007) 1045–1055
132. M. Dornbusch, S. Kirsch, C. Henzel, C. Deschamps, S. Overmeyer, K. Cox, M. Wiedow, U. Tromsdorf, M. Dargatz, U.Meisenburg, Prog. Org. Coat. 89 (2015), 332–343
133. Q. Zhou, Y. Wang, G. P. Bierwagen, Corros. Sci. 55 (2012) 97–106
134. Q. Zhou, Y. Wang, Prog. Org. Coat. 76 (2013) 1674–1682
135. M. Dornbusch, C. Henzel, C. Deschamps, S. Overmeyer, K. Cox, M. Wiedow, What Happens During the Swelling of a Coating ?, FATIPEC, ETTC, Cologne, 2014
136. R. Kiefer, R. Stilke, E. Boese, A. Heyn, M. Engelking, R. Hillert, Galvanotechnik (2012) 1904–1914
137. C. Langer, W. Wendland, K. Honold, L. Schmidt, J. Gutmann, M. Dornbusch, Corrosion Analysis of Decorative Microporous Chromium Plating Systems in Concentrated Aqueous Electrolytes, Engineering Failure Analysis 91 (2018) 255–274 and C. Langer, M. Dornbusch, Corrosion Behavior of Decorative Chromium Layer Systems in Concentrated Aqueous Electrolytes, NACE Corrosion 2017, New Orleans, 2017
138. Thesis (MA), L. Schmidt, Elektrochemische Untersuchung des Korrosionsmechanismus an galvanischen Chrom- und Nickelschichten, Niederrhein University of Applied Sciences, Krefeld, 2016
139. O. b. Suarez, J. b. Olaya, M. b. Suarez, S. Rodil, J. Chil. Chem. Soc. 57, 1 (2012) 977–982
140. B. Li, A. Lin, X. Wu, Y. Zhang, F. Gan, J. Alloys Compd. 453 (1–2) (2008) 93–101
141. L. Leita, A. Margon, A. Pastrello, I. Arcon, M. Contin, D. Mosetti, Environ Pollut. (Barking, Essex 1987) 157, 6 (2009) 1862–1866.
142. B. Pfeiffer, J. W. Schultze, J. Appl. Elecrochem. 21 (1991) 877–884
143. B. D. Bammel, Met. Finish. 101, 5 (2003) 38–43
144. H. M. Leister, Organic Coatings and plastics Chemistry 43 (1980) 486–491
145. G. Kortüm, D. Oelkrug, Naturwissenschaften 53, 23 (1966) 600–609
146. G. Kortüm, G. Schreyer, Z. Naturforschg. 11 a (1956) 1018–1022
147. S. Zhou, L. Wu, M. Xiong, Q. He, G. Chen, J. Dispersion Sci. Technol. 25, 4 (2004) 417–433
148. L. G. J. van der Ven, R. T. M. Leijzer, K. J. van den Berg , P. Ganguli, R. Lagendijk, Prog. Org. Coat. 58 (2007) 117–121
149. P. Mischke, Film Formation, Vincentz Network, Hanover, 2010
150. H. Herlinger, G. Stevens, Verfahren zur Herstellung lagerstabiler Polyamidimid-Harze und Überzugsmittel, die diese enthalten, DE10303635A1
151. K.-Y. Choi, Y.-T. Hong, D.-H. Suh, J.-C. Won, M.-H. Yi, Verfahren zur Herstellung von Polyamidimid-Harzen mit hohem Molekulargewicht, DE4440409
152. M. Dornbusch, U. Karl, H. Mizuno, Process for producing polyamideimides with the use of N-formylmorpholine, WO002015024824
153. M. Dornbusch, U. Karl, Polymerization in N-formylmorpholine, WO002015117857
154. T. Hirano, A. Koseki, M. Mizoguchi, M. Waki, H. Nakamura, Proceedings of the Electrical Electronics Insulation Conference, 17th 1985, New York, 189–191
155. Z. Rasheva, L. Sorochynska, S. Grishchuk, K. Friedrich, eXPRESS Polymer Letters 9, 3, (2015) 196–210
156. W. Zhang, R. Smith, C. Lowe, J Coat. Technol. Res. 6 (3) 315–328, 2009
157. N. J. Everall, JCT Coatings Tech 2005, 38–44

158. N. J. Everall, JCT Coatings Tech 2005, 46–52
159. R. Posner, A. M. Jubb, G. S. Frankel, M. Stratmann, H. C. Allen, Electrochim. Acta 83 (2012) 327–334
160. Thesis (PhD), M. Reichinger, Investigations of the direction-driven water and ion transport along interfaces and through polymer networks, University Paderborn, 2017
161. M. Reichinger, M. Dornbusch, Raman-SKP Study of the Aluminium/E-coat Interface, EUROCORR 2016 - Montpellier, France, 2016
162. Thesis (PhD), M. Santa, Combined *in situ* spectroscopic and electrochemical studies of interfacial and interphasial reactions during adsorption and de-adhesion of polymer films on metals, University Paderborn, 2010

4
Investigation of the Corrosion Rate

The investigation of the corrosion rate or corrosion performance of coated or uncoated metal substrates must be distinguished between scientific and industrial methods.

Scientific methods based on chemical, electrochemical or spectroscopic methods are useful to analyze an electrochemical or transport mechanism of the corrosion process. The results are chemical data of the surface, electrochemical data of the corrosion process such as the current density, surface capacity or the polarization resistance or physical data of the coating before and after the treatment such as the diffusion coefficient, amount of water in the coating or the resistance of the coating.

Industrial methods based on the treatment of samples in different corrosive environments are useful to evaluate a certain quality standard. The test conditions try to simulate the use of a metal part (car body, aircraft, ship, etc.) for a certain timeframe such as 5 or 10 years in a certain environment. The result is reduced to one or several parameters such as the delamination length of a sample or the marking of the surface after the treatment.

Of course nowadays the industry uses scientific methods and the academic field uses industrial tests but the separation is necessary to clarify what is the result of the experiment especially, if a mixture of both is used in an experimental series.

For industrial use the scientific and especially the electrochemical methods have to be standardized to generate comparable results. The interpretation of electrochemical impedance (EIS) data is complex and the measurements needs some hours. For industrial use the EIS measurement is to standardize in ISO 16773-1-3 but in DIN EN ISO 13129 alternative methods—easier and faster—such as current interrupter (CI) [1–3], relaxation voltammetry (RV) [4–6] and DC transient measurements [7, 8] have already be standardized.

4.1 Scientific Methods

4.1.1 Current-Density Potential Diagram and EIS

The electrochemical basics are already explained in Chap. 1.1 and the basics of the current-density potential diagram and the electrochemical impedance spectroscopy have been illustrated in Chap. 2.1.1 and 2.1.3/2.2.2, respectively.

With the current-density potential diagram often the corrosion protective properties of inhibitors in solution or conversion layers on the substrate are characterized. Inhibitors are predominantly used in cooling circuits but also in coatings especially to prevent flash rust in water based coatings. For financial and toxicological reasons the amount of inhibitors have to be minimized and therefore the efficiency of the inhibitor is a relevant parameter.

The inhibition efficiency (IE) could be investigated with the current density as follows [9–12]:

$$IE = \left(1 - \frac{i_{corr,Inhibitor}}{i_{corr,Substrate}}\right) \cdot 100 \qquad \text{Eq. 4.1}$$

With the use of EIS spectra the IE could be calculated from the resistance achieved by simulation of the data with an equivalent circuit (see Chap. 2.1.3) as follows [9, 13]:

$$IE = \left(1 - \frac{\frac{1}{R_{Inhibitor}}}{\frac{1}{R_{Substrate}}}\right) \cdot 100 \qquad \text{Eq. 4.2}$$

Finally the IE could be investigated by means of the weight loss of the samples (see Chap. 4.1.2). In Table 4.1 some data from literature are summarized. The polarization resistance R_{Pol} could be calculated from the current-density-potential diagram with the Stern-Geary equation (Eq. 1.48) or from EIS data by simulation of the spectra. There is some confusion with the unit of the resistance shown in Table 4.1. The corrosion potential is clear in [mV] and literal errors could easily be detected and the same is for the current density in [$\mu A{*}cm^{-2}$] and the capacity in [$\mu F{*}cm^{-2}$]. Only the resistance shows a broad range from $k\Omega{*}cm^2$ on metal substrate up to $G\Omega{*}cm^2$ on organic coatings and literal errors are hard to detect.

Table 4.1 Electrochemical data of some metal substrates with and without inhibitors in the electrolyte or conversion layers on the substrate. The E_{Corr} values from [11] and the capacity from [16] have been corrected regarding their unit. *: CPE (see Eq. 2.24), CF: Extract of the fruit *Capsicum frutescens* [9], ATMP: Aminotrimethylene phosphonic acid [14], Tween 80: oligomeric polyether with an unsaturated ester side group [14], UDI: 2-Undecyl-1,3-imidazoline [13], NI: 2-Nonyl-1,3-imidazoline [13], Metol: N-Methyl-p-aminophenol sulphate [11], PC: Prunus cerasus juice [16], Ce(dpp)3: Cerium diphenyl phosphate [15], Mm(dpp)3: "Mischmetal" diphenyl phosphate (Mischmetal: Ce, La, Nd) [15], CS: 1,1-(lauryl amido)-propyl ammonium chloride [12].

Substrate	Electrolyte	E_{corr} [mV]	i_{corr} [μA cm^{-2}]	R	C [μF cm^{-2}]	b_a [mV dec^{-1}]	b_c [mV dec^{-1}]	IE [%] Eq. 4.1	IE [%] Eq. 4.2	Lit.
Low Carbon Steel	1.0 M HCl	–521.6	888.7	94.03 Ω *cm²	77.09*					[9]
Low Carbon Steel	0.5 M H_2SO_4	–482.3	1534.9	63.1 Ω *cm²	51.2*					[9]
Low Carbon Steel	1.0 M HCl + 200 mg/l CF	–485.6	203.4	384.7 Ω *cm²	53.5*			77.1	75.6	[9]
Low Carbon Steel	1.0 M HCl + 1000 mg/l CF	–490.7	86.5	651.4 Ω *cm²	39.8*			90.3	85.6	[9]
Low Carbon Steel	0.5 M H_2SO_4 + 200 mg/l CF	–480.8	226.5	308.2 Ω *cm²	40.8*			85.2	79.8	[9]
Low Carbon Steel	0.5 M H_2SO_4 + 1000 mg/l CF	475.7	62.3	912.7 Ω *cm²	35.3*			95.9	93.1	[9]
St-37 Steel	1.0 M HCl	–480	546	32.8 Ω *cm²	78.4*	65	100			[16]
St-37 Steel	1.0 M HCl + 1 v/v% PC	–473	46	226.4 Ω cm²	26.7*	56	90	91.5	87.68	[16]
St-37 Steel	1.0 M HCl + 4 v/v% PC	–461	33.9	448.8 Ω cm²	25.0*	62	92	94.1	92.69	[16]

Table 4.1 contd. ...

...Table 4.1 contd.

Substrate	Electrolyte	E_{corr} [mV]	i_{corr} [μA cm^{-2}]	R	C [μF cm^{-2}]	b_a [mV dec^{-1}]	b_c [mV dec^{-1}]	IE [%] Eq. 4.1	IE [%] Eq. 4.2	Lit.
Mild Steel	1.0 M HCl	–475	41	30 Ω *cm^2	108					[11]
Mild Steel	0.5 M H_2SO_4	–450	28.9	110 Ω *cm^2	78					[11]
Mild Steel	1.0 M HCl + 0.08 M Metol	–525	9	140 Ω *cm^2	56			81.9	79	[11]
Mild Steel	0.5 M H_2SO_4 + 0.08 M Metol	–535	8	375 Ω *cm^2	45			77.4	71	[11]
304 SS	Groundwater	–76	10	5738 kΩ *cm^{-2}	81.68					[14]
304 SS	Groundwater + 50 ppm Zn^{2+}	–70	3.7	19825 kΩ *cm^{-2}	77.81			63		[14]
304 SS	Groundwater + 30 ppm ATMP	–1	2.5	32530 kΩ *cm^{-2}	67.7			75		[14]
304 SS	Groundwater + 150 ppm Tween 80	–14	5	15232 kΩ *cm^{-2}	79.99			50		[14]
304 SS	Groundwater + 50 ppm Zn^{2+} + 30 ppm ATMP	–10	0.62	78805 kΩ *cm^{-2}	59.32			93.8		[14]
304 SS	Groundwater + 50 ppm Zn^{2+} + 30 ppm ATMP + 150 ppm Tween 80	–27	0.1	393205 kΩ *cm^{-2}	35.1			99		[14]
Aluminium	1.0 M HCl	–902	4490	16.23 kΩ *cm^2	4.51*10^6	68	174			[13]
Aluminium	0.5 M H_2SO_4	–890	4160			56	120			[13]

Aluminium	1.0 M HCl + 500 ppm UDI	–898	89	247.03 kΩ *cm^2	0.29*10^6	60	170		98.02	[13]
Aluminium	1.0 M HCl + 500 ppm NI	–893	106			54	126		97.63	[13]
Aluminium	0.5 M H_2SO_4 + 500 ppm UDI	–884	320			50	84		92.3	[13]
Aluminium	0.5 M H_2SO_4 + 500 ppm NI	–879	479			40	110		88.47	[13]
Aluminium 99.99%	5.0 M HCl	–902	2591.31			180	188			[17]
Aluminium 99.99%	5.0 M HCl + 200 ppm Vanillin	–898	99.96			60	342	96.14		[17]
Aluminium 99.99%	5.0 M HCl + 2000 ppm Vanillin	–890	28.06			58	282	98.91		[17]
Aluminium 99.79%	1.0 M HCl	–1109	4.49	17.26 kΩ	4.51	190.7	432.5			[12]
Aluminium 99.79%	1.0 M HCl + 2*10^{-3} M CS	–966.8	0.67	112 kΩ	0.63	181.1	412.8	85	86	[12]
Aluminium 99.79%	1.0 M HCl + 1*10^{-2} M CS	–864	0.12	662 kΩ	0.10	180.4	411.6	97.3	97.7	[12]
Aluminium 2024-T3	0.1 M NaCl	–501	2.3							[15]
Aluminium 2024-T3	0.1 M NaCl + 2*10^{-4} M Ce(dpp)$_3$	–651	0.8							[15]

Table 4.1 contd. ...

...Table 4.1 contd.

Substrate	Electrolyte	E_{corr} [mV]	i_{corr} [μA cm^{-2}]	R	C [μF cm^{-2}]	b_a [mV dec^{-1}]	b_c [mV dec^{-1}]	IE [%] Eq. 4.1	IE [%] Eq. 4.2	Lit.
Aluminium 2024-T3	0.1 M NaCl + 2*10^{-4} M Mm(dpp)$_3$	–368	0.01							[15]
Magnesium AZ31B + Stannate based conversion layer	5% NaCl		11.4	4.4 Ω		201.6	271.1			[18]
Magnesium AZ31B + Cerium based conversion layer	5% NaCl		8.05	3.4 Ω		87.05	228.7			[18]
Magnesium AZ31B + Chromate based conversion layer	5% NaCl		5.06	4.5 Ω		103.3	106.7			[18]

Therefore the resistance data in Table 4.1 are cited similar to the original paper.

If a corrosion protective effect is visible, the current-density and the capacity of the surface decreases in contrast the resistance increases and the potential shifted regarding the inhibition effect (see Fig. 2.4).

The electrochemical data often combined with spectroscopic methods such as XPS [14], UV-VIS [14], infrared [14] or Raman [15] spectroscopy to give the complete picture which allows a design of inhibition by smart combinations of compounds (e.g., see the data from [14] in Table 4.1). Another approach is the quantum mechanical simulation of the interaction of the inhibitor with the metal substrate based on the Density-Functional-Theory DFT [9] and the correlation of the calculated with the experimental data to optimize the inhibition efficiency.

Besides the corrosion protective effect of an inhibitor or a conversion layer by decreasing the current density the electrochemical data give mechanistic information about the corrosion process on the surface (b-values).

The investigation of the relevant electrochemical parameters of pure metal substrates and metal coatings is standardized by DIN 50918 (see also DIN 50900).

4.1.2 *Weight Loss*

The easiest method to investigate the corrosion rate should be the weight of the sample after a corrosive treatment. The increase of the weight caused by the reaction of the metal substrate with oxygen should define the corrosion rate or the weight loss caused by the dissolution of metal ions or soluble metal compounds into the electrolyte.

Unfortunately the chemical composition of the corrosion products is complex as illustrated in Fig. 4.1 for steel in different environments (for zinc see Fig. 3.50). In a marine environment soluble cations formed and depending on the electrolyte akaganeite or lepidocrocite precipitated on the surface followed by a transformation in amorphous oxides and finally a slow generation of goethite and magnetite. In normal atmospheric conditions often iron (II) hydroxide precipitated on the surface followed by the transformation into lepidocrocite and finally magnetite. In soil the formation of the so called green rust ($GR(CO_3^{2-})$) occurs. This intermediate is a complex iron hydroxy-carbonate of Fe(II) and Fe(III) oxidized to Fe(III) green rust (exGr, Fe^{III}) and finally hematite and goethite could be detected on the corroded surface [19]. In carbonate rich electrolytes siderite is the dominant corrosion product [23] and in certain microbial conditions siderite and FeS precipitated and some hints of the formation of chukanovite formed from siderite have been reported [22].

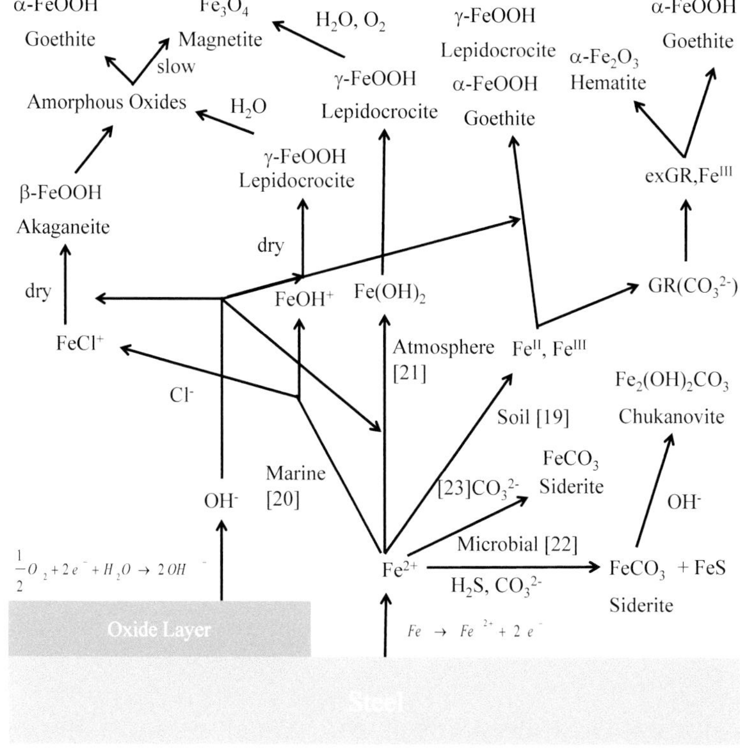

Fig. 4.1 Corrosion products on steel in different environments according to the cited literature.

In Table 4.2 some corrosion products mentioned in literature for the most important industrial substrates and solubility constants of metal compounds are summarized in Table 5.1 in the app.

The situation is more complex because the composition and the amount of corrosion products depend on the corrosive environment and with time. In Table 4.3 some corrosion products on zinc and steel at different places and different times of outdoor exposure are summarized. Even in small areas the corrosion products are different and the chronological change also is different.

Therefore the corrosion products have to be removed from the surface to investigate the mass loss and the corrosion rate. In Table 4.4 some methods are described to remove the corrosion products from the sample without dissolution of the substrate. For this challenge different methods depending on the metal substrate exist and some have been standardized (ASTM G1-90, ASTM G31-72, DIN EN ISO 8407 (2014) and DIN EN ISO 11474-2016).

Table 4.2 Corrosion products on metal substrates mentioned in literature. For iron and zinc see also Figs. 4.1 and 3.50, respectively.

Metal	Corrosion products (colour)	Comment	Lit.
Aluminium	AlOOH Boehmite (white)	Rarer Species [24, 25]	[24], [25]
	$Al(OH)_3$ Bayerite, Hydragillite, Gibbsite (white)		[24], [25], [26]
	Al_2O_3		[24], [25]
	Al_2O_3*$3H_2O$		[24], [25]
	$Al_x(OH)_y(SO_4)_z$	Rarer Species [24, 25]	[24], [25]
	$AlCl(OH)_2$*$4H_2O$	Rarer Species [24, 25]	[24], [25]
Copper	Cu_2O Cuprite (red-brown)		[24], [25]
	$Cu_4SO_4(OH)_6$ Brochantite (green)	Patina [24]	[24]
	$Cu_4SO_4(OH)_6$ *H_2O Posnjakite (light blue)	Patina [24]	[24], [25]
	$Cu_2SO_4(OH)_4$ Antlerite (green-dark green)	Patina [24]	[24], [25]
	$Cu_3SO_4(OH)_4$	Patina [24]	[24], [25]
	$Cu_2Cl(OH)_3$ Atacamite (green)	Rarer Species [24, 25]	[24], [25]
	$Cu_2CO_3(OH)_2$ Malachite (green)	Rarer Species [24, 25]	[24], [25]
	$Cu_2NO_3(OH)_3$	Rarer Species [24, 25]	[24], [25]
Iron	FeO Wustite (grey)		[27]
	$FeCO_3$ Siderite (yellow-brown)	Rarer Species [24, 25]	[24], [25]
	$Fe_2(OH)_2CO_3$ Chukanovite (brownish green)	See Fig. 4.1	[22]
	$FeCl_2$*$4H_2O$	In pits and cracks	[28]
	$FeSO_4$*$4H_2O$		[24], [25]
	α-FeOOH Goethite (light yellow-dark brown)	See Fig. 4.1	[24], [25]
	β-FeOOH Akaganeite (yellow brown)		[27]
	γ-FeOOH Lepidocrocite (dark red-dark brown)		[24], [27]
	δ-FeOOH Feroxyhite (brown)		[27]
	Green Rust (green)		[19]

Table 4.2 contd. ...

...Table 4.2 contd.

Metal	Corrosion products (colour)	Comment	Lit.
	$Fe(OH)_3$	On pits and cracks	[28]
	Fe_2O_3	On the surface besides pits and cracks [28]	[24], [25]
	α-Fe_2O_3 Hematite (steel-grey–black)		[27]
	γ-Fe_2O_3 Maghemite (brown)		[27]
	5 Fe_2O_3*$9H_2O$ Ferrihydrite (buffy)		[27]
	$Fe_x(OH)_yCl_z$	Rarer Species [24, 25]	[24], [25]
	Fe_3O_4 Magnetite (black)		[27]
	$Sn(OH)_2$		[29]
	SnO Romarchite (black)		[29]
Tin	$Sn(OH)_4$		[30], [29]
	SnO_2*H_2O		[30]
	SnO_2 Cassiterite (brown-black)		[29]
	$Zn(OH)_2$	Rarer Species [24, 25]	[24[,]31]
	ZnO Zincite (colorless–greenish yellow)	Zincite or Zinc oxide [32]	[24], [25]
	$Zn_5(OH)_6(CO_3)_2$ Hydrozincite (white–grey)		[24], [25], [31], [32]
	$ZnCO_3$ Smithsonite (white–greenish)		[24], [25]
	$ZnSO_4$	Rarer Species [24, 25]	[24], [25]
Zinc	$NaZn_4Cl(OH)_6SO_4$*H_2O Chlorohydroxysulphate	See Fig. 3.50	[32]
	$NaZn_4Cl(OH)_6SO_4$*$6H_2O$		[31]
	$Zn_4Cl_2(OH)_4SO_4$*$5H_2O$		[31]
	$ZnO(SO_4)_2$ Oxysulphate	See Fig. 3.50 Detected if $c(Cl^-) < c(SO_4^{2-})$	[32]
	$Zn_4SO_4(OH)_6$ Hydroxysulphate		[32]
	$Zn_4SO_4(OH)_6$*nH_2O		[31]
	$Zn_5Cl_2(OH)_8$*H_2O Simonkolleite (colourless)	Rarer Species [24, 25], detected if $c(Cl^-) > c(SO_4^{2-})$ [31]	[24], [25], [32], [31]

Table 4.3 Corrosion products after outdoor exposure on carbon steel, weathering steel and zinc.*: all places in ref. [33] in the idero-american region, ** all places in ref. [32] are on Santa Cruz de Tenerife (Canary Islands, Spain).

Metal	Detected corrosion products	Outdoor exposure [year]	Place	Lit.
Carbon Steel	Lepidocrocite, Goethite	1	Arties*	[33], [34]
	Lepidocrocite, Goethite	2		
	Lepidocrocite	1	Iguaza	[33], [34]
	Lepidocrocite	2		
	Lepidocrocite	1	Brasflia	[33], [34]
	Lepidocrocite, Goethite	2		
	58% Goethite, 3% Lepidocrocite, 39% Maghemite	16	Test sites	[27]
Weathering Steel	81% Goethite, 14% Lepidocrocite, 5% Maghemite	16	Test sites	[27]
	60% Goethite, 3% Lepidocrocite, 37% Maghemite		Bridge, high time of wetness	[27]
	37% Goethite, 63% Akaganeite		Bridge, high chloride content	[27]
Zinc 99.77%	Simonkolleite, Oxysulphate	1	Meteorologico**	[32]
	Simonkolleite, Oxysulphate	2		
	Hydrozincite	1	Botanico	[32]
	Hydrozincite, Oxysulphate	2		
	Zincite, Hydrozincite	1	Montaneta	[32]
	Hydrozincite, Oxysulphate	2		
	Hydrozincite, Simonkolleite, Chlorohydroxysulphate, Oxysulphate	1	Unelco Granadilla	[32]
	Simonkolleite, Chlorohydroxysulphate	2		
	Simonkolleite, Chlorohydroxysulphate, Oxysulphate, Hydroxysulphate	1	Valle Gran Rey	[32]
	Chlorohydroxysulphate, Oxysulphate	2		

Table 4.4 Chemical and electrochemical methods for the removal of corrosion products from corroded metal samples prior to weighing. Currents are cathodic. G1: ASTM G1-90, G31: ASTM G31-72, DIN1: DIN EN ISO 8407 (2014), DIN2: DIN EN ISO 11474-2016. For the ASTM standards see also [37].

Metal	Treatment		Comment	Lit.
	Solution	Conditions		
Al	Conc. HNO_3 + 50 g/l CrO_3	5–10 min. at 20°C		[35]
Al	Conc. HNO_3 + 5% Chromic acid	30 min. below 30°C	Loss weight of aluminium at 15°C is 0.3 g/m^2	[36]
Al	Conc. HNO_3	1–5 min. at 20–25°C		[G1], [DIN1]
Al	5% H_3PO_4 + 2% CrO_3	30 min. at 80–100°C with 1 A/dm^2, [G1, DIN2] 5–10 min. at 90°C, [DIN1] 80°C		[35], [G1], [DIN1], [DIN2]
Al	10% I_2 in Ethanol	10 hours at 50°C	Metal dissolves, oxide left	[35]
Al	5% Acetic acid	warming		[36]
Al	35 ml/l 85% H_3PO_4 + 20 g/l CrO_3	5 min. in boiling solution		[36]
Al	28 ml/l H_3PO_4 + 20 g/l $K_2Cr_2O_7$	70–80°C		[36]
Cu	5% H_2SO_4 + pickling inhibitor	5 min. at 75°C with 10 A/dm^2		[35], [36]
Fe and Steel	HCl (density 1.16) + 20 g/l Sb_2O_3 + 20 g/l (50 g/l [G1]) $SnCl_2$	15–20 min. at room temperature (up to 25 min. [36, DIN1]) or 20–25°C [G1, DIN1]	Clarke´s solution. Sb^{3+} acts as inhibitor and Sn^{2+} reduces Fe^{3+} to Fe^{2+}	[35], [36], [G1], [DIN1]
Fe and Steel	10–20% NaOH	5–10 min. at 60–100°C	Contact with zinc or cathodic current (1–10 A/dm^2) reduces metal loss	[35]
Fe and Steel	20% NaOH + 30 g/l zinc dust ([G1] granulated zinc)	5 min. in boiling solution	The zinc dust is added after the sample is immersed	[36]

Table 4.4 contd. ...

...Table 4.4 contd.

Metal	Treatment		Comment	Lit.
	Solution	Conditions		
Fe and Steel	20% H_2SO_4 + 0.05% Di-orthotolylthiourea [DIN1]: 5 g/l 1,3-Di-n-Butyl-2-thiourea	1 hour at 60°C, [DIN1] 1–5 min. at 20–25°C	Loss weight of steel is 13 g/m^2	[36], [DIN1]
Fe and Steel	20% Ammonium citrate solution	20 min. at 75–80°C	30 litre of the solution per m^2 metal surface. Loss weight of steel is < 3 g/m^2	[36]
Fe and Steel	Saturated citric acid	Long time at 20°C but below 60°C	Cathodic treatment	[36]
Fe and Steel	5–10% NaCN	3 hours at 20°C with 1.5 A/dm^2	Loss weight of steel is < 1 g/m^2	[36]
Fe and Steel	10% Ammonium citrate	1 A/dm^2		[36]
Stainless Steel	10–30% HNO_3	5 min. at 20–60°C		[35]
Stainless Steel	30% HNO_3	20°C		[36]
Stainless Steel	10% HNO_3	60°C, [DIN1] 20 min. at 60°C	Chloride free solution	[36], [DIN1]
Mg	200 g/l CrO_3 + 10 g/l Ag_2CrO_4 + 1 g/l $BaCrO_4$, [DIN1] 200 g/l CrO_3 + 10 g/l $AgNO_3$ + 20 g/l $Ba(NO_3)_2$	5 min. in boiling solution, [DIN1] 1 min. at 20–25°C	Ag^+ and Ba^{2+} remove Cl^- and SO_4^{2-}, respectively. Dissolves conversion coatings also	[35], [DIN1]
Mg	20% Chromic acid + 0.5–1.0% Ag_2CrO_4 + 2% $BaCrO_4$	5 min. in boiling solution	$AgNO_3$ could be used instead of Ag_2CrO_4	[36]
Mg	100 g/l $(NH_4)_2CrO_4$ at pH = 8–10	24 hours at room temperature with air agitation	Coatings not dissolved	[35], [36]
Ni	50% HCl, [DIN1] 150 ml/l HCl	Some minutes at room temperature, [DIN1] 1–3 min. at 20–25°C		[36], [DIN1]

Table 4.4 contd. ...

...Table 4.4 contd.

Metal	Treatment		Comment	Lit.
	Solution	Conditions		
Ni	10% H_2SO_4	Some minutes at room temperature, [DIN1]: 1–3 min. at 20–25°C		[36], [DIN1]
Sn	5% HCl	Up to 50 min. at 15°C, [DIN1] 10 min. at 20°C		[36], [DIN1]
Sn	15% Na_3PO_4	10 min. in boiling solution		[36], [DIN1]
Zn	Saturated solution of ammonium acetate, [DIN1]: 100 g/l of ammonium acetate	2 hours at room temperature, [DIN1]: 2–5 min. at 70°C		[35], [36], [DIN1]
Zn	200 g/l CrO_3 + 1 g/l $BaCrO_4$	1 min. at 80°C		[35]
Zn	I. 10% NH_4Cl II. 5% Chromic acid + 1% $AgNO_3$ III. Hot water	I. Several minutes at 60–80°C II. 15–20 s in boiling solution III. Rinsing		[36], [DIN1]
Zn	20% Chromic acid	1 min. at 80°C	Loss weight of zinc is 0.4 g/m^2	[36], [DIN2]
Zn	3% Citric acid + 10% KCN	15 A/dm^2	Loss weight of zinc is 0.4 g/m^2 day	[36]
Zn	I. 5% Na_3PO_4 II. 5% Na_3PO_4	I. Overnight II. 8 A/dm^2		[36]
Zn	I. 10% $(NH_4)_2S_2O_8$ II. 10% KCN	I. Immersion II. Cathodic treatment		[36]

The inhibition efficiency IE of an inhibitor or a conversion layer could be calculated with the weight loss as follows [9, 11, and 12]:

$$IE = \left(1 - \frac{\Delta w_{Inhibitor}}{\Delta w_{Substrate}}\right) \cdot 100 \qquad \text{Eq. 4.3}$$

Corrosion rates of industrially used metals in different environments are summarized in Table 5.4 in the app. There are different units to describe the mass loss of the samples. Common units are [µm/year], [ipy] (inch per year), [mpy] (mills per year = 0.0254 mm/year), [gmd] (g m^{-2} day^{-1}) and of course [mA/cm^2].

4.1.3 Immersion Test with ICP-OES

To avoid the dissolution of the corrosion products from the corroded surface especially in a screening of different corrosion protection concepts an alternative way to investigate the corrosion rate has been presented in ref. [38].

Example 4.1

The setup of the experiment is illustrated in Fig. 4.2. The samples immersed in 3% NaCl solution (samples of steel and zinc coated steel) or in 3% NaCl at pH = 3 (acetic acid) for aluminium for one week. For a constant volume the samples stored in a desiccator flushed with 100% humid air with the advantage that the CO_2 and O_2 content is also constant. The samples removed after 24, 72, 96 and 168 hours and washed with the immersion solution before the samples are stored in a fresh solution. To the used solution conc. HCl is added to solve solid corrosion products. The solution is analyzed by ICP-OES (inductively coupled plasma-optical emission spectroscopy) to investigate the metal content in the solution. The cumulative data gain a linear corrosion rate (average of at least two samples) in [mol l^{-1} h^{-1} cm^2]. In addition to the investigation of the metal

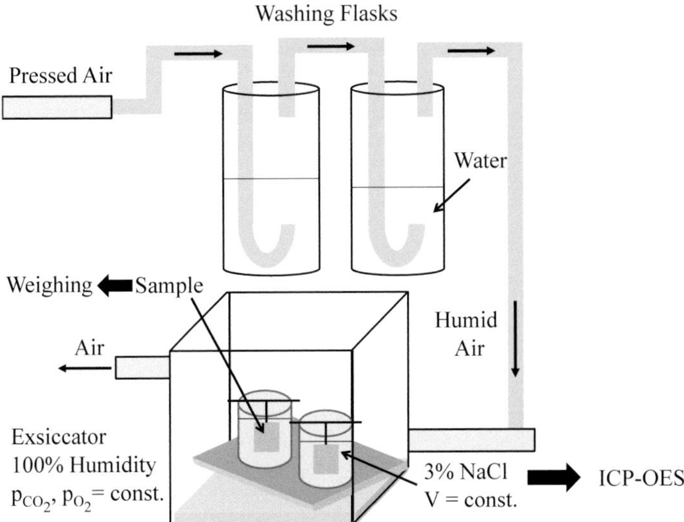

Fig. 4.2 Setup of the immersion test according to [38].

content in the electrolyte the samples could be weighed to define the mass loss or the mass increase caused by corrosion products in the solution or on the surface.

Some data achieved with this method are summarized in Table 4.5. The advantage of the method is that all substrates could be analyzed with one method and there is no need for different and often toxic solutions to remove the corrosion products from the surface (see Table 4.4). The method is useful for the investigation of corrosion rates on metal substrates coated with passive or conversion layers with a certain amount of corrosion. For organic coated samples the corrosion rate is too small to define within one week.

Table 4.5 Corrosion rates measured with the Immersion Test according to [38].

Substrate	Corrosion Rate [mol l^{-1} h^{-1} cm^{-1}]
Zinc coated steel	8.14
Phosphated zinc coated steel	6.17
Organic modified molybdenum based conversion layer on zinc coated steel	4.58

4.1.4 *Cyclic Test with EIS*

Organic coated metal substrates often tested in industrial corrosion tests such as the salt spray test (see Chap. 4.2.2). The test duration of high performance coatings depends on the market. For coil coating application often 1000 hours and in the aerospace market up to 4000 hours test duration are necessary to evaluate the corrosion protective performance of organic coatings. Since decades fast corrosion tests have been invented to reduce the test duration especially in the development of new corrosion protective coatings. One approach is illustrated in the following example.

Example 4.2

Bierwagon developed a test for one sample treated with an electrolyte such as Harrison solution (see Table 2.2) with thermal cycling for one week [39, 40]. The sample analyzed by means of EIS and some modifications of the setup mentioned in literature, too (e.g., in ref. [41]). This concept has been expanded to more samples and shortened to a thermal cycling (Fig. 4.4) for 36 hours claimed in ref. [42] and the setup is illustrated in Fig. 4.3.

The resulting impedance at 0.1 Hz shows the performance of the organic coated substrate, i.e., the higher the resistance the higher the corrosion protective properties. In [43] the impedance at 0.1 Hz has been used to interpret the EIS data and show a good correlation with the industrial test. The advantage of the method is that there is no need

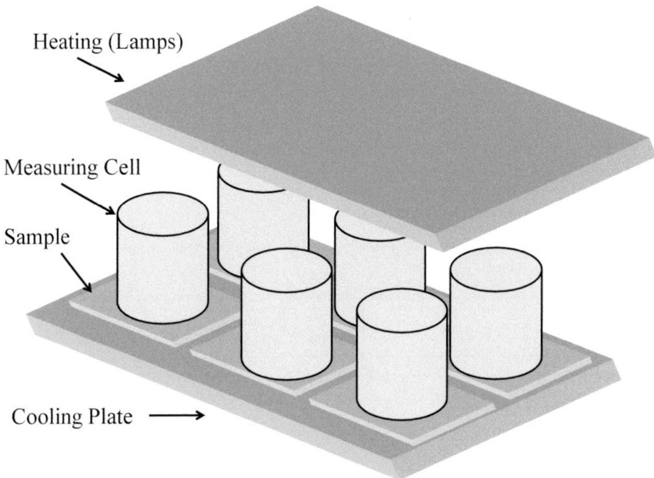

Fig. 4.3 Setup of the cyclic test according to [42].

to simulate the spectra with an equivalent circuit and the resulting uncertainty of the simulated data and without the challenge to choose the right circuit. The results correlate with standard test methods such as VDA 621-415 or ASTM B 287 [42].

The most important parameter of industrial tests is the delamination rate of a scratched sample. Therefore the setup in Fig. 4.3 has been used to investigate the delamination rate of organic coated samples [44]. The measurement of the delamination rate could be possible by means of simulating the EIS data with the Randles equivalent circuit and calculating the capacity C of the corroded surface [45, 46] to investigate the free and corroded surface area A according to the following equation (compare Eq. 2.10):

$$C = \varepsilon_0 \, \varepsilon_r \frac{A}{d} \wedge \varepsilon_0 = 8.85 \cdot 10^{-14} \, F\,cm^{-1} \qquad \text{Eq. 4.4}$$

The delamination rate in industrial tests is defined in one direction in [mm] therefore the following correlation should be possible [44]:

$$\sqrt{C} \approx \text{Delamination rate} \qquad \text{Eq. 4.5}$$

The data in [44] correlate with 1/delamination rate because of two reasons: First the thickness d of the passive layer has to be constant to correlate the capacity with the delamination rate and second the dielectric constant ε_r of the layer should be constant, i.e., the corrosion products and the composition of the corrosion products have to be constant, too. Both

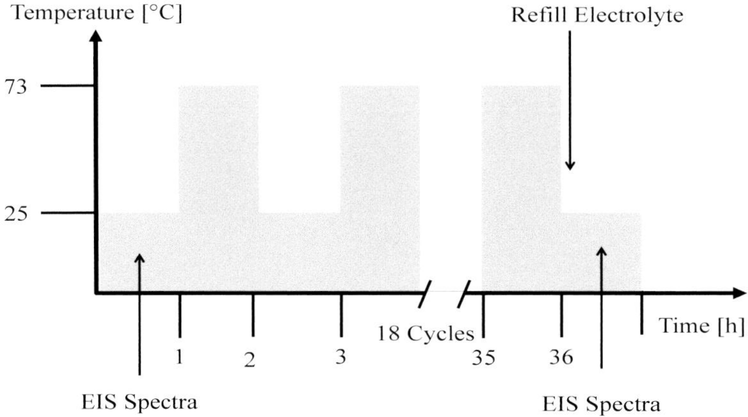

Fig. 4.4 Thermal cycling according to [42].

are not the case (see Figs. 3.50 and 3.51). For further discussion see the next chapter.

4.1.5 Delamination/Barrier Properties by Means of EIS

There are many ways to perform an EIS measurement on organic coated substrates. To achieve comparable data standardization should be usefully realized in DIN EN ISO 16773-2 and ASTM G106-89. The proposed electrochemical cell is similar to Fig. 2.46 with a two- or a three-electrode-setup and the experiment performed in a Faraday cage (see Chap. 2.2.2). To test the setup a confidence test, i.e., a measurement with a test circuit shown in Fig. 4.5 performed before the sample measurement started. The confidence test defines the inaccuracy of the method and this is necessary to compare data from different laboratories.

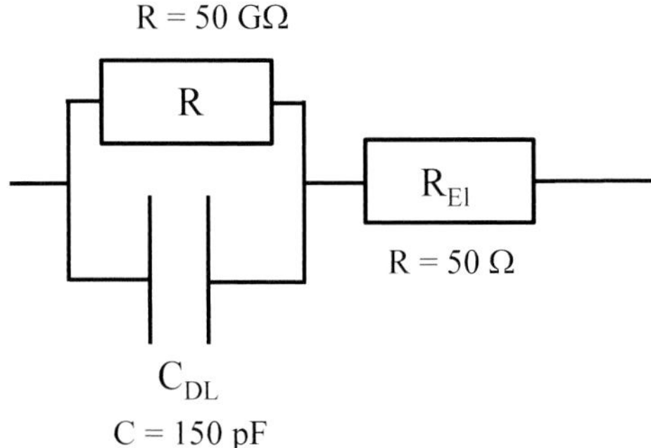

Fig. 4.5 Test circuit according to DIN EN ISO 16773-2 and ASTM G106-89.

The average of at least three samples has to be investigated and the thickness of the organic coatings should be as constant as possible. The frequency range of the experiment is proposed from 0.01 to 100 kHz with amplitude of 20 mV but higher amplitudes could be necessary for very high impedances, i.e., higher coating thicknesses. For more details see DIN EN ISO 16773-2.

Break-Point Frequency Method

With the assumption that during a corrosive treatment of an organic coated sample without a defect the coating delaminates or disbonds from the metal surface, the disbonded or delaminated area should be a parameter to investigate the barrier properties of the organic coating. In Fig. 4.6 a Bode plot of a corrosive treated coating is illustrated with relevant frequencies. The break-point frequency could be defined at the point of the spectrum where the phase angle is below 45° in the high frequency range. Based on the equivalent circuit of phase III in Fig. 2.45 the frequencies could be calculated based on the following parameters [47, 48]:

Coating capacity: $C_{Coat} = \varepsilon_r \, \varepsilon_0 \dfrac{A}{d}$ A: surface area Eq. 4.6

Coating resistance: $R_{Coat} = \dfrac{R^0_{Coat}}{A_d} = \dfrac{\rho \, d}{A_d}$ Eq. 4.7

ρ: Coating resistance d: Coating thickness A_d: Delaminated area

Double layer capacitance: $C_{DL} = C^0_{DL} \, A_d$ Eq. 4.8

Double layer resistance: $R_{CT} = \dfrac{R^0_{CT}}{A_d}$ Eq. 4.9

Delaminated area fraction: $D = \dfrac{A_d}{A}$ Eq. 4.10

The break-point frequency f_{break} and the minimum frequency f_{min} [49–51] with the phase angle Φ_{min} could be correlated to the delaminated area as follows:

$$f_{break} = \frac{1}{2} \pi \, R_{Coat} \, C_{Coat} = \frac{1}{2} \pi \, R^0_{Coat} \, C^0_{Coat} \, \frac{A_d}{A} = \frac{D}{2 \pi \, \varepsilon_r \, \varepsilon_0 \, \rho} = f^0_{break} \, D \qquad \text{Eq. 4.11}$$

$$f_{\min} = \sqrt{\frac{1}{4} \pi^2 \, C_{Coat} \, C_{DL} \, R^2_{Coat}} = \sqrt{\frac{D}{4 \pi^2 \, \varepsilon_r \, \varepsilon_0 \, C^0_{DL} \, \rho^2 \, d}} \qquad \text{Eq. 4.12}$$

$$\tan\Phi_{\min} = \sqrt{\frac{4C_{Coat}}{C_{DL}}} = \sqrt{\frac{4\varepsilon_r\,\varepsilon_0}{C_{DL}^0\,d\,D}} \qquad \text{Eq. 4.13}$$

As already shown (see Chap. 2.2.2) during the corrosive treatment with an electrolyte the coating receives water and/or electrolyte and therefore ε_r and ρ will change during the treatment. Furthermore the coating thickness increases during the water uptake. Finally f_{break}^0 is not a constant parameter but will change with time [48, 52]. If the coating is saturated with water or electrolyte, i.e., the capacity of the coating is nearly constant, the delaminated area fraction D could be calculated as follows [48, 52]:

$$\frac{f_{break}}{f_{\min}} = \sqrt{\frac{C_{DL}}{C_{Coat}}} = \sqrt{\frac{C_{DL}^0\,D}{C_{Coat}^0}} \qquad \text{Eq. 4.14}$$

This assumption is reasonable, if a blister occurs during the treatment (see Fig. 1.10) but without a visible delamination the interpretation of the delaminated area would be difficult. In ref. [53] a good correlation between the visible delamination and the investigation of the delaminated area by means of the break-point frequency method is illustrated with a large quantity of different E coats.

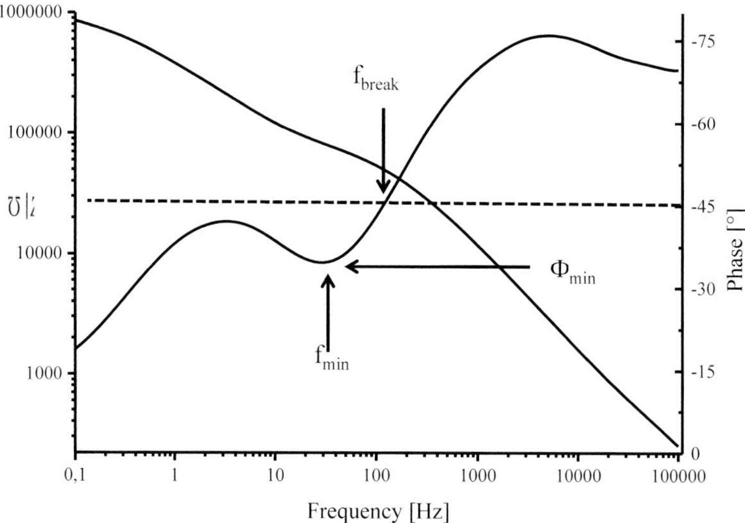

Fig. 4.6 Parameter of the break-point frequency method in a Bode plot of a corrosive treated organic coating without a defect.

For comparison of similar coatings with a variation of the corrosion protective pigment (zinc chromate, zinc phosphate, etc.) the break-point frequency method is also useful [54]. Considering the water uptake of the organic coating the method has been used to compare EIS data from coating systems (primer + top coat) after two years outdoor exposure with data from immersion test in 0.5% NaCl [52] with promising results. The advantage of the method is that there is no need to simulate the spectra with an equivalent circuit and the resulting uncertainty of the simulated data but the method based on the equivalent circuit of phase III. Therefore the method is limited to high performance coatings with thick layers and during the treatment no Warburg impedance could be detected (see [48]). If a Warburg impedance is visible as already mentioned by Haruyama [55], the equivalent circuit of phase IV–V in Fig. 2.45 has to be used to describe the process. Furthermore the samples have no defect such as a scratch because the model is based on an intact surface and delamination under the coating caused by wet deadhesion.

AC/DC/AC Method

To measure the break-point frequency weeks to months of treatment are necessary to investigate the data from the samples. For high performance epoxy resin based coatings with a thickness of 10 to 150 μm test periods of more than two years are necessary to detect the corrosion phenomenon or the decrease of barrier properties by means of EIS [56]. Industrial tests such as the salt spray test need around 1000 hours and climate chamber test in the automotive industry 10 weeks (see Chap. 4.2.5) therefore the EIS method with an immersion treatment is not as fast as the industrial methods. Because of that faster tests to investigate the corrosion protective performance have been developed. One method is the AC/DC/AC method illustrated in Fig. 4.7. The cathodic DC treatment forces formation of pores and delamination at the interface caused by a cathodic reaction [57] on organic coated samples without a defect.

Garcia and Suay [57] compared EIS measurements on E coats with different applications after 200 days immersion in 3.5% NaCl solution with the AC/DC/AC method (Fig. 4.7) for 24 hours. Both treatments show the same behaviour in the resulting EIS spectra of the coatings cured at different temperatures and applied with different voltages. Furthermore Olivier and Poelman [58] used the method on E coats with different conversion layers on different metal substrates with the same similar behaviour. The critique on the method is the use of the DC treatment because it is not the case in outdoor exposure or field tests and the use of defect free samples (see above).

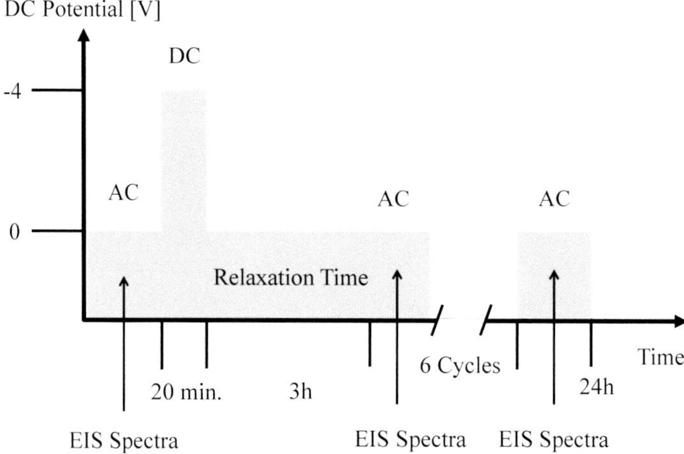

Fig. 4.7 AC/DC/AC test cycle according to [57].

Investigation of the Delamination by means of EIS

The challenge to measure the delamination rate of scratched samples by means of EIS has been mentioned in Chap. 4.1.4 caused by the simultaneous increase of the thickness and the surface area of the corrosion product layer. To avoid this problem Amirudin, Jernberg and Thierry [59] developed a method illustrated in Fig. 4.8. With the application of the counter electrode by vapour deposition on the organic coating the delamination from a scratch could be investigated by means of the break-point frequency method. Furthermore the results have been proved with IR thermography as a second independent method. The reasonable results show a linear

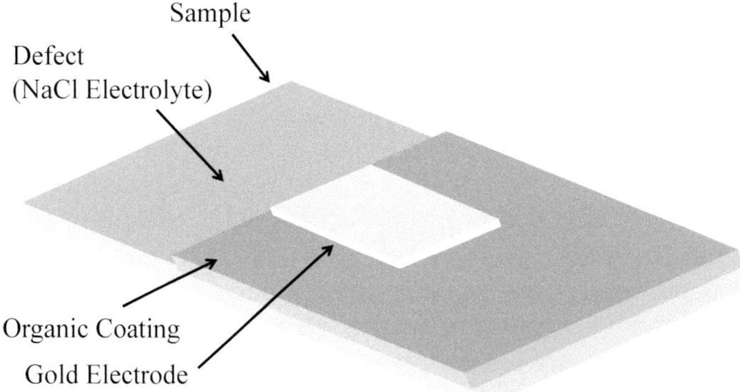

Fig. 4.8 Setup for the measurement of the delamination rate by means of EIS according to [59].

delamination rate similar to results achieved with the scanning Kelvin probe (see Chap. 2.2.3).

The disadvantage is the application of the gold electrode (400 nm) by vapour deposition because it is not useful for screening series.

Example 4.3

An easy approach to measure the delamination by means of EIS is illustrated in Fig. 4.9 and the thickness-area-problem is present with this setup. Because the metal substrate dominates the spectra a three-electrode setup has been used in a 3% NaCl electrolyte.

To test the setup two different coatings on steel have been used. One model coating with a very low wet adhesion and therefore a high delamination rate and an E coat with a very low delamination rate has been applied on steel substrates. The resulting Nyquist plots are illustrated in Figs. 4.10 and 4.11 for the model coating and the E coat, respectively. The corrosion layer is more than a barrier with a capacity visible in the high frequency range but at lower frequencies a capacity parallel with a resistance (equivalent circuit of phase III in Fig. 2.45) could be assumed as proved by the high phase angle and two time constants in the Bode plots (Figs. 4.12 and 4.13). Therefore the capacity and the resistance have been investigated from the Nyquist spectra according to Fig. 2.23. The combination of Eqs. 4.6 and 4.7 achieves at least a correlation between the product of the capacity with the resistance and the delaminated area shown in Fig. 4.14. A linear behaviour could be detected in the first 24 hours and the range could be proved by the investigation of the delamination

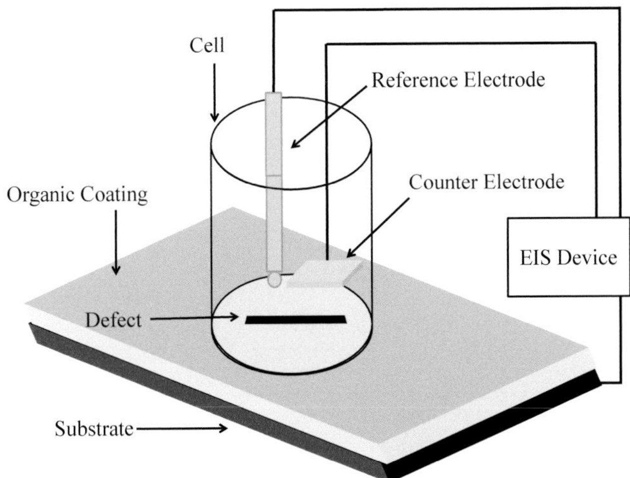

Fig. 4.9 Setup for the measurement of the delamination rate by means of EIS.

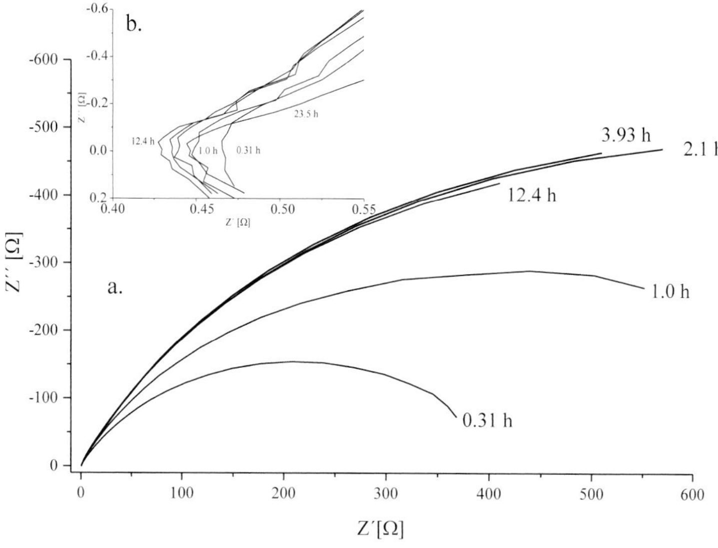

Fig. 4.10 Nyquist plots of the scratched model coating measured for 24 hours in 3% NaCl with the setup of Fig. 4.9. a: complete spectra, b: high frequency range.

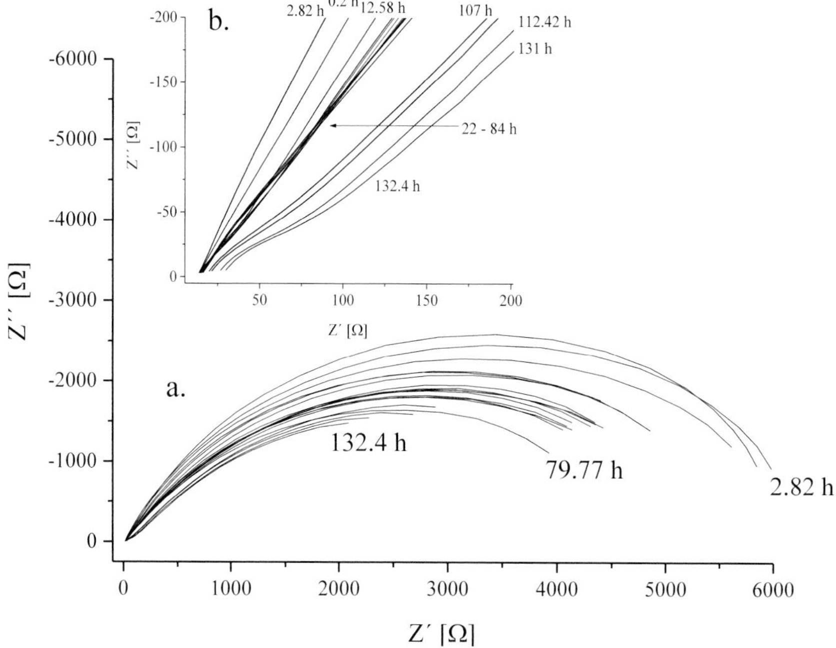

Fig. 4.11 Nyquist plots of the scratched E coating measured for 140 hours in 3% NaCl with the setup of Fig. 4.9. a: complete spectra, b: high frequency range.

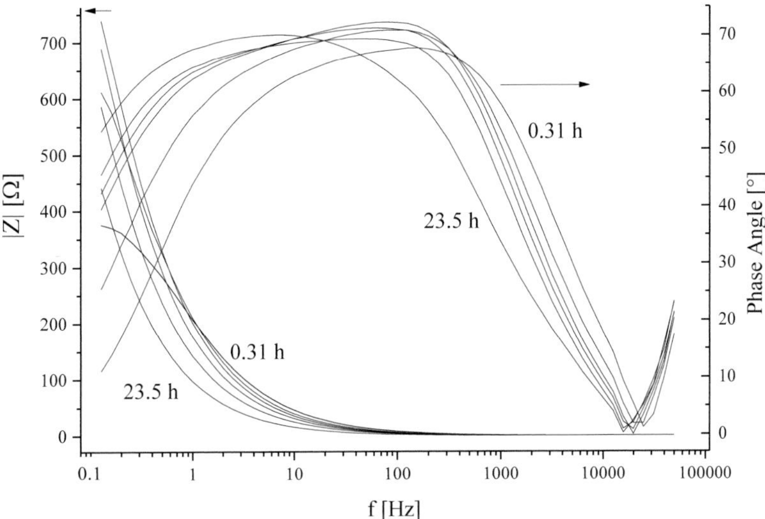

Fig. 4.12 Bode plots of the scratched model coating measured for 24 hours in 3% NaCl with the setup of Fig. 4.9.

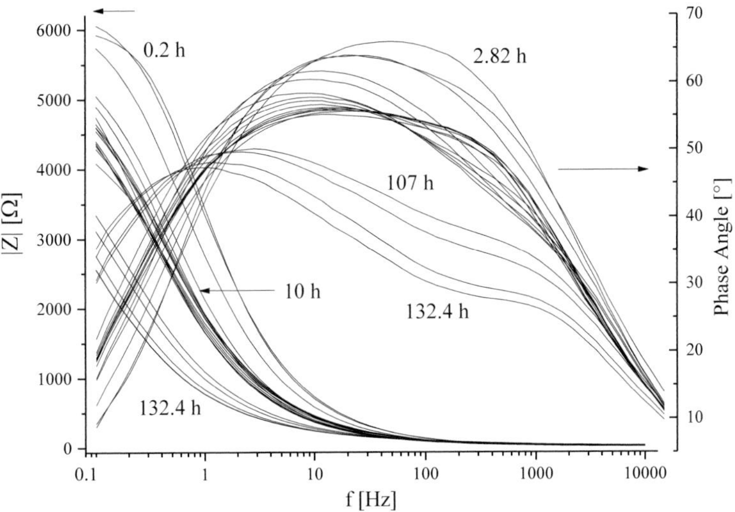

Fig. 4.13 Bode plots of the scratched E coating measured for 140 hours in 3% NaCl with the setup of Fig. 4.9.

rate of the samples with the naked eye. The linearity ends for the model coating because the delamination area achieves the border of the cell. The E coat samples show a passive range between 22–84 hours followed by the similar linear behaviour as in the first 22 hours. Although the data

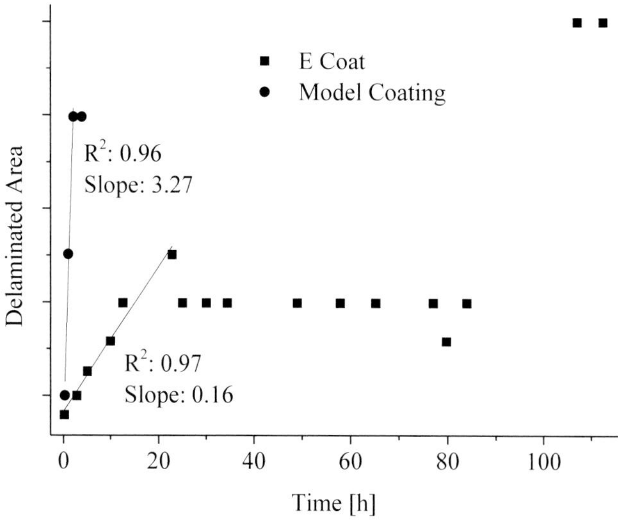

Fig. 4.14 Delamination rates calculated with the product of the resistance and the capacity from the Nyquist plots of Figs. 4.10 and 4.11 and based on the treatment illustrated in Fig. 4.9.

are based on several samples, more data are necessary to evaluate the method but these are at least promising results for further work because the thickness-area problem could be ignored or at least it did not disturb the measurement, if the goal is a comparison between different coatings and not absolute values.

Electrochemical Noise

All mentioned methods need to apply a potential/current on the samples to investigate data for the corrosion process or the corrosion protective properties and therefore the measurement disturbs the system in principle. The Electrochemical Noise method (EN) avoids it by measuring the current and potential noise between two identical electrodes whereas the current between both identical electrodes is measured with a zero resistance ammeter and the potential versus a reference electrode [60]. The noise resistance R_{noise} could be calculated with the standard deviations of the current σ_I and the potential σ_V data as follows [60]:

$$R_{noise} = \frac{\sigma_V}{\sigma_I} \qquad \text{Eq. 4.15}$$

The resistance could be correlated with the polarization resistance from EIS data for pure iron in different electrolytes with and without an inhibitor [61] and a good correlation of the EN data with EIS spectra has been found on iron in 0.5 M H_2SO_4 [60]. Furthermore with these methods organic coated samples have been analyzed and the data correlate in principle with other methods [62].

The EN data, potential and current fluctuations, could be transformed into the frequency domain by means of Fast Fourier Transformation (FFT) [60, 63]. The so called spectral noise resistance R_{SNoise} is calculated with the ratio of the potential and current FFTs at each frequency and the resulting limiting value R^0_{SNoise} could be correlated with R_{Noise} [60, 63]:

$$R_{SNoise}(f) = \frac{V_{FFT}(f)}{I_{FFT}(f)} \wedge R^0_{SNoise} = \lim_{f \to 0}\left(R_{SNoise}(f)\right) \qquad \text{Eq. 4.16}$$

Xiao and Mansfeld showed [63] that there is a good correlation between R_{Noise} and R^0_{SNoise} on organic coated metal substrates in 0.5 M NaCl electrolyte for 130 days but the magnitude of the resistances is different in comparison with the EIS data. There are more ways to use the EN data such as the calculation of the Power Spectral Density (PSD) from the FFT or by means of the maximum entropy method (MEM) [60, 64] and more complex calculations used in literature [65]. A comparison between PSD and MEM is illustrated in ref. [64] on stainless steel.

The weakness of the method is the missing of the fundamental correlation of the noise resistance with the corrosion rate. Furthermore the need of two identical electrodes for the experiment is hard to achieve. Jamali et al. [66] described a single electrode setup illustrated in Fig. 4.15 for the investigation of EN data.

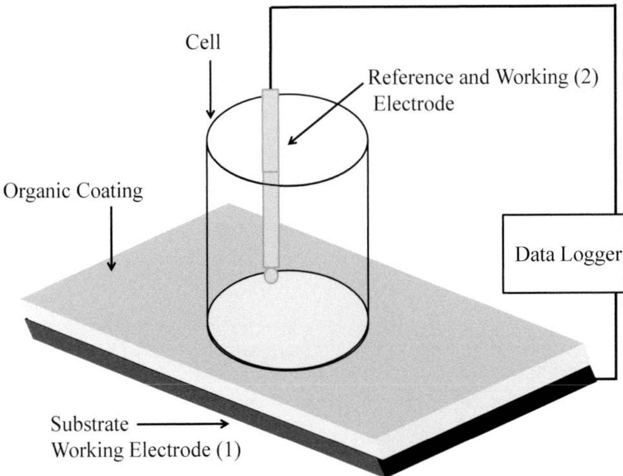

Fig. 4.15 Setup for the EN measurement with one electrode according to [66].

The available EN data (R_{Noise}, R^0_{SNoise}, PSD and MEM) correlate with corresponding EIS data and therefore promising results to use the EN for the investigation of corrosion rates.

4.2 Industrial Methods

The challenge to describe corrosion processes on real parts in real environments is caused by the long period necessary to detect the corrosion phenomenon and the fact that corrosion starts with a local process. Paik et al. [67] detected a Weibull distribution of the corrosion rate of organic coated ballast tanks in ships over a period of more than 10 years with a broad distribution caused by the uncertainties to detect the corrosion process. For instance in ref. [68] the outdoor exposure of organic coated steel samples has to extend from two to five years (Berlin 1970) to detect differences even on low performance coating systems in contrast to the two-year exposure in maritime environment. Therefore fast corrosion tests have been developed in every industry. Some test methods of the general and the automotive industry are summarized in Table 4.6. The term "fast" is very relative in accelerated corrosion tests because a normal Salt Spray Test (SST) takes 1000 hours, i.e., six weeks, and the most alternating climate tests in the automotive industry takes from 10 up to 30 weeks. In the aerospace industry specifications for the salt spray test duration of 3000 hours (ISO 9227) are common and up to 9000 hours (ISO 72523) are possible, i.e., 53 weeks. At least the tests are faster than the outdoor-exposure for years. This is the main reason for the development of really fast corrosion tests based on electrochemical methods mentioned above. The different industrial test methods are discussed and illustrated in the following chapters.

4.2.1 *Outdoor Exposure*

It should be the best option to test the corrosion protective properties of organic coatings or pure metal substrates in an outdoor exposure test because the accelerated corrosion tests have all a limited correlation to the real treatment in the field (see below). In DIN 55665 and ASTM G50 the major issues for outdoor exposure are summarized and exemplified in the following:

Test Sites

Of course the corrosion rate depends on the climate of the test site and therefore in all norms is distinct between rural, industrial and marine environments for the outdoor exposure. An impression of the environmental effect of the corrosion rate on different metal substrates is given in Table 5.4 in the app. In addition in Table 4.3 the effect is illustrated of places that are close to each other with significant differences of the corrosion mechanism.

Position

The position to the sun and the exposure angle to the horizontal define the amount of UV light, the wet/dry periods and the amount of water on the surface. In DIN 55665 angles of 5° or 20° are recommended and in ASTM G50 the angle to the horizontal of 30° or in ASTM D6675-01 (2006) and DIN EN ISO 11474 of 45° (oriented to the equator) is preferred. Finally for coil coating parts the European Coil Coating Association (ECCA) developed a procedure standardized in DIN EN 13523-19:2014 and DIN EN 13523-21:2017 with different angles of 5° (oriented to the south), 45° (oriented to the south) and 90° (oriented to the north) to optimize the simulation of the use of coil coating parts in the field.

Ground

Because of the fact that different grounds could generate a micro-environment for the test samples the distance has to be 0.45 m (DIN55665-2007), 0.75 m (DIN EN ISO 11474-2014) or 0.76 m (ASTM G50-2003) between the samples and the ground.

Test Duration

In DIN 55665 a treatment for at least 12 months is suggested. The duration could be reduced by an acceleration of the outdoor exposure to 6 months (see below).

Defect on the Surface of Organic Coated Metal Substrates

To investigate the delamination rate under the organic coating at a defect the samples are scratched in different ways illustrated in Fig. 4.16 (for a detailed discussion of the scratch see Chap. 4.2.2).

Accelerated Outdoor Exposure

The outdoor exposure could be accelerated by frequently spraying NaCl solution on the samples standardized in DIN EN ISO 11474 and ASTM D6675-01 (2006). In intervals (between 3 to 4 days in ASTM) a 5% NaCl (ASTM) or a 3% NaCl (DIN) solution sprayed over the samples to accelerate the corrosion process.

Analysis after the Test

The investigated parameters after the test are elucidated in the following chapter.

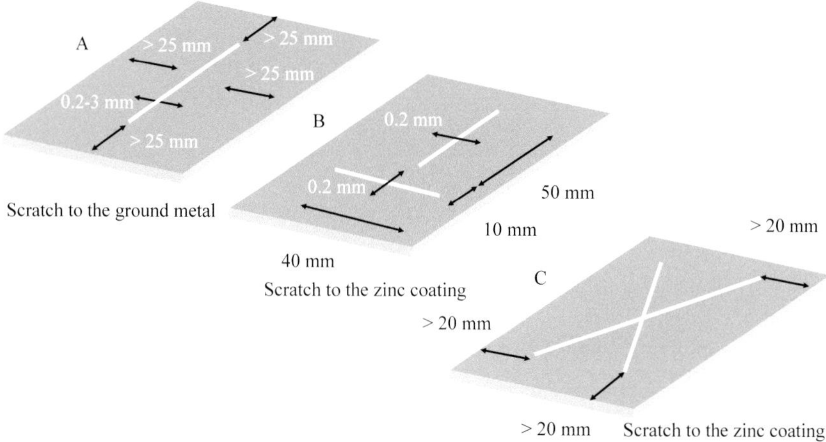

Fig. 4.16 Preparation of a defect on the test samples according to A: DIN 55665, B: DIN EN 13523-19 and C: DIN EN 13523-9(2014).

4.2.2 Salt Spray Test

The salt spray test according to ISO9227 or ASTM B-117-85 (and some more, see Table 4.6) performs a constant treatment of the samples with a NaCl electrolyte by means of spraying the solution on the samples for different test durations often between 120 and 1440 hours (DIN EN ISO 12944-6). For instance interior coatings for steel pipelines have to pass 480 hours in the salt spray test (ISO 15741). For conversion layers such as Cr(III) based layers on aluminium the duration of the salt spray test (without organic coating) depends on the performance level between 72 and 168 hours according to DIN EN 4729-2017.

Besides the pure salt spray test, further tests have been developed:

1. Salt spray test (SST)
   - ASTM B-117-85, DIN-EN-ISO9227 (former DIN 50021)

The test is often called Neutral Salt Spray test (NSS) because of the pH range of 6.5 to 7.2 and is used for all types of metal substrates coated with organic coatings and/or conversion layers.

2. Acetic acid accelerated salt spray test (ESS)
   - ASTM B287, DIN-EN-ISO9227 (former DIN 50021)

The test is used to determine the corrosion protective properties of organic coated or uncoated aluminium alloys and anodized aluminium.

Furthermore electrodeposited chromium or nickel surfaces on metal substrates could be analyzed with this test.

3. Copper accelerated salt spray test (CASS)
   - ASTM B 368, DIN-EN-ISO9227 (former DIN 50021)

This test is used on the same substrates or coatings as the ESS.

In all tests the samples placed in an angle of 15–20° from the vertical position and the salt fog sprays homogeneously over all samples.

Defect on the Surface

To simulate a defect on the organic coated surface a scratch is applied. There are many ways to apply defects on the surface (see Fig. 4.16). According to DIN EN ISO 9227 the width of the vertical scratch is between 0.1 and 1 mm and according to DIN EN ISO 17872 an X- or T-shaped scratch with a minimum width of 0.2 mm has to be applied on the surface. For organic coatings with a high thickness a scratch with a width of 2 mm is applied on the surface to the base metal (ISO 20340 a). An overview of different scratch methods and the results after corrosive treatment is illustrated in ref. [69].

As already mentioned (see Chap. 1.1.2) the cathodic process is diffusion controlled and therefore the corrosion rate is affected by the surface ratio of the anode and cathode area. Because of this the delamination or corrosion rate of small scratches is often lower as on broader defects [69].

Another challenge is the depth of the scratch on metal coated substrates. In Fig. 4.17 the different situations illustrated depending on the depth of the scratch on a metal coated substrate. If the scratch depth reaches the base metal, the metal coating generates a cathodic protection to the substrate. The coverage of the cathodic protection is in the mm-range and therefore the scratch width is very sensitive to the final result. Furthermore the shape of the scratch, i.e., trapezoid or rectangular, influenced the corrosion rate because it defines the anode/cathode ratio in the scratch (see Chap. 1.1.2). If the scratch only removes the organic coating

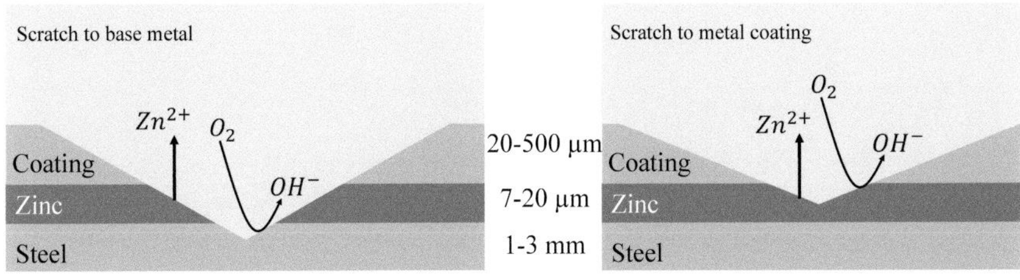

Fig. 4.17 Scratch on an organic and metal coated substrate.

as required in DIN EN 13523-8 for coil coating samples, the metal coating acts as a substrate until the coating is completely dissoluted during the corrosive treatment. Therefore in the beginning of the corrosive treatment the metal coating thickness is evaluated.

Because of the fact that the metal coating is very thin in comparison to the organic coating a defined scratch is difficult to produce by hand.

Analysis after the Test

After the treatment several parameters could be investigated to characterize the corrosion protective performance and adhesion of the organic coating as follows:

Degree of Blistering

The amount of blisters (see Chap. 1.2.2) is evaluated directly after the treatment according to DIN EN ISO 4628-2 or ASTM D 714 whereas distinguished between the amount and the size of the blisters by means of grades based on standards defined in the norm.

Rust Grade

The amount of corrosion visible on the organic coated surface is evaluated directly after the treatment according to DIN EN ISO 4628-3 or ASTM D 610 by means of grades based on standards defined in the norm.

Grade of Disbonding

In DIN EN ISO 4628-8 the corrosive disbonding is subdivided into deadhesion and corrosion, whereas deadhesion is called delamination. In scientific papers the corrosive disbonding is often also called delamination (see Chap. 2.2.3). The disbonded organic coating has to be removed carefully with a knife or scalpel to investigate the disbonded area. The disbonding rate is evaluated directly after the treatment or after 24 hours with the following equations [69]:

$$d = \frac{d_1 - w}{2} \qquad \text{Eq. 4.17}$$

d : grade of delamination in [mm]

w : width of the applied scratch in [mm]

d_1 : arithmetical average width of the delaminated area in [mm]

$$c = \frac{w_c - w}{2} \qquad \text{Eq. 4.18}$$

c : grade of corrosion in [mm]

w : width of the applied scratch in [mm]

w_1 : arithmetical average width of the corroded area in [mm]

The differentiation between delamination and corrosion occurs by the visible appearance of the surface, i.e., the area with corrosion products on the surface is corroded (Eq. 4.18) and the blank metal substrate is delaminated (Eq. 4.17).

Adhesion

Another consequence of the corrosive treatment could be the loss of adhesion of the organic coating far away from the defect. To investigate the residual adhesion a cross hatch test according to ISO 2409 or ASTM D 3359-07 (see also ref. [70]) has been performed normally 24 hours after the treatment.

Comparison of the NSS with Field Tests

The salt spray test has been criticized since decades because the test results have no correlation with outdoor exposure tests and do not mention field test results. The reason for that could be summarized as follows [70]:

Constant Humid Atmosphere

The humid atmosphere, i.e., a constant wet surface, avoids the formation of passive layers especially on zinc surfaces and therefore the corrosion rates of zinc coated metal substrates or zinc rich coatings are higher than in the field. Furthermore the composition of corrosion products on the surface (see Figs. 3.50 and 4.1) is also different. The constant wet surface forces a constant water/electrolyte uptake of the organic coating without dry periods, therefore the interface between the organic coating and the metal substrate is permanently treated with water/electrolyte and this is not the case under field conditions.

Chloride Concentration

As already mentioned the chloride content defines the composition or the formation of passive layers on the metal substrate especially on zinc surfaces. Therefore the high amount of chloride in the test avoids the formation of passive layers or forces the production of a completely different composition of the passive layers in comparison to field conditions.

Temperature

The constant high temperature (see Table 4.6) normally accelerates the corrosion and transport processes more than under field conditions.

Therefore the salt spray tests are not able to simulate real conditions in the field. To simulate real conditions alternating climate tests have been developed (see Chap. 4.2.5). To prove a certain quality level of an organic coating the salt spray tests are very useful, cheap and easy to proceed but only for one kind of substrate and one kind of organic coating. The comparison of the quality level of an organic coating on steel and on zinc coated steel is impossible and the comparison of a zinc rich primer with barrier coatings pigmented with inert pigments is also impossible. Nevertheless the salt spray test could separate different quality levels. In DIN EN ISO 12944-6 organic coatings for corrosion protection is distinguished in different protection levels according to different corrosion stress levels. The advantage of the direct correlation of a quality level to a certain corrosion stress level includes the risk that alternative organic coatings cannot be used because the stability in the salt spray test is not achieved but under real conditions the coating protects against corrosion. Therefore DIN EN ISO 12944-6 always suggests the outdoor exposure test in addition to the salt spray test.

Finally the standard deviation of the NSS is very high, i.e., for low performance coatings around 20–30% and for high performance coatings up to 50% because the delamination value is low and the error increases. Therefore more than one sample has to be used for an experiment at least three better still 10 samples per type of organic coating or metal substrate are meaningful for test series.

Example 4.4

The conversion layer based on phytic acid already presented in Exs. 2.5, 3.7 and 3.13 with promising results for the corrosion protective performance [71]. The use of a conversion layer is the combination with an organic coating and therefore the combination has been prepared and tested in the salt spray test (Fig. 4.18). The experiment has been interrupted after 68 hours because of the high delamination rate of the scratched samples. All conversion layers accelerate the delamination at the defect in comparison to the untreated steel substrate whereas the conversion layer applied at pH = 2 show the best results and could be improved with the addition of a wetting agent (V) and molybdate as corrosion inhibitor (K) (for details see Ex. 2.5).

The reason for the disappointing results is illustrated in Fig. 4.19. The contact angle of the conversion layers to water is very low, i.e., the surface is very hydrophilic (for details to investigate the contact angle/surface

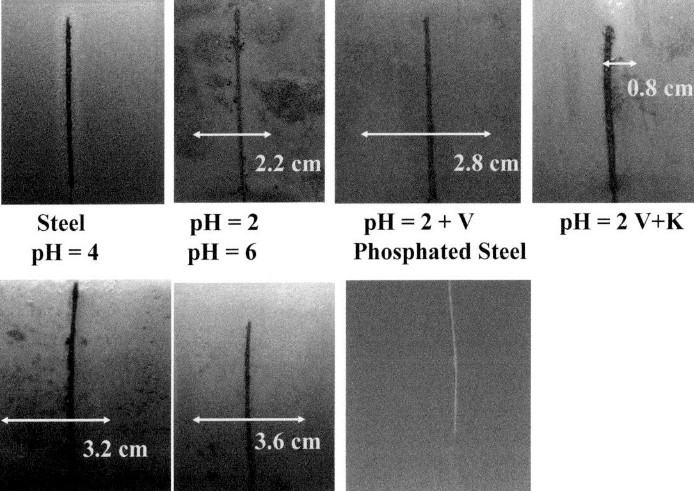

Fig. 4.18 Pictures after 68 hours salt spray test of steel samples coated with phytic acid based conversion layers covered with a clear coat and steel and phosphated zinc coated steel samples for comparison from [71]. With permission.

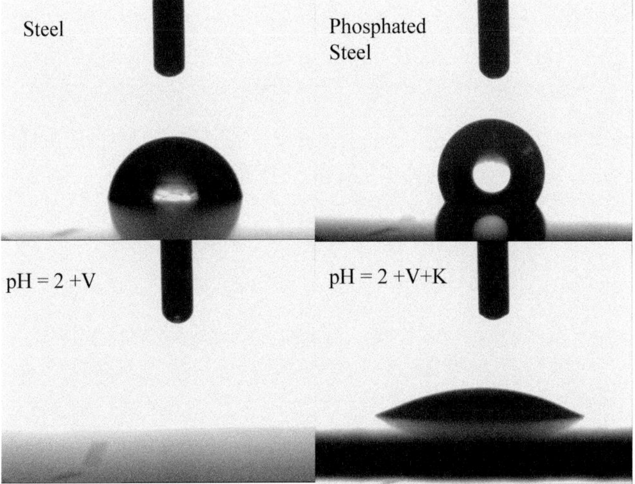

Fig. 4.19 Pictures of the investigation of the contact angle against water from the phytic acid based conversion layers and steel for comparison from [71]. With permission.

tension see ref. [72]). With the wetting agent the water drop absorbs immediately after application and the contact angle cannot be measured.

The promising electrochemical and spectroscopic data of the phytic acid based conversion layer cannot describe the corrosion protective properties of the complete coating system. The complete coating system has to be prepared and analyzed and the NSS is a good model treatment at a first look.

Example 4.5

The completely different behaviour of chromate containing and chromate free coil coating primer regarding the EIS data is illustrated in Ex. 2.12. The impedance spectroscopy shows the different barrier properties of the coatings and the SKP the delamination rate at a defect and in this example it is hard to measure because of the high performance level of the coil coating primers (see Ex. 2.13) [73].

The salt spray test results in Fig. 4.20 show more or less no difference of both primers, i.e., considering the standard deviation of the NSS no difference in the delamination rate at the scratch (in the middle of the samples, the uncoated edges are not analyzed according to the norms). Therefore the quality level of both coating systems is equal but the protective mechanism of the chromate free and chromate containing coating is different and cannot be analyzed by means of the NSS but with electrochemical and spectroscopic methods. Therefore the combination of industrial tests with electrochemical or spectroscopic methods generates the complete picture.

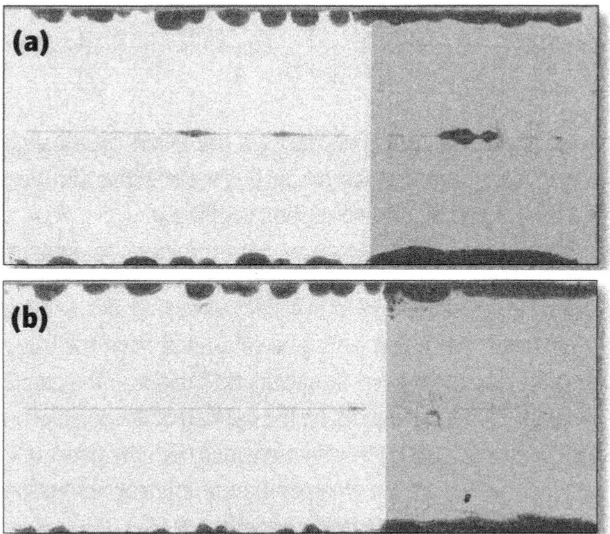

Fig. 4.20 Coil coating samples (left: primer and top coat, right: primer only) after 500 hours salt spray test. (a): Chromate containing primer with a chromate containing pre-treatment. (b): Chromate free primer and conversion layer from [73]. With permission.

4.2.3 *Constant Climate Tests*

In this paragraph all tests with a constant climate are summarized without the use of NaCl solution during the treatment. The constant climate tests mentioned in Table 4.6 are used to investigate the corrosion protective performance of organic coated or pure metal substrates or are part of special corrosion tests.

Constant Climate Test

The test standardized in EN ISO 6270-2-2005 is used to define a certain quality level according to DIN EN ISO 12944-6 (see Table 4.6).

CORR Test

For metal coatings based on Cu-Ni-Cr or Ni-Cr layers on steel the CORR test have been developed and standardized in DIN EN ISO 4541. The test uses a paste containing Cu(II), Fe(III) salts and ammonium chloride with kaolin as matrix. The paste applied on the surface and dried followed by a constant climate test with 80–90% humidity, i.e., no condensation of water on the surface, at 38°C for 16 hours. This cycle could be repeated several times. The resulting corrosion phenomenon on the surface such as blister formation (see Fig. 1.10) or pitting corrosion (see Fig. 1.9) are analyzed, e.g., according to ISO 1462. The modified CORR test (DIN 50958:2012) uses 100% humidity, i.e., condensation of water on the surface similar to the constant climate test, at 40°C and the corrosion areas on the surface uncovered after the test by immersion of the samples for 24 hours in water or 4 hours in the salt spray test (see Chap. 4.2.2).

Filiform Corrosion

Filiform corrosion is an anodic delamination (see Fig. 1.11) with the formation of filaments under the coating and look like worms under the coating. To initiate the anodic delamination chloride anions have to be placed at the defect by means of different methods as follows:

Immersion in NaCl Electrolyte (DIN EN ISO 4623-1:2002)

For organic coated steel substrates the scratched samples likewise as shown in Fig. 4.16 (B) immersed into a NaCl solution for 30–60 seconds followed by a constant climate test with 80% humidity at 40°C.

Salt Spray Test (DIN EN ISO 4623-1:2002)

For organic coatings with a higher corrosion protective performance on steel the scratched samples treated in the salt spray test (see Chap. 4.2.2) for maximum of 24 hours followed by the constant climate test with 80% humidity at 40°C.

On steel the anodic delamination is more likely at humidity below 85% and above often cathodic processes are observed.

Treatment with HCl (DIN EN 3665)

For organic coated aluminium substrates (e.g., cladded A2024-T3) used in the aircraft industry the initiation is achieved with HCl solution. In a closed chamber filled with conc. HCl (some cm) the scratched samples placed (1–24 hours) perpendicular with a certain distance above the solution so that the HCl steam initiated the corrosion at the scratch. After the initiation the samples treated in a constant climate test with 80% humidity at 40°C for normally 1000 hours. For a detailed discussion of the test see ref. [74].

After the test the number M and the length L of the filaments have to be investigated. Because of the fact that filiform corrosion is a very local phenomenon the arithmetic average of M and L of the left and the right side of the scratch are investigated to calculate an arithmetic average from both sides (DIN EN ISO 4628:10-2016). Finally the filiform corrosion of a sample is defined as L4/M3, i.e., the average length of all filaments is 4 mm and the average distance between two filaments at the scratch is 3 mm.

In ref. [75] outdoor exposure tests compared with the filiform corrosion test according to DIN EN 3665 on organic coated aluminium substrates (AlMg1, AlMgSi0.5). The results show that the corrosion test forces more filiform corrosion as generated in the outdoor exposure but with a principal correlation, i.e., if a coating system generates low corrosion in the test, the same result is visible in the outdoor exposure. On the other hand, if the climatic conditions of the outdoor exposure create cathodic delamination at the scratched samples, there is of course no correlation with the filiform test. Therefore to get the complete picture of the corrosion protective performance of a coating system, different tests have to be performed generating different kinds of corrosion mechanism.

4.2.4 *Immersion Tests*

In this paragraph all tests with an immersion of the samples into a corrosive solution are summarized.

Immersion in Water

The easiest treatment for an uncoated or coated metal substrate is the immersion into water standardized in DIN EN ISO 2812-2. The samples stored in a water tank with an angle of 0°–20° vertical and parallel to the flow direction at 40°C with or without a scratch in the organic coating according to ISO 15711. The duration of the test varies from hours to month for instance in DIN EN ISO 12944-6 the organic coating has to pass the immersion test between 2000 and 3000 hours and for interior coatings for steel pipelines 480 hours are required (ISO 15741). For coil coating systems (DIN EN 13523-9) the samples are scratched as illustrated in Fig. 4.16 method C and immersed in water with an angle between 15° to 20° at 40°C.

Contact Corrosion

Contact corrosion (see Chap. 1.2.1 and Fig. 1.9) based on the conductive join of two different metals could be analyzed according to DIN 50919:2016. The relevant values such as E_{cor}, E_a, E_c and i_{cor} (compare Fig. 1.4) could be investigated with an electrochemical cell similar to Fig. 2.1 but with reference electrodes for both metal electrodes and a conductive connection between the samples. The resulting weight loss (see Chap. 4.1.2) could be defined based on the anodic current density i_a with the Faraday law (DIN 50919-2016):

$$i_a = \frac{z \cdot F \cdot v}{M} \qquad \text{Eq. 4.19}$$

z : Charge of the metal cation

v : Mass loss rate [$g\ m^{-2}\ h^{-1}$]

F : Faraday constant, 96485 $C\ mol^{-1}$

M : Molar mass [$g\ mol^{-1}$]

As already mentioned the current density depends on the polarization resistance and the conductivity of the electrolyte according to Eq. 1.45.

Intergranular Corrosion

Intergranular corrosion (see Fig. 1.9) is an important corrosion mechanism on stainless steel and aluminium surfaces. The most important standards to test the intergranular corrosion on stainless steel surfaces are EN ISO 3651-2 and ASTM A 262 (compare ref. [76]) but of course there are more standards such as ref. [77]. The methods described in the standards based on temperature stress in acidic solutions and partially combined with an electrochemical treatment. For instance in method A in ASTM A 262 the samples are immersed in an oxalic acid solution and treated with a current density of 1 mA/cm^2 for 1.5 minutes. After the treatment the samples are analyzed under a microscope to investigate the intergranular corrosion qualitatively. A quantitative result could be generated with method B of the same standard by immersing the sample for 120 hours in a boiling solution of diluted H_2SO_4 and $Fe_2(SO_4)_3$ and investigating the corrosion rate by weight loss (see Chap. 4.1.2). For further reading see refs. [76, 78].

Pitting and Crevice Corrosion

The amount of pitting and crevice corrosion (see Chap. 1.2.1 and Fig. 1.9) on stainless steels could be investigated with the procedures described in ASTM G48.

Both corrosion mechanisms could be analyzed by immersion into an iron (III) chloride solution with or without additional HCl at a certain temperature for 75 hours. The tendency for crevice corrosion could be investigated with a model crevice illustrated in Fig. 4.21. A detailed discussion about the method is given in ref. [76]. Besides the crevice model according to ASTM G48 the automotive industry defined their own setups for instance the Renault crevice panel [79], the Volvo crevice panel [80] or spot welded samples (for further reading see refs. [81–83]).

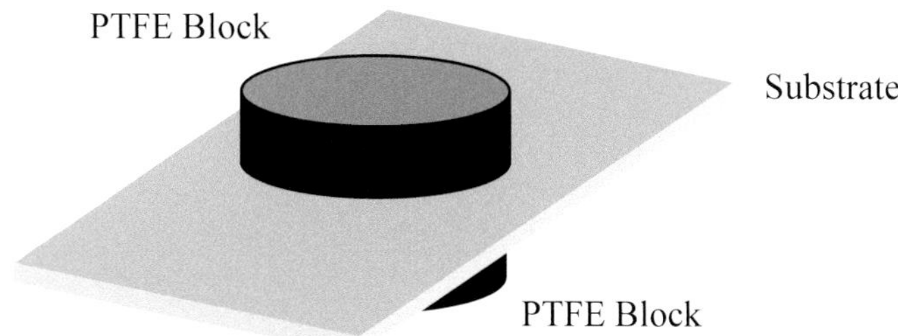

Fig. 4.21 Scheme of the sample preparation for the investigation of crevice corrosion according to ASTM G48.

4.2.5 Alternating Climate Tests

The aim of the alternating climate tests, i.e., alternate different treatments with or without NaCl solution in different humidities and temperatures, is to simulate the real conditions such as in the outdoor exposure or during the use of the coated metal part. In all alternating climate tests (see Table 4.6) a change from a wet to a dry surface occurs several times during the test cycle. During the drying phase a thin electrolyte layer takes place and accelerates the corrosion or delamination rate on the sample illustrated in Ex. 2.13. Otherwise during the dry period passive layers could be generated at the scratch or defects reducing the corrosion rate. Therefore the wet/dry cycles are important to simulate real conditions [70]. The challenge of the development of an alternating climate test is the huge effect of small climate changes under real conditions shown in Table 4.3. The resulting test is therefore the average of all climates or the most aggressive climate the coated metal part is used in. The resulting tests are different as illustrated in Table 4.6 and all tests are permanently optimized to increase the correlation of the test results with the field test.

Local effects such as the high calcium chloride content on the coated surface during the winter in Moscow generated a new corrosion mechanism and cannot be included into the standard alternating climate test. If the corrosion mechanism was analyzed as shown in Exs. 2.3, 2.6 and 3.19 for the Russian mud corrosion, a new corrosion test has to be developed to simulate these conditions. In the consequence every OEM developed a lot of tests to simulate the general corrosive attack and some local phenomenon to investigate the corrosion protective performance of uncoated and coated metal parts. Therefore Table 4.6 is just a very small overview of different alternating climate tests used in the industry. A broader overview about alternating climate test is given in refs. [70, 82, 84–86].

The conditions could change during the decades and also the corrosion tests. One example is the corrosion protective property of uncoated and coated metal substrates against sulphur dioxide containing atmosphere often called Kesternich test (for details see DIN 50018, ASTM B605 and DIN EN ISO 6988). The test has been developed because of the acid rain effect caused by the high sulphur dioxide content in exhaust emission for instance of lignite-fire power plants without flue gas desulphurization. In the last decades the test became less relevant because flue gas desulphurization is used worldwide (more or less) and the amount of sulphur dioxide is reduced (compare ref. [70]).

The examination of the test results are often similar to the methods described in Chap. 4.2.2.

The correlation of field tests with outdoor exposure and accelerated corrosion tests such as the alternating climate tests has been discussed for decades in literature and some results presented in the following paragraph.

Nowak et al. [87] compared different pre-treatment/passive layers (zinc phosphating, Ta_2O_5, TiO_2, Cr_2O_3 and Al_2O_3) on steel and zinc coated steel covered with an organic coating (primer and top coat). The passive layers increase the polarization resistance of the surface up to two orders of magnitudes investigated with electrochemical methods. Especially the Ta_2O_5 layer decreases the delamination area in comparison to the standard zinc phosphatation on steel but only in the Mazda test. More or less no effect could be detected in the GM 9540P test and the delamination rate of the zinc phosphated samples differs up to 300% between the industrial tests.

Zapponi et al. [88] compared Cr(IV) pre-treatment with chromium free systems on zinc coated steel covered with a polyester based organic coating (coil coating). The outdoor exposure and marine environments generate the same corrosion products in comparison to the Prohesion G85 test but only on the Cr(VI) pre-treatment samples, i.e., the classical system.

Table 4.6 Accelerated corrosion tests from the automotive and general industry used and illustrated in the mentioned literature. Abbreviations: min: minute, h: hour, d: day, w: week, DIN I: DIN EN ISO 11130-2017, DIN II: DIN EN ISO 12944-6.

Test	Treatment							Cycle duration	Test duration	
	Salt spray				Climate cycle					
	NaCl content [%]	pH	Amount [ml/h]	Cycle and Temp.	Relative humidity [%]	Duration	Temp. [°C]			Lit.
Alternating Climate Tests										
Renault ECCI D172028	1	4	5	30 min./d, 3.5 h/w, 35°C	20	1 h 35 min.	35		42 d	[83]
					55	2 h 40 min.	35			
					90	1 h 20 min.	35			
Renault 3C Test	5			24 h, 35°C	95	8 h	40	1 w	7 cycles	[89]
					Laboratory	16 h				
					64	48 h	20			
Volvo VICT VCS1027, 149	1	4	120	3*15 min. twice a w, 1.5 h/w, 45°C	50	4 h	45		42 d	[83]
					95	4 h	35			
Volvo VICT-2	1			1 h twice a week before (I) (Mo, Fr)	(I) 90	7 h	35		12 w	[70]
					(II) 90–45	1.5 h	35			
					(III) 45	2 h	35			
					(IV) 45–90	1.5 h	35			

VDA621-415 (Salt Spray DIN50021, Climate Cycling DIN 50017)	5	6.5–7.2	1.5	24 h/w, 40°C	100	4*8 h/w	40	1 w	70 d	[83], [70]
					50	4*16 h/w	18–28			
					50	2*48 h/w	23			
Volkswagen PV1210	5	6.5–7.2	1.5	4 h/d (5 days), 20 h/w, 23°C	50	4 h	23		42 d	[83]
					100	16 h	40			[83]
					Laboratory	16		1 w	3 cycles	[89]
General Motors GM9540P (method B)	NaCl 0.9 + $CaCl_2$ 0.1 + $NaHCO_3$ 0.1	6–9		4*15 min./d, 14 h/w, RT alternated with drying				24 h	40 cycles up to 72 cycles [87]	[83], [87]
					100	8 h	49			
					20	8 h	60			
GM Scab Test	5 Immersion			15 min., RT	85	22.5 h	60	24 to 25.5 h	20 cycles	[89], [87]
					Laboratory	75 min.				
						30 min. (once a w)	23			
						1 h	60			
Daimler Chrysler KWT DC	1	6.5–7.2	2	2 h/d (4 d), 8 h/w at 25°C	50–100		–15–50		42 d	[83]

Table 4.6 contd. ...

...Table 4.6 contd.

Test	Treatment							Cycle duration	Test duration	Lit.
	Salt spray				Climate cycle					
	NaCl content [%]	pH	Amount [ml/h]	Cycle and Temp.	Relative humidity [%]	Duration	Temp. [°C]			
Alternating Climate Tests										
Mazda	5	6.5–7.2		6 h at 35°C	97			24 h	8 cycles [87]	[87]
					20–30	3 h	50			
					95	14 h	50			
					Laboratory	1 h				
Ford Test	5			15 min., 23°C	100	8 h	25	1 w	10 cycles	[89]
					Laboratory	48 h				
Prohesion Test ASTM D 5894	5 + 3.5 $(NH_4)_2SO_4$	5.2		1 h at 23°C	Drying at 35°C			1 h	1400 h	[70]
Kesternich DIN 50018 (Volkswagen), ASTM B605	0 + SO_2 (1 or 2 l with 2 l water in a 300 l chamber)			8 h at 40°C	100 with bedewing the samples			24 h		[90], [70]
					Laboratory	16 h				

DIN EN ISO 6988	SO_2 (0.1 l with 2 l water in a 300 l chamber)			8 h at 40°C	50			24 h		[94]
					Laboratory	16 h				
Alternate Immersion Test in Salt Solution DIN EN ISO 11130-2010	3.5% (Immersion)	6.0–7.0		10 min. at 25°C				1 h		[94]
					45, 50 [DIN I]	50 min.	27, 70 [DIN I]			
Condensation Climate with Alternating Humidity and Air Temperature (AHT) DIN EN ISO 6270-2-2005	0	Dist. Water			100 with bedewing the samples	8 h	40	24 h		[94]
					Below 100	16 h	18–28			
Condensation Climate with Alternating Air Temperature (AT) DIN EN ISO 6270-2-2005	0	Dist. Water			100 with bedewing the samples	8 h	40	24 h		[94]
					Near 100	16 h	18–28			

Table 4.6 contd. ...

...Table 4.6 contd.

Test	NaCl content [%]	pH	Temp.	Relative humidity [%]	Test duration	Lit.
Salt Spray Tests						
Salt Spray DIN-EN-ISO9227	5 (15 ml/h)	6.5–7.2	35°C		28 d, 1440 h [92], 2160 h [92], 120–1440 h [DIN II]	[83]
Salt Spray ASTM B-117-85	5	6.5–7.2	35°C	97	2000 h	[89], [90]
Acetic Salt Spray DIN-EN-ISO9227, ASTM B287	5 + Acetic Acid	3.1–3.3	35°C	97	500 h [93]	[90]
Copper Accelerated Salt Spray DIN-EN-ISO9227, ASTM B 368	5 + $CuCl_2$ (1–4 g/l)	3.1–3.3 (adjusted with acetic acid)	35°C	97		[90]
Salt Spray PN-59/H-04603	3		6 h/d, 20°C			[91]

Constant Climate Tests						
Condensation Atmosphere with constant humidity DIN EN ISO 6270-2-2005	0	Dist. Water	40°C	100 with bedewing the samples	48–1440 h [DIN II]	[94]
CORR Test DIN EN ISO 4541	Samples coated with a kaolin paste composed of Cu(II), Fe(III) and Chloride	Dist. Water	38°C	80–90 without bedewing the samples	16 h	[95]
Modified CORR Test DIN 50958	Samples coated with a kaolin paste composed of Cu(II), Fe(III) and Chloride	Dist. Water	40°C	100 with bedewing the samples	16 h	[95]
Filiform Test DIN EN ISO 4623-1:2002, DIN EN 3665	Initiation with Chloride (NaCl or HCl)	Dist. Water	40°C	80	1000 h	[74]

The new approach of a water based chromium free conversion layer differs between the outdoor exposure and the alternating climate test.

LeBozec and Thierry [82] analyzed the effect of different conditions (wet/dry, temperature, salt spray duration, etc.) in the alternating climate test in comparison with outdoor exposure (marine environment) and a field test (fixed samples under a bus). Steel and zinc coated steel samples with zinc phosphate conversion layer covered with an E coat and top coats have been used in the test series. Furthermore Renault crevice samples have been prepared to analyze the tendency of crevice corrosion (see Chap. 4.2.4). Only one setup of the different alternating climate test correlates with the outdoor exposure but hardly with the field test.

LeBozec et al. [83] compared six different alternating climate tests (Renault 172028, Volvo VICT, VDA 621-415, VW P1210, GM 9540P and Daimler-Chrysler KWT-DC) with the salt spray test (ISO 9227), outdoor exposure (marine environment) and field tests (samples fixed on vehicles). Uncoated steel and zinc coated steel sheets and for the Renault test crevice samples covered with zinc phosphate and an E coat have been used. The corrosion rates of the uncoated samples differ on steel from 50 μm (GM 9540P) up to 200 μm (VDA 621-415) and on zinc coated steel from 8 μm (GM 9540P) and 90 μm (ISO 9227). The same differences could be detected on the organic coated samples. A comparison with the field test results was only possible with the alternating climate tests using a low NaCl content during the cycle such as KWT-DC, GM 9540P, Renault 172028 and Volvo VICT (compare Table 4.6). The alternating climate tests with a high NaCl content generate higher corrosion rates and the salt spray test definitely has no correlation with the field test as already mentioned (see Chap. 4.2.2).

The examples illustrate that different test conditions generate different results but alternating climate test are able to simulate field test [83]. The risk of the procedure is the optimization of the established coating system to the test and vice versa and new approaches may fail in the alternating climate test but pass the field test. Therefore the alternating climate test has to be proved periodically with field tests—a standard procedure in industry—and therefore several tests mentioned in this chapter have already been changed, i.e., optimized to the field and new coating systems.

4.3 Literature

1. H. Tanabe, T. Shinohara, M. Hoshino, Y. Sato; Application of Current Interrupter Method on the Estimation of Protective Coatings, Proceedings of INTERFINISH 80, p. 339–343 (1980) International Union for Surface Finishing (IUSF)
2. I. Sekine, M. Yuasa, K. Tanaka, T. Tsutsumi, F. Koizumi, N. Oda, H. Tanabe, M. Nagai, Relation of parameters obtained by various electrochemical measurements

for evaluation of abilities of protective coating films, J. Japan Soc. Colour Material 67, 424–430 (1994)
3. H. Tanabe, M. Nagai, H. Matsuno, M. Kano; Evaluation of protective coating by a current interrupter technique, Advances in Corrosion Protection by Organic Coatings II, The Electrochemical Society, Proceedings Volume 95-13, p.181–192 (1995)
4. G. Meyer, H. Ochs, W. Strunz, J. Vogelsang; 'Barrier coatings with high Ohmic resistance—comparison between Relaxation Voltammetry and Impedance Spectroscopy'; Proceed. of EMCR 1997, 25–29.08.1997, Trento Italy
5. H. Ochs, W. Strunz, J. Vogelsang; 'Relaxation voltammetry with organic barrier coatings on steel—experimental and theoretical approach'; Proceed. of EMCR 1997, Trento, Italy
6. W. Strunz, J. Vogelsang; 'Characterization and evaluation of organic coatings using relaxation voltammetry'; Proceed. of EUROCORR 1998, Utrecht, Netherlands
7. J. M. Sykes and H. E. M. Smith, Electrochemical measurements on polymer coated metal using a potential decay technique, Mater. Sci. Forum 44&45 (1989), 433 [Proceedings of the 3d3rd International Symposium on Electrochemical Methods in Corrosion Research, Zurich 1988]
8. J. D. B. Sharman, H. E. M. Smith and J. M. Sykes, Impedance of polymer coatings and polymer-coated steel measured by a potential decay technique—Advances in corrosion protection by organic coatings (Eds. M. Kendig and J.D. Scantlebury), Electrochem. Soc. 1989
9. E. E. Oguzie, K. L. Oguzie, C. O. Akalezi, I. O. Udeze, J. N. Ogbulie, V. O. Njoku, ACS Sustainable Chem. Eng. 1 (2013) 214–225
10. M. H. Wahdan, Mater. Chem. Phys. 49 (1997) 135–140
11. B. M. Praveen and T. V. Venkatesha, Int. J. Electrochem. Sci. 4 (2009) 267–275
12. S. S. Abd El Rehim, H. H. Hassan, M. A. Amin, Mater. Chem. Phys. 70 (2001) 64–72
13. M. A. Quraishi, M. Z. A. Rafiquee, S. Khan, N. Saxena, J. Appl. Electrochem. 37 (2007) 1153–1162
14. D. Gopi, S. Manimozhi, K. M. Govindaraju, P. Manisankar, S. Rajeswari, J. Appl. Electrochem. 37 (2007) 439–449
15. T. A. Markley, M. Forsyth, A. E. Hughes, Electrochim. Acta 52 (2007) 4024–4031
16. H. Ashassi-Sorkhabi, D. Seifzadeh, Int. J. Electrochem. Sci. 1(2006) 92–98
17. A. Y. El-Etre, Corros. Sci. 43 (2001) 1031–1039
18. A. R. Shashikala, R. Umarani, S. M. Mayanna, A. K. Sharma, Int. J. Electrochem. Sci. 3 (2008) 993–1004
19. L. Legrand, L. Mazerolles, A. Chauss´e, Geochim. Cosmochim. Acta 68, 17 (2004) 3497–3507
20. J. Alcantaram, D. de la Fuente, B. Chico, J. Simancas, I. Diaz, M. Morcillo, Materials 10 (2017) 406
21. J.-E. Hiller, Werkst. Korros. 11 (1966) 943–951
22. Thesis (PhD), H. Venzlaff, Die elektrisch mikrobiell beeinflusste Korrosion von Eisen durch sulfatreduzierende Bakterien Ruhr University Bochum, 2012
23. M. Odziemkowski, Spectroscop. Prop. Inorg. Organomet. Compd. 40 (2009) 385–449
24. S. SYED, Emirates Journal for Engineering Research 11, 1 (2006) 1–24
25. P. R. Roberge, Handbook of Corrosion Engineering, McGraw-Hill, USA, 1999, 58
26. F. Ostermann, Anwendungstechnologie Aluminium, Springer Verlag, 2007
27. D. C. Cook, Corros. Sci. 47 (2005) 2550–2570
28. E. Bardal, Corrosion and Protection, Springer Verlag, London, 2004
29. B. X. Huang, P. Tornatore, Y.-S. Li, Electrochim. Acta 46 (2000) 671–679
30. H. H. Hassan, K. Fahmy, Int. J. Electrochem. Sci. 3 (2008) 29–43
31. Y. Y. Chen, S. C. Chung, H. C. Shih, Corros. Sci. 48 (2006) 3547–3564
32. J. Morales, F. Diaz, J. Hernandez-Borges, S. Gonzalez, Corros. Sci. 48 (2006) 361–371
33. E. Almeida, M. Morcillo, B. Rosales, M. Marrocos, Mater. Corr. 51 (2000) 859
34. S. SYED, Emirates Journal for Engineering Research 11, 1 (2006) 1–24

35. G. Wranglen, Introduction to Corrosion and Protection of Metals Institute for Metallskydd, Stockholm, 1972
36. F. A. Champion, Corrosion Testing Procedures, Chapman and Hall, London, 1964
37. F. Blin, S. G. Leary, G. B. Deacon, P. C. Junk, M. Forsyth, J. Appl. Electrochem. 34 (2004) 591–599
38. M. Dornbusch, Lackschichtbildendes Korrosionsschutzmittel und Verfahren zu dessen stromfreier Applikation, DE102005023728
39. G. P. Bierwagon, L. Hi, J. Li, L. Ellingson, D. E. Tallmann, Prog. Org. Coat. 39 (2000) 67
40. G. P. Bierwagon, K. Allahar, B. Hinderliter, A. M. P. Simoes, D. Tallman, S. Croll, Prog. Org. Coat. 63 (2008) 250–259
41. H. Hayashibara, E. Tada, A. Nishikata, No. 11 ISIJ International 56, 11 (2016) 2029–2036
42. M. Fernanddez Gonzalez, M. Hickl, F. Hezel, T. Schauer, A. Miszcyk, Vorrichtung und Verfahren zur beschleunigten Durchführung von Korrosionstests, DE102004027792
43. G. Bierwagen, D. Tallman, J. Li, L. He, C. Jeffcoate, Prog. Org. Coat. 46 (2003) 148–157
44. M. Dornbusch, Prog. Org. Coat. 61 (2008) 240–244
45. B. R. Hinderliter, S. G. Croll, D. E. Tallman, Q. Su, G. P. Bierwagon, Elecrrochim. Acta 51 (2006) 4505–4515
46. F. Gui, R. G. Kelly, Corrosion 61 (2005) 119–129
47. S. Haruyama, M. Asari, T. Tsuru, Corrosion Protection by Organic Coatings, Eds M. Kendig, H. Leidhaiser, Proc. Vol. 87-2, 197–207, Electrochemical Society, Pennington, 1987
48. F. Mansfeld, J. Appl. Electrochem. 25 (1995) 187–202
49. F. Mansfeld, C. H. Tsai, Corrosion 47 (1991) 958
50. C. H. Tsai, F. Mansfeld, Corrosion 49 (1993) 726
51. Thesis (PhD), C. H. Tsai, University of Southern California, 1992
52. F. Mansfeld, C. H. Tsai, Determination of the Protective Properties of Polymer Coatings from High-Frequency Impedance Data, 12th International Corrosion Congress, Houston, September 1993, 128–150
53. J. M. McIntyre, Ha. Q. Pham, Prog. Org. Coat. 27 (1996) 201–207
54. A. Amiradin, C. Barreau, D. Thierry, Application of Electrochemical Impedance Spectroscopy to Study the Efficiency of Anti-corrosive Pigments in an Epoxy Resin, 12th International Corrosion Congress, Houston, September 1993, 114–127
55. S. Haruyama, M. Asari and T. Tsuru, Electrochem. Soc., Proc. 87-2 (1987) 197
56. J. N. Murray, H. P. Hack, Long-Term Electrochemical Characterizations of MIL-P-24441 Epoxy Coated Steel Using Electrochemical Impedance Spectroscopy (EIS), 12th International Corrosion Congress, Houston, September 1993, 151–156
57. S. J. Garcıa a, J. Suay, Prog. Org. Coat. 59 (2007) 251–258
58. M.-G. Olivier, M. Poelman (2012). Use of Electrochemical Impedance Spectroscopy (EIS) for the Evaluation of Electrocoatings Performances, Recent Researches in Corrosion Evaluation and Protection, Prof. Reza Shoja Razavi (Ed.), ISBN: 978-953-307-920-2, InTech
59. A. Amirudin, P. Jernberg, D. Thierry, Determination of Coating Delamination and Underfilm Corrosion during Atmospheric Exposure by means of Electrochemical Impedance Spectroscopy, 12th International Corrosion Congress, Houston, September 1993, 171–181
60. Encyclopaedia of Electrochemistry, Edited by A.J. Bard and M. Stratmann, 2007 Wiley-VCH Verlag GmbH & Co. KGaA, Weinheim, Vol. 4, Corrosion and Oxide Films, Chap. 7, G. S. Frankel, M. Rohwerder, Electrochemical Techniques for Corrosion, 687–723
61. F. Mansfeld, H. Xiao, J. Electrochem. Soc. 140, 8 (1993) 2205–2209
62. D. J. Mills, G. B. Bierwagen, B. Skerry, D. Tallman, 12th International Corrosion Congress, Houston, September 1993, 182–194
63. H. Xiao, F. Mansfeld, J. Electrochem. Soc. 141 (1994) 2332

64. M. G. Pujar, T. Anita, H. Shaikh, R. K. Dayal, H. S. Khatak, Int. J. Electrochem. Sci. 2 (2007) 301–310
65. Y. Shi, Zhao Zhang, J. Sua, F. Caoa, J. Zhang, Electrochim. Acta 51 (2006) 4977–4986
66. S. S. Jamali, D. J. Mills, J. M. Sykes, Prog. Org. Coat. 77 (2014) 733–741
67. J. Kee Paik, A. K. Thayamballi, Y. I. Park, J. Sung Hwang, Corros. Sci. 46 (2004) 471–486
68. J. Sickfeld, D. Wapler, Werkst. Korros. 21, 2 (1970) 77–85
69. A. Rudolf, Farbe&Lack (2010) 4
70. A. Forsgren, Corrosion Control through Organic Coatings, CRC Press, Boka Raton, 2006, Chapter 8, Corrosion Testing–Practice
71. M. Dornbusch, T. Biehler, M. Conrad, A. Greiwe, D. Momper, L. Schmidt, M. Wiedow, JUnQ, 6, 2, 1–7, 2016
72. T. Brock, M. Groteklaes, P. Mischke, European Coatings Handbook, Vincentz Network, Hanover, 2010
73. M. Dornbusch, M. Hickl, K. Wapner, L. Jandel, Millennium Steel 2008, 231–236
74. J. Teutsch, G. Bockmair, Farbe & Lack 2001, 01, 136–138
75. J. Pietschmann, Jahrbuch der Oberflächentechnik, 52, 1996, 221–237
76. G. Posch, J. Tösch, Welding in the World, 49, 9/10, 2005, 58–67
77. Indian Standard, IS10461 (Part 1):1994 (Reaffirmed 2006)
78. H. S. Khatak, B. Raj, Corrosion of Austenitic Stainless Steels: Mechanism, Mitigation and Monitoring, Woodhead Publishing Series in Metals and Surface Engineering, Elsevier 2002
79. Thesis (PhD), F. Zhu, Department of Materials and Engineering, Royal Institute of Technology, Stockholm, 2000
80. M. Ström, G. Ström, Proceeding of Eurocorr, Maastricht, 2006
81. S. Fujita, D. Mizuno, Corros. Sci. 49 (2007) 211
82. N. LeBozec, D. Thierry, Mater. Corr. 61, 10 (2010) 845–851
83. N. LeBozec, N. Blandin, D. Thierry, Mater. Corr. 59, 11 (2008) 889–894
84. B. Goldie, Prot. Coat. Eur. 1 (1996) 23
85. B. Appelman, J. Coat. Technol. 62 (1990) 57
86. B. S. Skerry, A. Alavi, K. I. Lindgren, J. Coat. Technol. 60 (1988) 97
87. W. B. Nowak, H. E. Townsend, L. Li, Corrosion Behavior of Oxide Coated Cold-rolled and Electrogalvanized Sheet Steel, 12th International Corrosion Congress, Houston, September 1993, 114–127
88. M. Zapponi, T. Perez, C. Ramos, C. Saragovi, Corros. Sci. 47 (2005) 923–936
89. J.-C. Charbonnier, The book of steel, Chap. 3, Corrosion, 335–362
90. F. Altmaier, Met. Finish. 110, 9A, 2012, 558, 560–564
91. T. Biestek, Atmospheric corrosion testing of electrodeposited zinc and cadmium coatings, Atmospheric Corrosion, W.H. Ailor, Ed. Wiley, New York, 1982, 631–643
92. M. Dornbusch, R. Rasing, U. Christ, Epoxy Resins, Vincentz Network, Hanover, 2016
93. P. Premendra, H. Terryn, J. M. C. Mol, J. H. W. de Wit, L. Katgerman, Mater. Corr. 60, 6, (2009) 399–406
94. K.-A. van Oeteren, Korrosionsschutz durch Beschichtungsstoffe, Band 2, Carl Hanser Verlag, München, 1980
95. A. Kutzelnigg, Die Prüfung metallischer Überzüge, Bd.4, Schriftenreihe Galvanotechnik, Eugen G. Leuze Verlag, Saulgau/Württ., 2. Aufl.1965

5 Appendix

5.1 Survey of the Examples Ordered by Themes

1 Substrates and Metal Coatings

Steel

Polarization Diagram: Examples 2.2 and 2.5
CV: Examples 2.4 and 2.5
SEM/EDX: Examples 3.5 and 3.6

Aluminium

Polarization Diagram: Examples 2.1 and 2.2
EIS: Example 2.7
IR: Example 3.14

Chromium/Nickel

Polarization Diagram: Example 2.3
EIS: Example 2.6
UV-VIS: Example 3.19

Zinc Coated Steel

Polarization Diagram: Example 2.2
XPS: Example 3.1, Example 3.4
SEM/EDX: Example 3.5
UV-VIS: Example 3.8
IR: Example 3.11, Example 3.14
Raman: Example 3.15, Example 3.16
Immersion Test: Example 4.1

Copper

Polarization Diagram: 2.1
EIS: Example 2.7

2 *Conversion Layers*

Phosphated Steel

Polarization Diagram: Example 2.1
CV: Example 2.4
EIS: Example 2.7
SEM/EDX: Example 3.6
Immersion Test: Example 4.1

Zirconium Based Conversion Layers

XPS: Example 3.2, Example 3.3
UV-VIS: Example 3.9, Example 3.10
IR: Example 3.12

Phytic Acid Based Conversion Layers

Polarization Diagram: Example 2.5
CV: Example 2.5
SEM/EDX: Example 3.7
IR: Example 3.13
Salt Spray Test: Example 4.4

Molybdenum Based Conversion Layers

Immersion Test: Example 4.1

3 *Organic Coatings*

Zinc Flake Coatings

CV: Example 2.8
EIS: Example 2.9

Coil Coating

EIS: Example 2.12
SKP: Example 2.13
Salt Spray Test: Example 4.5
Raman: Example 3.24

Autophoretic Coatings

UV-VIS: Example 3.20

Wire Enamel

IR: Example 3.22

Acrylic Dispersion Based Coatings

EIS: Example 2.10
UV-VIS: Example 3.18

E Coat

SKP: Example 2.14, Example 2.15
SEM/EDX: Example 3.17
IR: Example 3.21
Raman: Example 3.23
Delamination by Means of EIS: Example 4.3

Can Coatings

EIS: Example 2.11

5.2 Electrochemical Data

5.2.1 *Solubility Constants*

Experimental data vary with the method and because of systematic errors in the experimental setup. Some data are very constant since decades in literature and some show very broad deviations. To stress both the very accurate and the inaccurate data, several values from literature are summarized in the tables. Literature with similar data collected in one bracket and the order of the values is similar to the order of the literature.

Table 5.1 Solubility constants of important corrosion products.

Compound	Solubility product K_L/ [Equilibrium constant K]	Lit.	ΔG [kJ/mol]	Lit.	Reaction for K_L
Aluminium					
γ-AlOOH Boehmite	$10^{-31.7}$, $1.9*10^{-33}$, $10^{-34.02}$	[8], [149], [16]	–917.82 ± 1.9	[13]	$AlOOH + H_2O \rightleftharpoons Al^{3+} + 3OH^-$
amorphous $Al(OH)_3$	$10^{-31.2}$, $10^{-32.34}$	[8], [16]			$Al(OH)_3 \rightleftharpoons Al^{3+} + 3OH^-$
γ-$Al(OH)_3$ Bayerite	$10^{-33.0}$, $10^{-35.52}$	[8], [16]			$Al(OH)_3 \rightleftharpoons Al^{3+} + 3OH^-$
α-$Al(OH)_3$ Hydrargillite	$10^{-33.8}$, $10^{-36.30}$	[8], [16]	–1154.89	[13]	$Al(OH)_3 \rightleftharpoons Al^{3+} + 3OH^-$
α-Al_2O_3 Corundum	$10^{-33.45}$	[16]	–1582.3	[2, 4, 13]	
Chromium					
Cr_2O_3			–1058.1, –1058.07	[2], [13]	
active $Cr(OH)_3$	10^{-30}, $6.7*10^{-31}$	[8], [149]	–844.40	[13]	$Cr(OH)_3 \rightleftharpoons Cr^{3+} + 3 OH^-$
$Cr(OH)_3 * 3H_2O$			–1612.46	[13]	
γ-CrOOH			–661.90	[13]	
α-CrOOH			–667.40	[13]	

Table 5.1 contd. ...

...Table 5.1 contd.

Compound	Solubility product K_L/ [Equilibrium constant K]	Lit.	ΔG [kJ/mol]	Lit.	Reaction for K_L
			Copper		
CuO Tenorite	$10^{-7.65}$	[10]			$CuO + 2H^+ \rightleftharpoons Cu^{2+} + H_2O$
active CuO	$10^{-19.7}$	[8]	–128.29, –129.7	[13], [2, 4]	$CuO + H_2O \rightleftharpoons Cu^{2+} + 2OH^-$
inactive CuO	$10^{-20.5}$, $10^{-20.12}$	[8], [16]			
Cu_2O	$10^{-14.7}$	[8]	–144.68, –146.0	[8], [2, 4]	$Cu_2O + H_2O \rightleftharpoons 2Cu^+ + 2OH^-$
inactive $Cu(OH)_2$	$10^{-18.8}$, $10^{-18.81}$, $10^{-19.4}$, $5.6*10^{-20}$	[8], [16], [10], [149]	–359.02	[13]	$Cu(OH)_2 \rightleftharpoons Cu^{2+} + 2OH^-$
$Cu(OH)_{1.5}Cl_{0.5}$	$10^{-17.3}$	[8]			
$CuCO_3$	$1.37*10^{-10}$	[149]			$CuCO_3 \rightleftharpoons Cu^{2+} + CO_3^{2-}$
$Cu_2(OH)_2CO_3$ Malachite	$10^{-5.80}$	[10]			$Cu_2(OH)_2CO_3 + 2H^+ \rightleftharpoons 2Cu^{2+} + 2H_2O + CO_3^{2-}$
$Cu_3(OH)_2(CO_3)_2$ Azurite	$10^{-18.0}$	[10]			$Cu_3(OH)_2(CO_3)_2 + 2H^+ \rightleftharpoons 3Cu^{2+} + 2H_2O + 2CO_3^{2-}$
			Iron		
α-FeOOH Goethite	$10^{-40.4}$, $10^{-37.4}$	[10], [11]	–489.8 ± 1.2	[13]	$FeOOH + H_2O \rightleftharpoons Fe^{3+} + 3OH^-$
γ-FeOOH Lepidocrocite			–480.1 ± 1.4	[13]	
$FeCO_3$ Siderite	$3,13*10^{-11}$, $10^{-10.4}$	[1], [10]	–666.7	[2]	$FeCO_3 \rightleftharpoons Fe^{3+} + CO_3^{2-}$
$Fe(OH)_2$	$4,87*10^{-17}$, $10^{-14.3}$, $1.8*10^{-15}$, $10^{-13.29}$, from $HFeO_2^-$ $10^{-18.3}$	[1], [10], [3], [5], [5]	–483.9, –500.16	[3], [13]	$Fe(OH)_2 \rightleftharpoons Fe^{2+} + 2OH^-$

	$[10^{11.72}]$	[10]			$Fe(OH)_2 + 2H^+ \rightleftharpoons Fe^{2+} + 2H_2O$
	$[10^{-2.14}]$	[10]			$Fe(OH)_2 + 2H^+ + 2e^- \rightleftharpoons Fe + 2H_2O$
active $Fe(OH)_2$	10^{-14}	[8]	–446.13	[13]	$Fe(OH)_2 \rightleftharpoons Fe^{2+} + 2OH^-$
inactive $Fe(OH)_2$	$10^{-15.1}$	[8]			$Fe(OH)_2 \rightleftharpoons Fe^{2+} + 2OH^-$
$Fe(OH)_3$	$6*10^{-38}$, $10^{-38.7}$, $2,79*10^{-39}$	[3], [10], [1]	–695.0	[3]	
	$[10^{4.35}]$	[10]			$Fe(OH)_3 + H^+ + e^- \rightleftharpoons Fe(OH)_2 + H_2O$
	$[10^{16.07}]$	[10]			$Fe(OH)_3\ 3H^+ + e^- \rightleftharpoons Fe^{2+} + 3H_2O$
	$[10^{3.02}]$	[10]			$Fe(OH)_3\ 3H^+ \rightleftharpoons Fe^{3+} + 3H_2O$
active $Fe(OH)_3$	$10^{-38.7}$	[8]			$Fe(OH)_3 \rightleftharpoons Fe^{3+} + 3OH^-$
inactive $Fe(OH)_3$	$10^{-39.1}$	[8]			$Fe(OH)_3 \rightleftharpoons Fe^{3+} + 3OH^-$
FeO Wustite	$10^{-16.5}$	[11]	–244.5	[3]	
α-Fe_2O_3 Hematite	$10^{-42.7}$	[8, 10]	–742.2, –741.5, –744.27 ± 1.25	[2, 4], [3], [13]	$Fe_2O_3 + 3H_2O \rightleftharpoons 2Fe^{3+} + 6OH^-$
γ-Fe_2O_3 Maghemite			–718.5	[3]	
Fe_3O_4 Magnetite			–1015.23, –1015.4, –1014.9	[13], [2, 4], [3]	
$FeCO_3$	$2.11*10^{-11}$	[149]			$FeCO_3 \rightleftharpoons Fe^{2+} + CO_3^{2-}$

Table 5.1 contd. ...

...Table 5.1 contd.

Compound	Solubility product K_L/ [Equilibrium constant K]	Lit.	ΔG [kJ/mol]	Lit.	Reaction for K_L
			Magnesium		
MgO			–569.3, –569.43	[2], [4]	
$MgCO_3$	$6,82*10^{-6}$	[1]			
$MgCO_3 \cdot 3H_2O$	ca. 10^{-5}, $2,38*10^{-6}$	[149], [1]			
$MgCO_3 \cdot 5H_2O$	$3,79*10^{-6}$	[1]			
$Mg(OH)_2$	$5.5*10^{-12}$, $5,61*10^{-12}$	[149], [1]			
active $Mg(OH)_2$	$10^{-9.2}$	[8]			$Mg(OH)_2 \rightleftharpoons Mg^{2+} + 2OH^-$
inactive $Mg(OH)_2$	$10^{-10.9}$	[8]			$Mg(OH)_2 \rightleftharpoons Mg^{2+} + 2OH^-$
			Nickel		
$NiCO_3$	$1,42*10^{-7}$	[1]			
β-$Ni(OH)_2$	$5,48*10^{-16}$	[1]	–450.77	[13]	
active $Ni(OH)_2$	$10^{-14.7}$	[8, 149]	–388.77	[13]	$Ni(OH)_2 \rightleftharpoons Ni^{2+} + 2OH^-$
inactive $Ni(OH)_2$	$10^{-17.2}$	[8]			$Ni(OH)_2 \rightleftharpoons Ni^{2+} + 2OH^-$
NiO			–211.1	[13]	
			Tin		
SnO	$10^{-26.2}$	[8]	–251.9	[2]	$SnO + H_2O \rightleftharpoons Sn^{2+} + 2OH^-$
amorphous SnO_2	10^{-58}	[8]			

active SnO_2	10^{-61}	[8]	–515.8	[2]	$SnO_2 + 2H_2O \rightleftharpoons Sn^{4+} + 4OH^-$
inactive SnO_2	$10^{-64.5}$	[8]			$SnO_2 + 2H_2O \rightleftharpoons Sn^{4+} + 4OH^-$
$Sn(OH)_2$	$5*10^{-26}$, $5,45*10^{-27}$	[149], [1]	–491.6	[2]	$Sn(OH)_2 \rightleftharpoons Sn^{2+} + 2OH^-$
$Sn(OH)_4$	10^{-56}	[149]			$Sn(OH)_4 \rightleftharpoons Sn^{4+} + 4OH^-$
H_2SnO_2	$6*10^{-18}$	[149]			$H_2SnO_2 \rightleftharpoons H^+ + HSnO_2^-$
Zinc					
$ZnCO_3$ Smithsonite	$1,46*10^{-10}$, $10^{-10.78}$, $6*10^{-11}$	[1], [10], [149]	–731.5	[2]	$ZnCO_3 \rightleftharpoons Zn^{2+} + CO_3^{2-}$
$ZnCO_3 \cdot H_2O$	$5,42*10^{-11}$	[1]			
α-$Zn(OH)_2$	$1.43*10^{-16}$	[15, 16]			
ε-$Zn(OH)_2$	$10^{-16.47}$, $10^{-16.92}$, $3*10^{-17}$, $8.41*10^{-18}$	[8], [9], [1], [15, 16]	–553.5, –555.93 ±0.21	[2], [13]	$Zn(OH)_2 \rightleftharpoons Zn^{2+} + 2OH^-$
γ-$Zn(OH)_2$	$1.40*10^{-17}$, $10^{-16.70}$, $10^{-16.26}$	[15, 16], [9], [8]			$Zn(OH)_2 \rightleftharpoons Zn^{2+} + 2OH^-$
amorphous $Zn(OH)_2$	$10^{-15.95}$, $1.66*10^{-16}$, $2*10^{-16}$, $10^{-16.4}$	[9], [15, 16], [6, 8], [10]			$Zn(OH)_2 \rightleftharpoons Zn^{2+} + 2OH^-$
β_1 $Zn(OH)_2$	$1.90*10^{-17}$, $2.2*10^{-17}$, $10^{-16.7}$	[15, 16], [6], [8, 9]			$Zn(OH)_2 \rightleftharpoons Zn^{2+} + 2OH^-$
β_2 $Zn(OH)_2$	$10^{-16.20}$, $7*10^{-17}$	[8], [6, 7]			$Zn(OH)_2 \rightleftharpoons Zn^{2+} + 2OH^-$
ZnO Zincite	$10^{-11.33}$	[10]			$ZnO + 2H^+ \rightleftharpoons Zn^{2+} + H_2O$
active ZnO	$8*10^{-17}$, $10^{-16.66}$, $10^{-16.89}$	[6], [8], [9, 10]	–320.5, –320.91 ±0.25, –318.3	[2], [13], [4]	$ZnO + H_2O \rightleftharpoons Zn^{2+} + 2OH^-$

Table 5.1 contd. ...

...Table 5.1 contd.

Compound	Solubility product K_L/ [Equilibrium constant K]	Lit.	ΔG [kJ/mol]	Lit.	Reaction for K_L
		Zinc			
inactive ZnO	$1.3*10^{-17}$, $10^{-16.1}$ to $10^{-16.43}$, $10^{-16.83}$	[6], [9], [8]			$ZnO + H_2O \rightleftharpoons Zn^{2+} + 2OH^-$
$Zn_5(OH)_8Cl_2$ Simonkolleite	$3*10^{-15}$, $10^{-14.2}$	[6], [8]			
$Zn(OH)_{1.71}Cl_{0.29}$	$6*10^{-16}$	[6]			
$Zn_5(OH)_6(CO_3)_2$ Hydrozincite	$10^{-14.4}$	[8]			

5.2.2 Thermodynamic Data

Table 5.2 Standard enthalpy of formation, standard potential and pH dependence of substrate metals. (Literature with similar data collected in one bracket and the order of the values is similar to the order of the literature.)

Compound	Standard enthalpy of formation ΔG [kJ/mol]	Lit.	ΔE [V]	Lit.	Reaction	Reaction enthalpy ΔG [kJ/mol]/ΔE pH dependence [V]/ Equilibrium constant K	Lit.
H_2O	–237.1, –237.13, –237.14, –237.4, –237.5	[2], [4], [13], [3], [10]	1.229, 1.23	[17], [10, 20]	$O_2 + 4\,H^+ + 4\,e^- \rightleftharpoons 2\,H_2O$	$E = 1.228 - 0.0591\,pH + 0.0147 \log p_{O2}$	[16]
	160	[10]	–0.8277, –0.828, –0.83	[17], [22], [10, 20]	$2\,H_2O + 2\,e^- \rightleftharpoons H_2 + 2\,OH^-$	$E = -0.83 + 0.0591\,pOH - 0.0296 \log p_{H2}$	calc.*
	–154	[10]	0.40, 0.401	[10, 20], [17, 22]	$O_2 + 2\,H_2O + 4\,e^- \rightleftharpoons 4\,OH^-$	$E = 0.40 + 0.0591\,pOH + 0.0147 \log p_{O2}$	calc.
			0	def.	$2H^+ + 2e^- \rightleftharpoons H_2$	$E = -0.0591\,pH - 0.0296 \log p_{H2}$	calc.
OH^-	–157.24	[4]					
Aluminium							
Al^{3+}	–481.2, –485.0, –487.2 ± 2.3	[4], [12], [13]	–1.66, –1662, –1.67, –1.68	[3, 22], [17], [21], [20]	$Al^{3+} + 3\,e^- \rightleftharpoons Al$	$E = -1.663 - 0.0197 \log [(Al^{3+})]$	[16]
$Al(OH)^{2+}$	–694.1, –696.0 ± 2.5	[12], [13]					

Table 5.2 contd. ...

...Table 5.2 contd.

Compound	Standard enthalpy of formation ΔG [kJ/mol]	Lit.	ΔE [V]	Lit.	Reaction	Reaction enthalpy ΔG [kJ/mol]/ΔE pH dependence [V]/ Equilibrium constant K	Lit.
Aluminium							
$Al(OH)^{2+}$	–900.2 ± 2.3	[13]					
γ–AlOOH Boehmite	–917.82 ± 1.9	[13]					
$Al(OH)_3$			–2.31	[17]	$Al(OH)_3 + 3\,e^- \rightleftharpoons Al + 3\,OH^-$	E = –2.31 + 0.0591 pOH	calc.
α–$Al(OH)_3$ Hydrargillite	–1154.89	[13]					
Al_2O_3	–1576.5	[18]			$2Al + 1.5\,O_2 \rightleftharpoons Al_2O_3$		
	–1574.9	[18]			$2Al + 3H_2O \rightleftharpoons Al_2O_3 + 3H_2$		
			–1.550	[16]	$2Al + 3H_2O \rightleftharpoons Al_2O_3 + 6H^+ + 6e^-$	E = –1.550 – 0.0591 pH	[16]
α-Al_2O_3 Corundum	–1582.3	[2, 4, 13]					
$[Al(OH)_4]^-$	–1305.3, –1305.7 ± 2.1	[12], [13]	–2.328	[17]	$Al(OH)_4^- + 3\,e^- \rightleftharpoons Al + 4\,OH^-$	E = –2.328 + 0.0788 pOH + 0.0197 log[$(Al(OH)_4^-)$]	calc.
AlO_2^-	–830.9	[12]	–1.262	[16]		E = –1.262 + 0.0788 pH – 0.0197 log [(AlO_2^-)]	[16]
$H_2AlO_3^-$			–2.33	[17]	$H_2AlO_3^- + H_2O + 3\,e^- \rightleftharpoons Al + 4\,OH^-$	E = –2.33 + 0.0788 pOH + 0.0197 log[$(H_2AlO_3^-)$]	calc.

Chromium							
Cr^{2+}			–0.90, –0.91, –0.913	[20, 21], [22], [17, 16]	$Cr^{2+} + 2e^- \rightleftharpoons Cr$	$E = -0.193 + 0.0295 \log[(Cr^{2+})]$	[16]
Cr^{3+}	–215.48	[13]	–0.407, –0.424	[17, 16], [21]	$Cr^{3+} + e^- \rightleftharpoons Cr^{2+}$	$E = -0.407 + \log[(Cr^{3+})/(Cr^{2+})]$	[16]
			–0.74, –0.744	[3, 20, 22], [17]	$Cr^{3+} + 3e^- \rightleftharpoons Cr$	$E = -0.744 + 0.0197 \log[(Cr^{3+})]$	[16]
$Cr(OH)^{2+}$	–430.96	[13]	1.258	[16]	$Cr_2O_7^{2-} + 12H^+ + 6e^- \rightleftharpoons 2CrOH^{2+} + 5H_2O$	$E = 1.258 - 0.1182\,pH - 0.0098 \log C$	[16]
			1.402	[16]	$CrO_4^{2-} + 7H^+ + 3e^- \rightleftharpoons CrOH^{2+} + 3H_2O$	$E = 1.402 - 0.1379\,pH$	[16]
$Cr(OH)_2^+$	–628.15	[13]	1.135	[16]	$Cr_2O_7^{2-} + 10H^+ + 6e^- \rightleftharpoons 2Cr(OH)_2^+ + 3H_2O$	$E = 1.135 - 0.0985\,pH - 0.0098 \log C$	[16]
			1.279	[16]	$CrO_4^{2-} + 6H^+ + 3e^- \rightleftharpoons Cr(OH)_2^+ + 2H_2O$	$E = 1.279 - 0.1182\,pH$	[16]
$Cr(OH)_3$	–844.40	[13]	–1.48	[17]	$Cr(OH)_3 + 3e^- \rightleftharpoons Cr + 3OH^-$	$E = -1.48 + 0.0591\,pOH$	calc.
$Cr(OH)_3 * 3H_2O$	–1612.46	[13]					
γ–CrOOH	–661.90	[13]					
α–CrOOH	–667.40	[13]					
Cr_2O_3	–1047.3, –1058.07, –1058.1	[18], [13], [2]		[17]	$2Cr + 1{,}5\,O_2 \rightleftharpoons Cr_2O_3$		

Table 5.2 contd. ...

...Table 5.2 contd.

Compound	Standard enthalpy of formation ΔG [kJ/mol]	Lit.	ΔE [V]	Lit.	Reaction	Reaction enthalpy ΔG [kJ/mol]/ΔE pH dependence [V]/ Equilibrium constant K	Lit.
Chromium							
	–1045.7	[18]	–0.654, –0.579	[16], [16]	$2Cr + 3H_2O \rightleftharpoons Cr_2O_3 + 3H_2$	E = –0.654 – 0.0591 pH	[16]
CrO_2			1.48, 1.556	[17], [16]	$CrO_2 + 4 H^+ + e^- \rightleftharpoons Cr^{3+} + 2H_2O$	E = 1.556 – 0.2364 pH – 0.0591 log $[(Cr^{3+})]$	[16]
$Cr(OH)_4^-$	–1007.75	[13]					
CrO_2^-			–1.2	[17]	$CrO_2^- + 2 H_2O + 3 e^- \rightleftharpoons Cr + 4 OH^-$	E = –1.2 + 0.0788 pOH + 0.0197 log$[(CrO_2^-)]$	calc.
			–0.213	[16]	$CrO_2^- + 4 H^+ + 3 e^- \rightleftharpoons Cr + 2 H_2O$	E = –0.213 – 0.0788 pH + log $[(CrO_2^-)]$	[16]
CrO_4^{2-}	–727.8	[12]	1.350	[17, 16]	$HCrO_4^- + 7 H^+ + 3 e^- \rightleftharpoons Cr^{3+} + 4 H_2O$	E = 1.350 – 0.1379 pH	[16]
			–0.13	[17]	$CrO_4^{2-} + 4 H_2O + 3 e^- \rightleftharpoons Cr(OH)_3 + 5 OH^-$	E = –0.13 + 0.0985 pOH + 0.0197 log$[(CrO_4^{2-})]$	calc.
$Cr_2O_7^{2-}$	–1301.1	[12]	1.36, 1.38	[17], [20, 21]	$Cr_2O_7^{2-} + 14 H^+ + 6 e^- \rightleftharpoons 2 Cr^{3+} + 7 H_2O$	E = 1.37 – 0.1379 pH + 0.0099 log$[(Cr_2O_7^{2-})/(Cr^{3+})^2]$	calc.

Copper							
Cu^+	49.98, 50	[4], [10]	0.52, 0.520, 0.521, 0.522	[10, 20], [16], [3, 17, 22], [19]	$Cu^+ + e^- \rightleftharpoons Cu$	$E = 0.520 + 0.0591 \log [(Cu^+)]$	[16]
Cu^{2+}	65.04, 65.27, 65.49, 66	[13], [8], [4], [10]	0.15, 0.153, 0.159, 0.160, 0.167	[10], [17, 16, 22], [21], [20], [19]	$Cu^{2+} + e^- \rightleftharpoons Cu^+$	$E = 0.153 + 0.0591 \log [(Cu^{2+})/(Cu^+)]$	[16]
			0.34, 0.340, 0.3419, 0.342, 0.337	[20], [21], [17], [22], [3, 19, 16]	$Cu^{2+} + 2 e^- \rightleftharpoons Cu$	$E = 0.337 + 0.0295 \log [(Cu^{2+})]$	[16]
$Cu(OH)^+$	–116.00	[13]					
inactive $Cu(OH)_2$	–359.02	[13]	–0.222	[17]	$Cu(OH)_2 + 2 e^- \rightleftharpoons Cu + 2 OH^-$	$E = -0.222 + 0.0591\ pOH$	calc.
			–0.08	[17]	$2 Cu(OH)_2 + 2 e^- \rightleftharpoons Cu_2O + 2 OH^- + H_2O$	$E = -0.08 + 0.0591\ pOH$	calc.
					$Cu + 2H_2O \rightleftharpoons Cu(OH)_2 + 2H^+ + 2e^-$	$E = 0.609 - 0.059\ pH$	[19]
					$Cu_2O + 3H_2O \rightleftharpoons 2Cu(OH)_2 + 2H^+ + 2e^-$	$E = 0.747 - 0.059\ pH$	[19]
CuO			0.570, 0.609	[19, 16], [16]	$Cu + H_2O \rightleftharpoons CuO + 2H^+ + 2e^-$	$E = 0.570 - 0.059\ pH$	[19, 16]
			0.669, 0.747	[19, 16], [16]	$Cu_2O + H_2O \rightleftharpoons 2CuO + 2H^+ + 2e^-$	$E = 0.669 - 0.059\ pH$	[19, 16]

Table 5.2 contd. ...

...Table 5.2 contd.

Compound	Standard enthalpy of formation ΔG [kJ/mol]	Lit.	ΔE [V]	Lit.	Reaction	Reaction enthalpy ΔG [kJ/mol]/ΔE pH dependence [V]/ Equilibrium constant K	Lit.
Copper							
			0.620	[16]	$Cu^+ + H_2O \rightleftharpoons CuO + 2H^+ + e^-$	$E = 0.620 - 0.1182\ pH - 0.0591 \log[(Cu^+)]$	[16]
active CuO	–128.29, –129.7	[13], [2, 4]					
inactive CuO							
Cu_2O	–144.68, –146.0	[8], [2, 4]	–0.36	[17]	$Cu_2O + H_2O + 2\ e^- \rightleftharpoons 2\ Cu + 2\ OH^-$	$E = -0.36 + 0.0591\ pOH$	calc.
	–145.9	[18]			$2Cu + 0.5\ O_2 \rightleftharpoons Cu_2O$		
	–145.4	[18]			$2Cu + H_2O \rightleftharpoons Cu_2O + H_2$		
			0.471	[19, 16]	$2Cu + H_2O \rightleftharpoons Cu_2O + 2H^+ + 2e^-$	$E = 0.471 - 0.059\ pH$	[19, 16]
Iron							
Fe^{+2}	–78.9, –79, –85, –90.50 ± 1.0	[4, 12], [10], [3], [13]	–0.41, –0.440, –0.447	[10], [3, 20, 21, 22], [17]	$Fe^{2+} + 2\ e^- \rightleftharpoons Fe$	$\log K = 13.86$	[10]
						$E = -0.44 + 0.0296 \log[(Fe^{2+})]$	calc.
						$pH = 0.5[\log K - \log(a_{Fe(II)})]$	[10]

Fe^{+3}	–4.7, –10.6, –16.23 ± 1.1	[4, 12], [3], [13]	–0.036, –0.037, –0.04	[3], [17], [20]	$Fe^{3+} + 3e^- \rightleftharpoons Fe$	$E = -0.037 + 0.0197 \log[(Fe^{3+})]$	calc.
	–74	[10]	0.77, 0.771	[10, 20], [17, 16, 21, 22]	$Fe^{3+} + e^- \rightleftharpoons Fe^{2+}$	logK = 13.06	[10]
						$E = 0.771 + 0.0591 \log [(Fe^{3+})/(Fe^{2+})]$	[16]
$FeOH^+$	–274.18, –277.4	[13], [12]	0.16	[17]	$Fe_2O_3 + 4H^+ + 2e^- \rightleftharpoons 2FeOH^+ + H_2O$	$E = 0.16 - 0.1182\,pH - 0.0591\log[(FeOH^+)]$	calc.
$FeOH^{+2}$	–229.4, –234.1	[12], [3]	0.914	[16]	$Fe^{2+} + H_2O \rightleftharpoons FeOH^{2+} + H^+ + e^-$	$E = 0.914 - 0.0591\,pH + 0.0591 \log [(FeOH^{2+})/(Fe^{2+})]$	[16]
$Fe(OH)_2^+$	–438.0, –444.7	[12], [3]	1.194	[16]	$Fe^{2+} + 2H_2O \rightleftharpoons Fe(OH)_2^+ + 2H^+ + e^-$	$E = 1.194 - 0.1182\,pH + 0.0591 \log [(Fe(OH)_2^+)/(Fe^{2+})]$	[16]
FeOOH	–489.8 ± 1.2	[13]	2.08	[17]	$HFeO_4^- + 4H^+ + 3e^- \rightleftharpoons FeOOH + 2H_2O$	$E = 2.08 - 0.0788\,pH + 0.0197 \log[(HFeO_4^-)]$	calc.
α-FeOOH Goethite	–65	[10]	0.67	[10]	$a\text{-}FeOOH + 3H^+ + e^- \rightleftharpoons Fe^{2+} + 2H_2O$	$E = 0.67 - 0.1773\,pH - 0.0591 \log [(Fe^{2+})]$	calc.
γ-FeOOH Lepidocrocite	–480.1 ± 1.4	[13]					
$FeCO_3$ Siderite	–666.7	[2]					
$Fe(OH)_2$	–483.9, –500.16	[3], [13]			$Fe(OH)_2 \rightleftharpoons Fe^{2+} + 2OH^-$	logK = 11.72	[10]

Table 5.2 contd. ...

...Table 5.2 contd.

Compound	Standard enthalpy of formation ΔG [kJ/mol]	Lit.	ΔE [V]	Lit.	Reaction	Reaction enthalpy ΔG [kJ/mol]/ΔE pH dependence [V]/ Equilibrium constant K	Lit.
			Iron				
						pH = 0.5[logK – log $(a_{Fe(II)})$]	[10]
active $Fe(OH)_2$	–446.13	[13]					
$Fe(OH)_3$	–695.0	[3]	–0.56	[17]	$Fe(OH)_3 + e^- \rightleftharpoons Fe(OH)_2 + OH^-$	E = –0.56 + 0.0591 pOH	calc.
					$Fe(OH)_3 \rightleftharpoons Fe^{3+} + 3OH^-$	logK = 3.02	[10]
						pH = 1/3 [log K – log$(a_{Fe(II)})$]	[10]
	–92	[10]	0.95	[10]	$Fe(OH)_3 + 3H^+ + e^- \rightleftharpoons Fe^{2+} + 3H_2O$	E = 0.95 – 0.1773 pH – 0.0591 log[(Fe^{2+})]	calc.
FeO Wustite	–244.5	[3]					
Fe_2O_3 Haematite	–741.5, –742.2, 744.27 ± 1.25	[3], [2, 4], [13]					
Fe_2O_3			2.09	[17]	$2\ HFeO_4^- + 8\ H^+ + 6\ e^- \rightleftharpoons Fe_2O_3 + 5\ H_2O$	E = 2.09 – 0.0788 pH + 0.0197 log[$(HFeO_4^-)$]	calc.
			0.16	[17]	$Fe_2O_3 + 4\ H^+ + 2\ e^- \rightleftharpoons 2\ FeOH^+ + H_2O$	E = 0.16 – 0.1182 pH – 0.0591 log[$(FeOH^+)$]	calc.
γ-Fe_2O_3 Maghemite	–718.5	[3]					

Fe_3O_4	–1007.5	[18]			$3Fe + 2O_2 \rightleftharpoons Fe_3O_4$		
	–1005.3	[18]	–0.085	[16]	$3Fe + 4H_2O \rightleftharpoons Fe_3O_4 + 4H_2$ ([19] or $8H^+ + 8e^-$)	E = –0.085 – 0.0591 pH	[16]
Fe_3O_4 Magnetite	–1014.9, –1015.23, –1015.4	[3], [13], [2, 4]					
$Fe(OH)_4^-$	–842.67	[13]					
$HFeO_4^-$			2.07	[17]	$HFeO_4^- + 7 H^+ + 3 e^- \rightleftharpoons Fe^{3+} + 4 H_2O$	E = 2.07 – 0.1379 pH + 0.0197 log[($HFeO_4^-$)/(Fe^{3+})]	calc.
FeO_4^{2-}			1.7, 1.9, 2.2	[16], [22], [17]	$FeO_4^{2-} + 8 H^+ + 3 e^- \rightleftharpoons Fe^{3+} + 4 H_2O$	E = 1.700 – 0.1580 pH + 0.0197 log[(FeO_4^{2+})/(Fe^{3+})]	[16]
Magnesium							
Mg^{2+}	–454.8	[4, 12]	–2.356, –2.36, –2.363, –2.37	[21], [20], [16], [3, 22]	$Mg \rightleftharpoons Mg^{2+} + 2e^-$	E = –2.363 + 0.0295 log[(Mg^{2+})]	[16]
$MgOH^+$	–626.7	[12]					
MgO	–569.3, –569.43	[2], [4]	–1.722, –1862	[16], [16]	$Mg + H_2O \rightleftharpoons MgO + 2H^+ + 2e^-$	E = –1.862 – 0.0591 pH	[16]
Nickel							
Ni^{2+}	–43.9, –45.6	[13], [12]	–0.250, –0.257	[3, 16, 22], [17, 20, 21]	$Ni^{2+} + 2 e^- \rightleftharpoons Ni$	E = –0.250 + 0.0295 log[(Ni^{2+})]	[16]
$Ni(OH)^+$	–227.15, –227.6	[13], [12]					

Table 5.2 contd. ...

...Table 5.2 contd.

Compound	Standard enthalpy of formation ΔG [kJ/mol]	Lit.	ΔE [V]	Lit.	Reaction	Reaction enthalpy ΔG [kJ/mol]/ΔE pH dependence [V]/ Equilibrium constant K	Lit.
Nickel							
β-$Ni(OH)_2$	–450.77	[13]					
active $Ni(OH)_2$	–388.77	[13]					
$Ni(OH)_2$			–0.72	[17, 20, 21]	$Ni(OH)_2 + 2\,e^- \rightleftharpoons Ni + 2\,OH^-$	E = –0.72 + 0.0591 pOH	calc.
NiO	–209.7, –211.1	[18], [13]		[17]	$Ni + 0.5\,O_2 \rightleftharpoons NiO$		
	–209.1	[18]	0.110, 0.116	[16], [16]	$Ni + H_2O \rightleftharpoons NiO + H_2$	E = 0.110 – 0.0591 pH	[16]
NiO_2			1.59, 1.593, 1.68	[20], [16, 21], [22]	$NiO_2 + 2e^- + 4H^+ \rightleftharpoons Ni^{2+} + 2H_2O$	E = 1.593 – 0.1182 pH – 0.0295 log[(Ni^{2+})]	[16]
Tin							
Sn^{2+}	–27.2	[4, 12]	–0.136, –0.1375, –0.14	[3, 16, 22], [17], [20]	$Sn^{2+} + 2\,e^- \rightleftharpoons Sn$	E = –0.136 + 0.0295 log[(Sn^{2+})]	[16]
Sn^{4+}			0.15, 0.151	[20, 22], [17, 16]	$Sn^{4+} + 2\,e^- \rightharpoonup Sn^{2+}$	E = 0.151 + 0.0295 log[(Sn^{4+})/(Sn^{2+})]	[16]
$Sn(OH)^+$	–254.8	[12]	–0.194	[17]	$SnO_2 + 3\,H^+ + 2\,e^- \rightleftharpoons SnOH^+ + H_2O$	E = –0.194 – 0.0887 pH – 0.0296 log[($SnOH^+$)]	calc.

$Sn(OH)_3^+$			0.142	[17]	$Sn(OH)_3^+ + 3\ H^+ + 2\ e^- \rightleftharpoons Sn^{2+} + 3\ H_2O$	$E = 0.142 - 0.0887\ pH + 0.0296\ \log[(Sn(OH)_3^+)/(Sn^{2+})]$	calc.
$Sn(OH)_2$	–491.6	[2]	–0.096	[19]	$Sn + 2H_2O \rightleftharpoons Sn(OH)_2 + 2H^+ + 2e^-$	$E = -0.096 - 0.059/pH$	[19]
$Sn(OH)_4$			0.075	[19]	$Sn(OH)_2 + 2H_2O \rightleftharpoons Sn(OH)_4 + 2H^+ + 2e^-$	$E = 0.075 - 0.059/pH$	[19]
			0.088	[19]	$SnO + 3H_2O \rightleftharpoons Sn(OH)_4 + 2H^+ + 2e^-$	$E = 0.088 - 0.059/pH$	[19]
SnO	–251.9	[2]		[18]	$Sn + H_2O \rightleftharpoons SnO + 2H^+ + 2e^-$	$E = -0.104 - 0.059V/pH$	[19]
SnO_2	–515.8	[2]	–0.077, –0.094, 0.120	[16], [17], [16]	$SnO_2 + 4\ H^+ + 2\ e^- \rightleftharpoons Sn^{2+} + 2\ H_2O$	$E = -0.077 - 0.1182\ pH - 0.0295\ \log[(Sn^{2+})]$	[16]
			–0.106, –0.117	[16], [17]	$SnO_2 + 4\ H^+ + 4\ e^- \rightleftharpoons Sn + 2\ H_2O$	$E = -0.106 - 0.050/pH$	[19]
						$E = -0.106 - 0.0591/pH$	[16]
			–0.945	[17]	$SnO_2 + 2\ H_2O + 4\ e^- \rightleftharpoons Sn + 4\ OH^-$	$E = -0.945 + 0.0591\ pOH$	calc.
			–0.108	[19]	$SnO + H_2O \rightleftharpoons SnO_2 + 2H^+ + 2e^-$	$E = -0.108 - 0.050/pH$	[19]
			–0.121	[19]	$Sn(OH)_2 \rightleftharpoons SnO_2 + 2H^+ + 2e^-$	$E = -0.121 - 0.059/pH$	[19]
$HSnO_2^-$			–0.909	[17]	$HSnO_2^- + H_2O + 2\ e^- \rightleftharpoons Sn + 3\ OH^-$	$E = -0.909 + 0.0887\ pOH + 0.0296\ \log[(HSnO_2^-)]$	calc.

Table 5.2 contd. ...

...Table 5.2 contd.

Compound	Standard enthalpy of formation ΔG [kJ/mol]	Lit.	ΔE [V]	Lit.	Reaction	Reaction enthalpy ΔG [kJ/mol]/ΔE pH dependence [V]/ Equilibrium constant K	Lit.
Tin							
			0.333	[19]	$Sn + 2H_2O \rightleftharpoons HSnO_2^- + 3H^+ + 2e^-$	E = 0.333 – 0.0886/pH	[19]
$Sn(OH)_6^{2-}$			–0.93	[17]	$Sn(OH)_6^{2-} + 2\,e^- \rightleftharpoons HSnO_2^- + 3\,OH^- + H_2O$	$E = -0.93 + 0.0887\ pOH + 0.0296\ \log[(Sn(OH)_6^{2}/(HSnO_2^-)]$	calc.
Zinc							
Zn^{2+}	–147, –147.06, –147.1, –147.23	[10], [4], [12], [13]	–0.76, –0.7618, –0.7626, –0.763	[10, 20], [17], [21], [3, 16, 22]	$Zn^{2+} + 2\,e^- \rightleftharpoons Zn$	$E = -0.763 + 0.0295\ \log[(Zn^{2+})]$	[16]
$Zn(OH)^+$	–330.1, –334.65	[12], [13]	–0.497	[17]	$ZnOH^+ + H^+ + 2\,e^- \rightleftharpoons Zn + H_2O$	$E = -0.497 - 0.0296\ pH + 0.0296\ \log[(ZnOH^+)]$	calc.
$ZnCO_3$ Smithsonite	–731.5	[2]					
amorphous-$Zn(OH)_2$					$2\,Zn(OH)_2 + \rightleftharpoons 2ZnO$ active $+ 2H_2O$	ΔG = –2.468	[9]
					$2\,Zn(OH)_2 + \rightleftharpoons 2ZnO$ stable $+ 2H_2O$	ΔG = –6.569	[9]

ε-$Zn(OH)_2$	–553.5, –555.93 ± 0.21	[2], [13]			2 e $Zn(OH)_2$ + ⇌ 2ZnO active + $2H_2O$	ΔG = 4.100	[9]
					2 e $Zn(OH)_2$ + ⇌ 2ZnO stable + $2H_2O$	ΔG = 0	[9]
γ-$Zn(OH)_2$					2 γ $Zn(OH)_2$ + ⇌ 2ZnO active + $2H_2O$	ΔG = 2.970	[9]
					2 γ $Zn(OH)_2$ + ⇌ 2ZnO stable + $2H_2O$	ΔG = 1.130	[9]
$Zn(OH)_2$			–1.249	[17]	$Zn(OH)_2 + 2 e^- \rightleftharpoons Zn + 2 OH^-$	E = –1.249 + 0.0591 pOH	calc.
β_1 $Zn(OH)_2$					2 β_1 $Zn(OH)_2$ + 2 e^- ⇌ 2ZnO active + $2H_2O$	ΔG = 2.678	[9]
					2 β_1 $Zn(OH)_2$ + 2 e^- ⇌ 2ZnO stable + $2H_2O$	ΔG = 1.422	[9]
ZnO			–0.400, –0.439	[16], [16]	$Zn + H_2O \rightleftharpoons ZnO + 2H^+ + 2e^-$	E = –0.439 – 0.0591 pH	[16]
active ZnO	–318.3, –320.5, –320.91 ± 0.25	[4], [2], [13]	–1.260	[17]	$ZnO + H_2O + 2 e^- \rightleftharpoons Zn + 2 OH^-$	E = –1.260 + 0.0591 pOH	calc.
ZnO_2^{2-}			–1.211, –1215	[9], [17]	$ZnO_2^{2-} + 2 H_2O + 2 e^- \rightleftharpoons Zn + 4 OH^-$	E = –1.21 + 0.1182 pOH + 0.0296 log[(ZnO_2^-)]	calc.
			0.441	[16]	$Zn + 2H_2O \rightleftharpoons ZnO_2^{2-} + 4H^+ + 2e^-$	E = 0.441 – 0.1182 pH + 0.0295 log[(ZnO_2^{2-})]	[16]

Table 5.2 contd. ...

...Table 5.2 contd.

Compound	Standard enthalpy of formation ΔG [kJ/mol]	Lit.	ΔE [V]	Lit.	Reaction	Reaction enthalpy ΔG [kJ/mol]/ΔE pH dependence [V]/ Equilibrium constant K	Lit.
Zinc							
$[Zn(OH)_4]^{2-}$	–862.74, –863.30	[9], [13]	–1.199, –1.285, –1.29	[17], [21], [20]	$Zn(OH)_4^{2-} + 2e^- \rightleftharpoons Zn + 4OH^-$	ΔG = –233.68	[9]
						$E = -1.2 + 0.1182\ pOH + 0.0296 \log[(Zn(OH)_4^{2-})]$	calc.
					$Zn^{2+} + 4OH^- \rightleftharpoons [Zn(OH)_4]^{2-}$	$K = 1.4*10^{15}$	[9]
						ΔG = –86.61	[9]

calc.: calculated according to the Nernst equation with standard conditions

5.2.3 Practical Galvanic Series

The disadvantages of the data from literature are the often missing details of the alloy, the treatment of the sample and the electrolyte. Therefore the potential of tin varies from –0.44 to –0.18 V and for copper from –0.01 to –0.37 V. Therefore some values from literature are summarized in the table for comparison. Literature with similar data collected in one bracket and the order of the values is similar to the order of the literature.

Table 5.3 Practical galvanic series in flowing seawater.

Material	Potential [V]	Literature
Graphite	0.2 to –0.3	[22, 128]
Gold	0.243	[3]
Platinum	+0.3, 0.18 to –0.26	[132], [22, 128]
Silver	0.149, –0.05, –0.15 to –0.1	[3], [132], [22, 128]
Titanium	0.04 to –0.06, –0.111	[128], [3]
Nickel 99.6	0.046	[3]
Copper	0.01, –0.18, –0.37 to –0.30	[3], [132], [22, 128]
Silver Solder 4404	–0.015	[3]
Stainless Steel Type 326,317 (1.44xx)	–0.1 to 0.0	[128]
V2A Steel 1.4300	–0.045	[3]
Nickel 200	–0.1 to –0.2	[22]
Nickel	–0.27	[132]
Tin	–0.184, –0.33 to –0.30, –0.44	[3], [22, 128], [132]
Lead	–0.25 to –0.19, –0.47	[128], [132]
Stainless Steel Type 430 passive (1.4016)	–0.28 –0.20	[128]
Zinc 98.5	–0.284	[3]
Hard Chromium on Steel	–0.291	[3]
AlCuMg	–0.339	[3]
Stainless Steel 316, 317 active (1.44xx)	–0.46 to –0.33	[128]
Cadmium	–0.519, –0.74 to –0.70	[3], [128]
Stainless Steel Type 430 (1.4016)	–0.57 to –0.45	[128]
Low Alloy Steel	–0.63 to –0.57	[128]
Aluminium 99.5	–0.667	[3]
Mild Steel, Cast Iron	–0.72 to –0.60	[128, 132]
AlMgSi	–0.785	[3]
Zinc Coating (100 μm acidic) on Steel	–0.794	[3]
Zinc Coating (100 μm cyan.) on Steel	–0.806	[3]
Aluminium Alloys	–1.00 to –0.76	[128]
Zinc	–1.03 to –0.98, –1.06	[128], [132]
Magnesium	–1.63 to –1.60	[22, 128, 132]

5.2.4 Corrosion Rates

Table 5.4 Corrosion rates in µm/year of metals and alloys in different environments. (Literature with similar data collected in one bracket and the order of the values is similar to the order of the literature.)

Alloy	Atmosphere	Rate µm/year	Lit.
	Iron		
Carbon Steel	Flowing seawater	100–160	[131]
	Flowing seawater (1.2 m/s)	324	[131]
	Rural	5–10	[131]
	Urban	10–30	[131]
	Industrial	30–60	[131]
	Marine	10–40	[131]
	Arctic	4	[131]
Steel (0.3% Cu)	Seawater	116	[22]
	Industrial	7	[22]
	Marine	14	[22]
Steel (0.2% C, 0.02% Cu, 0.02% Ni, 0.02% Cr)	Urban/Industrial	12.2	[22]
Steel (0.1% C, 0.03% Ni, 1.1% Cr, 0.4% Cu)	Urban/Industrial	2.3	[22]
	Zinc		
Zinc	Rural	0.2–3, 0.7–2.8	[22], [131]
Zinc 99.9%	Rural	0.9	[22]
Zinc 99.0%	Rural	1.1	[22]
Zinc	Rural, tropical	0.14–4.3	[131]
Zinc	Urban, small cities	0.7–2.8	[131]
Zinc 99.9%	Urban/Industrial	5.1	[22]
Zinc 99.0%	Urban/Industrial	4.9	[22]
Zinc	Marine, coastal	0.5–8, 0.7–2.8	[22], [131]
Zinc 99.9%	Marine, coastal	1.6	[22]
Zinc 99.0%	Marine, coastal	1.8	[22]
Zinc	Marine, ships	6	[131]

Table 5.4 contd. ...

...Table 5.4 contd.

Alloy	Atmosphere	Rate µm/year	Lit.
	Zinc		
Zinc	Industrial	2–16, 4.3–14	[22], [131]
Zinc	Indoors	0.07–0.6	[131]
Zinc	Flowing seawater	20–85	[131]
	Aluminium		
Aluminium	Flowing seawater	1–5 except for pits	[131]
Aluminium	Urban/Industrial	0.8	[22]
Aluminium	Marine	0.7	[22]
Aluminium	Rural	0.03	[22]
	Copper		
Copper	Flowing seawater	13–90	[131]
Copper	Urban/Industrial	1.2	[22]
Copper	Marine	1.3	[22]
Copper	Rural	0.6	[22]
	Nickel		
Nickel	Flowing seawater	< 25 except for pits	[131]
Nickel	Urban/Industrial	3.3	[22]
Nickel	Marine	0.1	[22]
Nickel	Rural	0.2	[22]
	Tin		
Tin	Urban/Industrial	1.2	[22]
Tin	Marine	2.3	[22]
Tin	Rural	0.5	[22]

5.3 Spectroscopic Data

5.3.1 XPS Data

XPS data are often very accurate with a deviation of less than 0.5 eV. There is an exception for zinc and zirconium data and therefore several data from literature for metals and their compounds are summarized in the table for comparison. Literature with similar data collected in one bracket and the order of the values is similar to the order of the literature.

Table 5.5 XPS data of metal substrates, conversion layer compounds and common coating compounds.

Compound	[eV]	Literature	Comment
Water (adsorbed)	533.1, 533.4	[43], [19]	O1s
Oxide	530, 530.2, 530.3, 530.7	[19], [43], [31], [32]	O1s
Hydroxide	531.4, 531.7, 531.9, 531.8	[43], [19], [32], [33]	O1s
Aluminium Compounds			
Al	72.9	[75]	Al2p
	1419	[75]	$L_{23}M_1M_{23}$ (Al Kα)
Al_2O_3 Sapphire	74.4	[75]	Al2p
α-Al_2O_3	73.9	[75]	Al2p
γ-Al_2O_3	74.0	[75]	Al2p
Al_2O_3	74.2	[37]	Al2p
	531.5	[37]	O1s
AlOOH Boehmite	74.2	[75]	Al2p
	531.5	[75]	O1s
$Al(OH)_3$ Gibbsite	74.0	[75]	Al2p
	531.5	[75]	O1s
$Al(OH)_3$ Bayerite	74.2	[75]	Al2p
	531.4	[75]	O1s
Barium Compounds			
$BaSO_4$	779.9–780.8	[75]	Ba3d5/2
Calcium Compounds			
$CaCO_3$	346.5–347.2	[75]	Ca2p3/2
$CaMoO_4$	346.9–347.5	[75]	Ca2p3/2
$Ca_3Si_3O_9$	346.8–347.2	[75]	Ca2p3/2
Carbon Compounds			
Graphite	284.5	[75]	C1s
Carbon	284.60, 284.62	[33], [38]	C1s
Carbonate	290.0, 290.5, 291	[50], [40], [123]	C1s
CH	285.00	[39]	C1s
C–C(O)–O–C	285.4	[39]	C1s
O=C–O–**C**	286.3	[39]	C1s
O=**C**–O–C	288.7	[39]	C1s
$-CF_2$ (PTFE*2)	292.2	[42]	C1s
$-CF_3$(PTFE)	295.3	[42]	C1s

Table 5.5 contd. ...

...Table 5.5 contd.

Compound	[eV]	Literature	Comment
	Cerium Compounds		
CeO_2	111.7	[23]	Ce4d3/2
	108.2	[23]	Ce4d5/2
	881.8–882.4	[75]	Ce3d5/2
	529.4, 529.3	[23], [37]	O1s
Ce_2O_3	530.3	[37]	O1s
	Chromium Compounds		
Cr_2O_3	576.9	[75]	Cr2p3/2
	586.7	[75]	Cr2p1/2
	531.0	[75]	O1s
CrO_3	576.7	[51]	Cr2p3/2
	530.2	[51]	O1s
$Cr(OH)_3$	577.2	[51]	Cr2p3/2
	531.3	[51]	O1s
	Copper Compounds		
Cu	567.9	[19]	CuL_3MM (Al Kα)
	932.67	[19]	Cu2p3/2
Cu_2O	569.7	[19]	CuL_3MM (Al Kα)
	932.35	[19]	Cu2p3/2
CuO	568.7	[19]	CuL_3MM (Al Kα)
	933.6	[75]	Cu2p3/2
	953.5	[75]	Cu2p1/2
	529.6	[75]	O1s
$Cu(OH)_2$	570.2	[19]	CuL_3MM (Al Kα)
	934.5, 935.1	[19], [75]	Cu2p3/2
	531.2	[75]	O1s
$CuCO_3$	935.0	[75]	Cu2p3/2
	Iron Compounds		
Fe^{2+}	709.5	[51]	Fe2p3/2
Fe^{3+}	710.5	[51]	Fe2p3/2
α-FeOOH Goethite	711.0, 711.25	[75], [70]	Fe2p3/2
	530.1	[71]	O1s
	531.8	[71]	O1s

Table 5.5 contd. ...

...Table 5.5 contd.

Compound	[eV]	Literature	Comment
		Iron Compounds	
γ-FeOOH Lepidocrocite	712	[24]	Fe2p3/2
$Fe(OH)_2$	531.3	[75]	O1s
Fe^0	707	[24]	Fe2p3/2
FeO	709.0	[75]	Fe2p3/2
Fe_2O_3	710.7, 710.9	[32], [75]	Fe2p3/2
	724.5	[32, 75]	Fe2p1/2
Fe_3O_4 Magnetite	710.8	[72]	Fe2p3/2
		Magnesium Compounds	
Mg	48.6	[40]	Mg2p
MgO	531.2	[40]	O1s
$Mg(OH)_2$ Brucite	49.2–49.8, 50.8	[75], [40]	Mg2p
	532.2	[40]	O1s
		Molybdenum Compounds	
MoO_4^{2-}	232.2	[49]	Mo3d5/2
	235.4	[49]	Mo3d3/2
MoO_2	229.1, 229.0–229,5	[49], [75]	Mo3d5/2
	232.2	[49]	Mo3d3/2
	529.9–531.1	[75]	O1s
$MoO_3 * nH_2O$	233.0	[75]	Mo3d5/2
	530.4–531.6	[75]	O1s
		Nitrogen Compounds	
$R\text{-}NH_2$ Amine	399.4	[43]	N1s
$R\text{-}NH_3^+$ Ammonium	400.6	[43]	N1s
		Silicon Compounds	
Silica	103.2–103.9, 103.3	[75], [69]	Si2p3/2
	532.7	[68]	O1s
Silicates	102.0–103	[75]	
$Al_4[(OH)_8 \mid Si_4O_{10}]$ Kaolinite	103.0	[75]	Si2p3/2
$Mg_3Si_4O_{10}(OH)_2$ Talc	103.1	[75]	Si2p3/2

Table 5.5 contd. ...

...Table 5.5 contd.

Compound	[eV]	Literature	Comment
Silicon Compounds			
α-SiO2	103.3	[75]	Si2p3/2
Quartz	103.7	[75]	Si2p3/2
	533.2	[75]	O1s
Silanes	101.9–102.9	[75]	Si2p3/2
Methyl silicone Resin	102.9	[75]	Si2p3/2
	532.7	[75]	O1s
Phenyl silicone Resin	102.7	[75]	Si2p3/2
	532.6	[75]	O1s
Sulphur Compounds			
Ph-S (PPS*)	285.0	[42]	C1s
	164.0	[42]	S2p
SO_4^{2-} Sulphate	169.8	[42]	S2p
Tin Compounds			
Sn	484.6	[19]	Sn3d5/2
Sn(II)	485.6	[19]	Sn3d5/2
Sn(IV)	486.4	[19]	Sn3d5/2
SnO	486.0–486.9	[75]	Sn3d5/2
	530.1	[75]	O1s
SnO_2	486.5–486.9	[75]	Sn3d5/2
	530.6	[75]	O1s
Titanium Compounds			
TiO_2 Anatase	459.0	[66]	Ti2p3/2
TiO_2	457.7	[33]	Ti2p3/2
	463.5	[33]	Ti2p1/2
	529.7	[33]	O1s
Yttrium Compounds			
$Y(OH)_3$ amorphous	160–162.5	[36]	Y3d3/2
	157.8–160.5	[36]	Y3d5/2
	532–533.3	[36]	O1s
Y_2O_3	157.7–158.3	[36]	Y3d3/2
	155.6–157.1, 156.4–157.0	[36], [75]	Y3d5/2
	528.8–531.2	[36]	O1S

Table 5.5 contd. ...

...Table 5.5 contd.

Compound	[eV]	Literature	Comment
Yttrium Compounds			
YOOH	158.9–159.04	[36]	Y3d3/2
	156.6–158.6	[36]	Y3d5/2
	529.3–531	[36]	O1S
Zinc Compounds			
Zn	494.4	[43, 56, 57]	ZnLMM
	993	[50]	$ZnL_3M_{45}M_{45}$
	1021.60	[56, 57]	Zn2p3/2
Zn^{2+}	988.5	[50]	$ZnL_3M_{45}M_{45}$
ZnO Zincite	498.2, 498.5	o.d., [43, 56, 57, 122]	ZnLMM
	1021.75, 1022.05	[58], [56, 57, 122]	Zn2p3/2
	988.5	[58]	$ZnL_3M_{45}M_{45}$
	530.0, 530.3	[50], o.d.	O1s
$Zn_5(OH)_6(CO_3)_2$ Hydrozincite	499.5	[56, 57, 122]	ZnLMM
	1022.4, 1022.5	[50], [56, 57, 122]	Zn2p3/2
$Zn(OH)_2$	499.2, 500.1	o.d., [43, 56, 57, 122]	ZnLMM
	1022.2, 1022.80	[50], [56, 57, 122]	Zn2p3/2
	531.8, 532	[50], o.d.	O1s
$ZnCO_3$	499.2	[56, 57, 122]	ZnLMM
	1022.30, 1022.50	[122], [56, 57]	Zn2p3/2
Zirconium Compounds			
Zr	179.6	[53]	Zr3d5/2
ZrO_2	181.5, 182.0, 182.1, 182.2, 182.5, 182.0–182.5	[33], o.d., [54], [53, 136], [55], [75]	Zr3d5/2
	184.1, 184.4	[33], o.d.	Zr3d3/2
	529.7, 529.9, 530.1, 530.2–530.9	[33], [53], o.d., [75]	O1s
ZrO_2*nH_2O	181.8	[136]	Zr3d5/2
ZrO_x	180.9, 182.4	[134]	Zr3d3/2
$Zr(OH)_4$	182.8, 183.6, 183.7	o.d., [53], [134]	Zr3d5/2
	184.9	o.d.	Zr3d3/2
	531.1, 531.2	o.d., [53]	O1s
$Zr(OH)_4*nH_2O$	183.6	[136]	Zr3d5/2

o.d.: own data (see examples), *Polyphenylene sulphide (PPS)

5.3.2 *Vibrational Data*

Table 5.6 Vibrational data of corrosion products on metal substrates. (Literature with similar data collected in one bracket and the order of the values is similar to the order of the literature).

Compound	Infrared			Raman		
	[cm^{-1}]	Lit.	Comment	[cm^{-1}]	Lit.	Comment
Aluminium Compounds						
AlO_2^-	450–480	[108]				
	515–560	[108]				
	620–670	[108]				
	800–920	[108]				
Al_2O_3	560–580	[108]				
	650–660	[108]				
	840–850	[108]				
	1573	[62]				
$Al(OH)_3$ Gibbsite				177	[106]	
				242	[106]	
				322	[106]	
	667	[112]		381	[106]	
	804	[112]		538	[106]	
	970	[112]		569	[106]	
	1022	[112]		892	[106]	
	3348–3396	[112]	O-H st.	3361	[106]	
	3460	[112]	O-H st.	3431	[106]	
	3529	[112]	O-H st.	3520	[106]	
	3622	[112]	O-H st.	3615	[106]	
$Al(OH)_3$ Bayerite				240	[106]	
				250	[106]	
	630	[112]		299	[106]	
	770	[112]		325	[106]	
	972	[112]		525	[106]	
	3410	[112]	O-H st.	547	[106]	
	3470	[112]	O-H st.	3421	[106]	
	3550	[112]	O-H st.	3542	[106]	
	3655	[112]	O-H st.	3651	[106]	

Table 5.6 contd. ...

...Table 5.6 contd.

Compound	Infrared			Raman		
	[cm^{-1}]	Lit.	Comment	[cm^{-1}]	Lit.	Comment
Aluminium Compounds						
$NaAlCO_3(OH)_2$ Dawsonite	503, 539, 685	[143]		588	[144]	
	840, 928, 948	[143]		1090	[144]	
	1390, 1399	[143], [144]				
	1555, 1562	[143], [144]				
	3277, 3280	[144], [143]	O-H st.	3287	[144]	
Chromium Compounds						
$Cr(OH)_3$	555	[62]	Cr(III)-OH			
	594	[62]	Cr(III)-OH			
CrOOH	918–933	[62]				
$Cr_2O_3{*}2H_2O$	840	[62]	Cr(VI)-O			
	1427–1450	[62]				
CrO_4^{2-}	770–960	[108]		848, 770–960	[62], [108]	
$Cr_2O_7^{2-}$				200–290	[108]	
				350–380	[108]	
	730–800	[108]		515–600	[108]	
	840–900	[108]		901, 840–900	[62], [108]	
				900–1000	[108]	
Copper Compounds						
$Cu(OH)_2$	690	[101]	Cu-O	460, 500–650	[62], [93]	
	3300, 3320	[133], [101]	O-H st.			
	3570, 3576	[133], [101]	O-H st.			
Cu_2O				145	[93]	
				214	[93]	

Table 5.6 contd. ...

...Table 5.6 contd.

Compound	Infrared			Raman		
	[cm^{-1}]	Lit.	Comment	[cm^{-1}]	Lit.	Comment
Copper Compounds						
	617, 610–630	[64], [108]	Cu-O	644–649	[93]	
	880	[133]	Cu-O			
	1160	[133]	Cu-O			
CuO	416	[64]	Cu-O			
	441	[64]	Cu-O			
	453	[64]	Cu-O			
	513	[64]	Cu-O	525	[62]	Cu-O
	568	[64]	Cu-O	625	[62]	Cu-O
Iron Compounds						
$Fe(OH)_2$	737	[110]		420	[74]	
	3728	[109]	O-H st.	510, 550	[74], [91]	
$Fe(OH)_3$				303	[90]	
				387	[90]	
				698	[90]	
FeOOH	900	[133]				
	1300	[133]				
Green Rust				430	[153]	
				510	[153]	
γ-FeOOH Lepidocrocite				247–250, 250, 252	[153], [90, 121], [74]	
	795	[86]	O-H ben.	376, 379, 380–400	[90, 121], [74], [153]	
	887	[86]	O-H ben.	525, 528, 529	[121], [74], [153]	
	1018, 1021	[74], [86]	O-H- st.	650	[121]	
	3160	[74]	O-H- st.	1300	[121]	
	3620, 3525	[74]	O-H- st.			

Table 5.6 contd. ...

...Table 5.6 contd.

Compound	Infrared			Raman		
	[cm^{-1}]	Lit.	Comment	[cm^{-1}]	Lit.	Comment
Iron Compounds						
α-FeOOH Goethite				136–147	[153]	
				247–250, 251	[153], [93, 62]	
				298–300, 299, 300, 303	[153], [90], [74], [93, 62]	
	785, 795, 800	[102], [86, 74], [63]	O-H ben.	380–400, 386, 387, 390, 420	[153], [74], [90], [93], [62]	
	887, 890, 892	[86, 102], [63], [74]	O-H ben.	481, 482	[74], [93, 62]	
	1021	[86]	O-H ben.	549, 552, 554	[74], [93, 62], [90]	
	3140	[74]	O-H- st.	667–690, 682	[153], [93]	
	3660, 3484	[74]	O-H- st.	1003	[93]	
β-FeOOH Akaganeite				136–147	[153]	
				305–320, 314	[153], [74]	
				380, 380–400	[74], [153]	
				408–415	[153]	Shoulder
	795	[86]	O-H ben.	549	[74]	
	887	[86]	O-H ben.	710–725, 722	[153], [74]	
δ-FeOOH Feroxyhite	640	[63]				
	1200	[63]				
$5Fe_2O_3$* $9H_2O$ Ferrihydrite				495–502	[153]	

Table 5.6 contd. ...

...Table 5.6 contd.

Compound	Infrared			Raman		
	[cm^{-1}]	Lit.	Comment	[cm^{-1}]	Lit.	Comment
Iron Compounds						
α-Fe_2O_3 Hematite				293	[90]	
	325	[102]		299	[90]	
	465	[102]		412	[90]	
	557	[102]		613	[90]	
γ-Fe_2O_3 Maghemite				381	[74]	
				486	[74]	
	550–670, 510–610	[86], [108]	Fe-O	550	[62]	
				670, 676	[74], [121]	
	700	[133]		710, 710–725, 718	[62], [153], [74]	
				1360, 1450	[121], [62]	
Fe_3O_4 Magnetite	570 ± 5	[104]		532, 543, 550	[74], [93, 62], [90]	
				660, 667, 667–690, 670, 674, 676	[79], [74], [153], [90, 91], [93, 62], [121]	
$FeCO_3$*H_2O Siderite	195	[105]		509	[62]	
	230	[105]		734	[62]	
	378	[105]		1082	[62]	
Magnesium Compounds						
$Mg(OH)_2$ Brucite	370	[78]				
	460	[78]				
	3690	[78]	(s) O-H- st.			
Nickel Compounds						
α-$Ni(OH)_2$	463	[63]	Ni-O- st.			
β-$Ni(OH)_2$	444	[63]	Ni-O- st.			

Table 5.6 contd. ...

...Table 5.6 contd.

Compound	Infrared			Raman		
	[cm^{-1}]	Lit.	Comment	[cm^{-1}]	Lit.	Comment
Tin Compounds						
$Sn(OH)_2$	545	[126]		470	[126]	
	722	[126]		752	[126]	
				904	[126]	
$Sn(OH)_4$	560	[126]		580	[126]	
				750	[126]	
SnO	420	[126]		470	[126]	
	520	[126]				
	719	[126]		760	[126]	
SnO_2	410	[126]		470	[126]	
	606, 305	[29]	Sn-O- st.	636	[126]	
	615, 321	[29]	Sn-O- st.	776	[126]	
Zinc Compounds						
ZnO Zincite				100	[99, 125]	
	325, 350–600, 320–340	o.d., [99], [108]		330, 375	[99, 125], [100, 125]	
	414, 400–420, 460	o.d., [108], [78]		440	[99, 100, 121, 125]	
	428, 450–460	o.d., [108]		543, 550, 563–570, 570, 580	[77, 125], o.d., [25], [100], [121, 125]	
	500–520, 532	[108], o.d.		1070	[100]	
	560	[108]		1140	[100]	
	3442, 3450	o.d., [78]				

Table 5.6 contd. ...

...Table 5.6 contd.

Compound	Infrared			Raman		
	[cm^{-1}]	Lit.	Comment	[cm^{-1}]	Lit.	Comment
Zinc Compounds						
$ZnCO_3$ Smithsonite	202	[105]		140	[99, 125]	
	309	[105]		216	[99, 125]	
	360	[105]		383	[99, 125]	
	745	[99]		730	[99, 125]	
	873	[99]		1063	[99, 125]	
	1420	[99]	Carbonate ben.			
β-$Zn(OH)_2$/ε-$Zn(OH)_2$ *				170/150	[100]	
				210	[100]	
	410–486	[124]		250, 242–258	[100], [124]	
	514	[124]		300/370, 366–379	[100], [124]	
	547	o.d.		380, 385	[100], [125]	
	715, 717	[107], [124]		420/480, 475	[100], [124]	
	750, 772	[107], [124]		740/720, 711	[100], [124]	
	830, 847, 912	[107], [124], o.d.		800/760, 757	[100], [124]	
	1025, 1031	[107], [124]		1025, 1030	[124], [100]	
	1062, 1080	o.d. [107], [124]		1080, 1085	[100], [124]	
				1050	[100]	ε-$Zn(OH)_2$
				3100	[100]	β $Zn(OH)_2$
				3150/3190, 3189	[100], [124]	

Table 5.6 contd. ...

...Table 5.6 contd.

Compound	Infrared			Raman		
	[cm⁻¹]	Lit.	Comment	[cm⁻¹]	Lit.	Comment
Zinc Compounds						
	3146, 3260	[124], [107]	OH- st.	3250/3260	[100, 124]	
	3383	o.d.		3340, 3450	[100], [125]	β-$Zn(OH)_2$ Hydroxide
	3552	o.d.				
$Zn_5(OH)_8Cl_2$ Simonkolleite	280	[78, 124]		210, 213	[99, 100, 125], o.d.	
	370, 393	[78], [124]		240.5, 255, 256, 258, 260	[77], [25], [99, 125], o.d., [100]	
	460, 488	[78], [124]		390, 392.2, 394, 398, 399, 409.7	[100, 124], [77], [25, 125], [99], o.d., [77]	
	695, 711, 720	[107], [124], [78]		727, 730	o.d., [100]	
	897, 905	[124], [78]		910, 916	[100, 124], o.d.	
	1030, 1040, 1041	[107], [78], [124]		1030	[100]	
				1340	[77]	
	3400, 3480, 3490	[121], [107], [78]	OH- st.	3450, 3453, 3456, 3458	[100], [99, 125], o.d., [124]	
	3540, 3558	[107], [124]	OH- st.	3480, 3484, 3486, 3487	[100], [99, 125], [124], o.d.	

Table 5.6 contd. ...

...Table 5.6 contd.

Compound	Infrared			Raman		
	[cm^{-1}]	Lit.	Comment	[cm^{-1}]	Lit.	Comment
Zinc Compounds						
	3590	[107]	OH- st.	3580	[100]	
$Zn_5(OH)_6$ $(CO_3)_2$ Hydrozincite	317	[105]				
	370, 372	[78], [105]				
	396, 398	[105], [124]		385	[124]	
	460, 467, 475	[78], [105], [124]				
	513, 515	[105], [124]		705	[124]	
	835, 836, 838	[78], [124], o.d.	Carbonate ben.	734	[124]	
	1040, 1044, 1170	[78], [124], o.d.		1060	[124]	
	1380, 1382, 1390, 1428	[99], [124], [78], o.d.	Carbonate ben.	1369 and 1437	[124]	
	1500, 1501, 1507, 1510	[78], o.d., [124], [99]	Carbonate ben.	1545	[124]	
	3250, 3310, 3385	[99], [124], o.d.	OH-st.	3388	[124]	
$Zn(OH)_4^{2-}$ Zincate				285	[9]	
				322	[9]	
	430	[9]		430	[9]	
				484	[9]	

Sym.: symmetric, asym.: asymmetric, def.: deformation, st.: stretching, ben.: bending, o.d.: own data, *: lines of both types

Table 5.7 Vibrational data of conversion layers. (Literature with similar data collected in one bracket and the order of the values is similar to the order of the literature.)

Compound	Infrared			Raman		
	[cm^{-1}]	Lit.	Comment	[cm^{-1}]	Lit.	Comment
Cerium Compounds						
CeO_2	503	[103]	Ce-O-O st.	465	[23]	Fluorite structure
	728	[103]	Ce-O-O st.			
	855	[108]	Ce-O-O st.			
Molybdenum Compounds						
MoO_4^-				310–350	[108]	
				380–400	[108]	
	810–850	[108]		810–850	[108]	
	890–935	[108]		890–935	[108]	
$Mo_7O_{24}^{6-}$				924–934	[127]	
MoO_2				340	[127]	
				512	[127]	
				763	[127]	
α-MoO_3	370–390	[108]		448.8	[76]	
	550	[108]				
	860–870	[108]		804.8	[76]	
	1090–1110	[108]				
Phosphorus Compounds						
PO_4^{3-}	1005, 1000–1100	[87], [108]	Asym. st.	1000, 1000–1100	[90], [108]	Sym. st. (iron phosp. layer)
	737, 827, 951	[141]	Phytic acid			
HPO_4^{2-}	850, 825–920	[87], [108]				
	890	[87]				
	990	[87]				
	1075	[87]				

Table 5.7 contd. ...

...Table 5.7 contd.

Compound	Infrared			Raman		
	[cm^{-1}]	Lit.	Comment	[cm^{-1}]	Lit.	Comment
Phosphorus Compounds						
	1150–1350, 1371–1375	[108], [142]	P=O-st., [142] phytic acid	1150–1350	[108]	P=O-st.
	1600–2500, 1600	[108], [141]	Acid salts with P-OH group, [141] phytic acid	1600–2500	[108]	Acid salts with P-OH group
	2525–2725	[108]	Acid salts with P-OH group	2525–2725	[108]	Acid salts with P-OH group
$H_2PO_4^-$	875	[87]				
	940	[87]				
	1075	[87]				
	1155, 1150–1350	[87], [108]	P=O-st.	1150–1350	[108]	P=O-st.
	1600–2500	[108]	Acid salts with P-OH group	1600–2500	[108]	Acid salts with P-OH group
	2525–2725	[108]	Acid salts with P-OH group	2525–2725	[108]	Acid salts with P-OH group
H_3PO_4	890	[87]				
	1005, 910–1040, 972–1011–147	[87], [108], [140]	P-O-st., [140] phytic acid	910–1040	[108]	P-O-st.
	1175, 1150–1350	[87], [108]	P=O-st.	1150–1350	[108]	P=O-st.

Table 5.7 contd. ...

...Table 5.7 contd.

Compound	Infrared			Raman		
	[cm^{-1}]	Lit.	Comment	[cm^{-1}]	Lit.	Comment
Phosphorus Compounds						
	1250	[87]				
	1680	[108]	P=O	1680	[108]	P=O
	2100–2300	[108]	O-H-st.	2100–2300	[108]	O-H-st.
	2560–2700	[108]	O-H-st.	2560–2700	[108]	O-H-st.
Titanium Compounds						
TiO_2 Brookite				205	[62]	
				320	[62]	
				360	[62]	
				412	[62]	
TiO_2	320–360	[108]		320–360	[108]	
	460–525	[108]		460–525	[108]	
	660–700	[108]		660–700	[108]	
TiO_2 nano	466	[33]	Ti-O-Ti			
	511	[33]	Ti-O			
Yttrium Compounds						
amorphous $Y(OH)_3$	720	[36]	Y-OH ben.			
	980	[36]	Y-OH ben.			
	1390	[36]	OH-ben.			
	1510	[36]	OH-ben.			
	1630	[36]	OH-ben.			
	3300	[36]	OH-st.			
hexagonal $Y(OH)_3$	720	[36]	Y-OH ben.			
	980	[36]	Y-OH ben.			

Table 5.7 contd. ...

...Table 5.7 contd.

Compound	Infrared			Raman		
	$[cm^{-1}]$	Lit.	Comment	$[cm^{-1}]$	Lit.	Comment
Yttrium Compounds						
	2800	[36]	OH-st.			
	3600	[36]	OH-st.			
α-$Y(OH)_3$	675	[36]	Y-OH ben.			
	800	[36]	Y-OH ben.			
	1390	[36]	OH-ben.			
	1420	[36]	OH-ben.			
	1520	[36]	OH-ben.			
	3410	[36]	OH-st.			
	3500	[36]	OH-st.			
	3600	[36]	OH-st.			
Y_2O_3	375	[36]	Y-O st.			
	465	[36]	Y-O st.			
	560	[36]	Y-O st.			
YOOH	3600	[36]	OH st.			
	2800	[36]	OH st.			
	1630	[36]	OH ben.			
	1510	[36]	OH ben.			
	1390	[36]	OH ben.			
	980	[36]	Y-OH ben.			
	720	[36]	Y-OH ben.			
	475	[36]	Y-O st.			
	350	[36]	Y-O st.			
Zirconium Compounds						
ZrO_2	415, 422–423, 449, 470	o.d., [135], o.d. [137]		179	[138]	Zr-Zr

Table 5.7 contd. ...

...Table 5.7 contd.

Compound	Infrared			Raman		
	[cm^{-1}]	Lit.	Comment	[cm^{-1}]	Lit.	Comment
Zirconium Compounds						
	503–519, 517, 540	[135], o.d., [94]	Zr-O-Zr	190	[138]	Zr-Zr
	572, 576–580, 584	o.d., [135], [137]	Zr-O	334	[138]	Zr-Zr
	630	[137]	Zr-O-Zr	346	[138]	ZrO
	700, 710	[94], [33]	Zr-O-Zr	382	[138]	O-O, Zr-O
	707	[94]	tetragonal Zr	453, 476	[98], [138]	Zr-O, O-O [138]
	739, 746	[135], o.d.		557, 568	[138], [98]	Zr-O, O-O [138]
	870	[137]	tetragonal Zr	614	[138]	O-O
	1385	[33]	Zr-O	633, 640	[138], [89]	O-O [138]
$ZrO(OH)_2$* x H_2O Conversion Layer	409	o.d.	compare ZrO_2			
	548	o.d.	Zr-O-Zr			
	667	o.d.				
	842	o.d.				
	1407	o.d.	Zr-O-H			
	1524	o.d.				
	1646	o.d.	H-O-H-ben.			
	3420	o.d.	O-H-st.			
$Zr(OH)_4$	472	o.d.				
	648, 680.5, 681.6	o.d., [111]	Two bands [111]			

Table 5.7 contd. ...

...Table 5.7 contd.

Compound	Infrared			Raman		
	[cm^{-1}]	Lit.	Comment	[cm^{-1}]	Lit.	Comment
Zirconium Compounds						
	1005	[95]	Zr-OH			
	1115, 1120	[97], [96]	Zr-OH			
	1346, 1383	o.d.	Two bands			
	1437, 1453	[33]	Zr-O-H			
	1628	o.d.	H-O-H-ben.			
	3420, 3783	o.d., [111]	O-H-st.			
$Zr(OH)_2$* $2H_2O$	610–625	[137]	Zr-O-Zr			
	1535–1575	[137]	Zr-O-H ben.			
	1620–1635	[137]	O-H ben. in H_2O			
	3330–3350	[137]	O-H st.			
$Zr(OH)_2$* H_2O	615–630	[137]	Zr-O-Zr			
	584	[137]	Zr-O			
	1540	[137]	Zr-O-H ben.			
	1635	[137]	O-H ben. in H_2O (annealing up to 175°C)			
	3330–3370	[137]	O-H st.			

sym.: symmetric, asym.: asymmetric, def.: deformation, st.: stretching, ben.: bending, o.d.: own data

Table 5.8 Vibrational data of common organic coating compounds. (Literature with similar data collected in one bracket and the order of the values is similar to the order of the literature.)

Compound	Infrared			Raman		
	[cm^{-1}]	Lit.	Comment	[cm^{-1}]	Lit.	Comment
H_2O Water	3920	[139]	O-H-st.			
	3300, 3400, 3490	[84], [43], [139]	O-H-str.			
	1625, 1645, 1650	[78], [139], [84]	H-O-H-ben.	1637	[82]	298° K
	3710	[114]	in solution	228	[82]	77° K
	3100–3600	[114]	Crystal water			
	3200–3600	[114]	O-H in H-bridge to alcohols/ amines			
	2500–3200	[114]	O-H in H-bridge to carboxylic groups			
D_2O Water	2900	[139]	O-D-st.			
	2500, 2540	[84], [139]	O-D-st.	1203	[82]	298° K
	2450	[139]	O-D-st.			
	1200, 1215	[84], [139]	D-O-D-ben.	223	[82]	77° K
–OH	3600	[43]	st.			
	1260–1410	[114]	O-H-def.			
H_3O^+ (HCl)	2600–2700	[84]	sym. and asym. st.			
	1665	[84]	asym. ben.			
	1125	[84]	sym. ben.			
Aluminium Compounds						
Al_2O_3*	415–435	[108]				
$2SiO_2$*	460–475	[108]				
$2H_2O$ Kaolin	515–550	[108]				
	905–920	[108]				
	935–960	[108]				
	1090–1120	[108]				
	3645–3655	[108]	O-H-st.			

Table 5.8 contd. ...

...*Table 5.8 contd.*

Compound	Infrared			Raman		
	[cm^{-1}]	Lit.	Comment	[cm^{-1}]	Lit.	Comment
Barium Compounds						
$BaSO_4$	415	[108]		415	[108]	
	460	[108]		460	[108]	
	605–615	[108]	Doublet, asym. def. of sulphate	605–615	[108]	Doublet
	630–640, 631	[108], [133]		630–640	[108]	
	1070–1090	[108]	Doublet	1070–1090	[108]	Doublet
	1110–1130	[108]		1110–1130	[108]	
	1180–1200	[108]	Sharp	1180–1200	[108]	Sharp
	1470	[108]	Broad			
	3430	[108]	Broad	3430	[108]	Broad
Calcium Compounds						
$CaCO_3$ Calcite	98	[105]				
	110	[105]				
	228, 230	[105], [108]				
	319, 330	[105], [108]				
	360	[105]				
	695–705, 706	[108], [139]		695–705	[108]	
	715	[108]		715	[108]	
	845–860, 879	[108], [139]				
	1160	[108]				
	1410(14)–1495(92)	[108], [139]		1410–1495	[108]	
	1770–1815	[108]				
	2500–2530	[108]				

Table 5.8 contd. ...

...Table 5.8 contd.

Compound	Infrared			Raman		
	[cm^{-1}]	Lit.	Comment	[cm^{-1}]	Lit.	Comment
Calcium Compounds						
$CaCO_3$ Aragonite	110	[105]				
	215	[105]				
	263	[105]				
	695–705	[108]		695–705	[108]	
	715	[108]		715	[108]	
	845–860	[108]				
	1160	[108]				
	1410–1495	[108]		1410–1495	[108]	
	1770–1815	[108]				
	2500–2530	[108]				
Carbon Compounds						
Graphite				1575, 1582	[79], [25]	
				1355	[79]	
CO_3^{2-}	885, 800–890	[87], [108], 139]	ben.			
	1065, 1020–1100	[87], [108		1050, 1080, 1020–1100	[62], [89], [108]	st.
	1385, 1320–1530, 1430, 1410–1450	[87], [108], [113], [114, 139]		1320–1530	[108]	
HCO_3^-	845, 830–840	[87], [108]	ben.			
	1010, 990–1000	[87], [108]		990–1000	[108]	
	1310	[87]				
	1360, 1290–1370	[87], [108]		1290–1370	[108]	

Table 5.8 contd. ...

...Table 5.8 contd.

Compound	Infrared			Raman		
	[cm^{-1}]	Lit.	Comment	[cm^{-1}]	Lit.	Comment
Carbon Compounds						
CH_2	1430–1470	[114]	C-H-def.	1440–1480, 1450–1475	[108], [119]	
	2850	[26, 114]	sym. st.	2840–2870	[108]	
	2920, 2930, 2960	[26], [43], [114]	asym. st.	2915–2940	[108]	
CH_3	1470, 1430–1470	[43], [114]	ben.	1440–1465	[108]	
	2960	[26]	asym. st.	2950–2975	[108]	
R-OH				800–970	[119]	Sym. C-C-O-st.
	1040–1150	[114]	C-O-st.	1050–1150	[119]	Asym. C-O-st.
	1260–1410	[114]	O-H-def.	1350–1450	[119]	O-H in plane
	3200–3600	[114]	O-H in H-bridges	3300–3400	[119]	O-H-st.
	3590–3650	[114]	O-H free			
R-C$(CH_3)_2$-R	1380	[114]	Doublet	1335–1385	[108]	Two bands of almost equal intensity
R-NH-CO_2^- R-NH_3^+	3200–3000	[43]	Carbamate			
NH_2-CO-R				1400–1420	[119]	C-N-st. Amide III
	1650	[114]	Amide I	1660–1680	[119]	C=O-st. Amide I
	1640	[114]	Amide II			
R-NH-CO-R				1280–1310	[119]	Amide III
	1630–1680, 1640, 1650	[114], [30], [84, 133]	Amide I	1620–1680, 1640–1680	[27], [119]	Amide I
	1660	[81]	Amide in polyamideimide			

Table 5.8 contd. ...

...Table 5.8 contd.

Compound	Infrared			Raman		
	[cm^{-1}]	Lit.	Comment	[cm^{-1}]	Lit.	Comment
Carbon Compounds						
	1550, 1640	[30, 133], [114]	Amide II			
R-CO-NR-CO-R	1780	[81]	Imide in polyamideimide	1200–1360	[27]	Amide III
R-CO-NR-CO-R	1700/1770	[114]	Imide	1700/1770	[108]	five membered rings
R-O-CO-NR_2	1690–1740, 1700–1735	[114], [113]	Urethane	1680–1740	[108]	Alkyl urethanes
R_2N-CO-NR_2	1660	[113]	Urea, C=O st.	1635–1680	[108]	Solid phase
Polyether polyurethane (aromatic)	1110	[117]	C-O			
	1230	[117]	C-C-st.			
	1530	[117]	Coupling peak of –NH ben. and –CN st.			
	1600	[117]	st. of the double bonds in the aromatic ring			
	1640–1763	[117]	–C=O st. of the urethane			
	2850	[117]	sym. st. of the CH_2 group			
	2940	[117]	asym. st. of the CH_2 group			
	3200–3500	[117]	–NH/–OH st.			
Polyurethanes				860–865	[157]	C-O-C/C-C-C ben.
				1181	[157]	C-O-C st./arom. Ring

Table 5.8 contd. ...

...Table 5.8 contd.

Compound	Infrared			Raman		
	[cm^{-1}]	Lit.	Comment	[cm^{-1}]	Lit.	Comment
Carbon Compounds						
	1540–1560	[158]	Amide II	1521–1540	[157]	NH ben. CN st.
				1615	[157]	arom. Ring C=C
	1635–1645	[139]	C=O-st. (hydrogen-bonded urea carbonyl)	1650	[157]	C=O st. urethane
	1703–1710, 1705–1710	[158], [139]	C=O-st. (hydrogen-bonded urethane)			
	1730, 1730–1740	[139], [158]	C=O-st. (non-hydrogen-bonded urethane)			
	2860	[139]	sym. C-H-st.			
	2940	[139]	asym. C-H-st.			
	3260–3290	[158]	N-H st. (to ether)			
	3305–3320, 3315–3340	[139], [158]	N-H-st. (hydrogen bonded)			
	3445, 3445–3450	[139], [158]	N-H-st. (non-hydrogen bonded)			
trans-C=C	965	[27]		1665	[27]	
cis-C=C				1622–1655	[27]	
cis-H-C=C	3010	[34]				
Styrene				1413	[156]	C=CH_2 wag.
				1632	[156]	C=C st.
Unsaturated Polyester				1632	[156]	C=C st.
				1661	[156]	C=C st.

Table 5.8 contd. ...

...Table 5.8 contd.

Compound	Infrared			Raman		
	[cm^{-1}]	Lit.	Comment	[cm^{-1}]	Lit.	Comment
Carbon Compounds						
Polyacrylates	962	[139]	C-CH_3-ben. (PMMA)	800	[108]	PMMA
	1170	[108]	C-O-st.			
	1170	[139]	C-O-C-ben. (PMMA)			
	1260	[108]	C-O-st.			
	1729	[139]	C=O-st. (PMMA)			
Phenyl	830	[27]	often used as internal standard			
Isocyanurate	760	[35]				
	1690	[35]				
Melamine (HMMM)	815, 820	[115], [108]	Out of plane def. of the triazine ring			
	910	[115]	R-O-CH_3 def. of not reacted ether groups	914	[88]	Residual methoxy group
	1040	[108]	C-O-st.	984	[88]	Melamine ring
				1395	[88]	Melamine side chain C-O
	1545, 1560	[115], [108]	In-plane def. of the triazine ring	1557	[88]	Melamine side chain C-N
Polyester	725, 730	[115], [108]	C-H-torsion of the aromatic ring in aromatic based polyesters	730	[108]	
	1050–1330	[114]	Two bands, C-O-def.	953	[88]	
	1250–1310	[108]	Esters of aromatic acids (phthalates etc.)	1250–1310	[108]	Esters of aromatic acids (phthalates etc.)

Table 5.8 contd. ...

...Table 5.8 contd.

Compound	Infrared			Raman		
	[cm^{-1}]	Lit.	Comment	[cm^{-1}]	Lit.	Comment
Carbon Compounds						
	1725	[115]	C=O-st.	810	[88]	
Ph-COOR				1010–1050	[119]	C-O-C-st.
				1210–1310	[119]	
	1715–1730	[113]	C=O-st.			
R-COOR				1110–1140	[119]	C-O-C-st.
	1735–1750	[113, 114]	C=O-st.	1640–1685	[108]	
Ph-COOH	1680–1700	[113, 114]	C=O-st.			
R-COOH	1700–1725	[113, 114]	C=O-st.			
R-COO$^-$	1300–1420	[113, 114]	sym. st.			
	1550–1610	[113, 114]	asym. st.			
R-NCO Isocyanate	2250–2275, 2280	[113], [89]	sym. NCO	not visible	[89]	
				1400–1440, 1525	[119], [89]	sym. NCO
Epoxy (BPA based)	914	[27]		639–642	[154]	Epoxy-ring def.
				718–736	[154]	Epoxy-ring def.
				821	[154]	C-H wag.
				916–918	[154]	Epoxy-ring def.
				1012	[154]	arom. ring st.
				1026–1031	[154]	C-O st.
				1186–1188	[154]	C-H wag.
				1252–1258, 1256	[154], [27]	Epoxy-Ring St.
				1610	[154]	arom. Ring st.
				2930	[154]	Aliphatic C-H

Table 5.8 contd. ...

...Table 5.8 contd.

Compound	Infrared			Raman		
	[cm^{-1}]	Lit.	Comment	[cm^{-1}]	Lit.	Comment
Carbon Compounds						
				3070	[154]	arom. C-H
Epoxy (GMA)	850	[83]				
Epoxide	915–920	[116]	Ring vibration			
	1190	[116]	CH_2 out of plane	1256, 1230–1280, 1260–1280	[89], [108], [119]	
	3050	[116]	C-H-st.			
Epoxy Resin Cured (Zinc Rich Paint)	850	[77]				
	1250	[77]				
	1460	[77]				
	1510	[77]				
	1600	[77]				
Phenol-Formaldehyde Resins	3350	[108]	O-H-st.			
	1230	[108]	C-O-st.			
	760/820	[108]	C-H out of plane			
R-Cl	560–820, 560–830	[113], [114]	C-Cl-st.	505–760	[108]	C-Cl-st.
Ph-Cl	1030-1100	[108, 113, 114]	Ring and C-Cl-st.	1030–1100	[108]	Ring and C-Cl-st.
R-F	1120–1360, 1120–1365	[113], [114]	C-F-stretching	1000–1400	[108]	C-F-st.
Ph-F	1100-1270	[108, 113, 114]	Ring and C-F-st.	1100–1270	[108]	Ring and C-F-st.
α-PVDF	531	[80]				
	766	[80]				
β-PVDF	511	[80]				
	840	[80]				
γ-PVDF	430	[80]				

Table 5.8 contd. ...

...Table 5.8 contd.

Compound	Infrared			Raman		
	[cm^{-1}]	Lit.	Comment	[cm^{-1}]	Lit.	Comment
Carbon Compounds						
PPS* Uncured	481	[42]	S-S st.			
	554	[42]	Ph-S st.			
	742, 820	[42]	C-H ring out of plane			
	1076–1094	[42]	C-H ring in plane			
	1011	[42]	C=C ring in plane			
	1472	[42]	C=C ring st.			
	1390	[42]	C-C ring in plane ben.			
	1573	[42]	arom. ring st.			
PPS Cured	507(481)	[42]	S-S st. is reduced or disappeared			
	1904	[42]	C=O			
	1230	[42]	C-O			
	1180	[42]	S=O			
	710–630	[42]	C-S			
	1296	[42]	O=S=O asym. st.			
Magnesium Compounds						
$Mg_3Si_4O_{10}(OH)_2$ Talc	445–455	[108]				
	530–540	[108]				
	665–675	[108]				
	1005–1030	[108]				
	3660, 3675, 3685	[108]	Three bands, O-H-st.			

Table 5.8 contd. ...

...Table 5.8 contd.

Compound	Infrared			Raman		
	[cm^{-1}]	Lit.	Comment	[cm^{-1}]	Lit.	Comment
Nitrogen Compounds						
R-N-H	3300–3500	[114]	N-H-st.			
R-NH_2	1560–1650	[114]	N-H-def.	1660	[155]	Amine in epoxy based coating
R_2-N-H	1490–1580	[114]	N-H-def.			
R-NH_3^+	3030–3130	[114]	N-H-st.	3100–3350	[108]	
	1500/1600	[114]	primary/ secondary	1500/1600	[108]	
NH_4^+	1390–1485	[139]				
	3030–3335	[139]				
NO_3^-	800–860	[139]				
	1340–1410	[139]				
Phosphorus Compounds						
PO_4^{3-}	1005, (950)1000–1100, 1050	[87], [108, 114, 139], [113]	asym. st.	1000, 1000–1100	[90], [108]	sym. st. (iron phosphate layer)
	1180–1240	[114]	P-O-st.			
	1250–1300	[114]	P=O-st.			
P-O-H	2560–2700	[114]	associated OH			
Silicon Compounds						
Silica	466, 490	[85], [30]	O-Si-O ben.	556	[23]	Si-O
	800, 802, 806, 785–805	[83], [43], [30], [108]	Si-O-Si sym. st.			
	932, 950	[43], [83]	Si-OH st.			
	1038, 1050–1100, 1250–1000, 1100, 1075–1100	[85], [30, 133], [43], [83], [108]	Si-O-Si asym. str.	1100, 1000–1100	[23], [108]	Si-O

Table 5.8 contd. ...

...Table 5.8 contd.

Compound	Infrared			Raman		
	[cm^{-1}]	Lit.	Comment	[cm^{-1}]	Lit.	Comment
Silicon Compounds						
	1150–1175	[108]				
	1200–1225, 1230	[108], [133]				
	3439, 3200–3700	[85], [108, 139]	Si-O-H st.	3200–3700	[108]	Si-O-H st.
Silicate	900–1100	[108]		480	[89]	
				900–1100	[108]	
SiO_3^{2-}	450–500	[108]				
	750–790	[108]				
	960–1030	[108]				
SiO_4^{2-}	470–540	[108]				
	860–1180, 900–1100	[108], [139]				
Silicone	1249–1285, 1240–1290	[43, 139], [108]	Si-CH_3 sym. def.	1240–1290	[108]	Si-CH_3 sym. def.
	1423, 1410	[43], [108]	Si-CH_3 asym. ben., [55] three CH_3-groups	1410	[108]	Si-CH_3 asym. ben.
	1110	[139]	Si-C_6H_5-st.			
	1430	[139]	Si-C_6H_5-st.			
R-O-Si Alkoxy-silanes	945–990, 950	[108], [145]	Si-O-C-st.	945–990	[108]	Si-O-C-st.
	1000–1100, 1050–1110	[108], [139]	Si-O-C-stretching	1000–1100	[108]	Si-O-C-st.
Silanole	820–870	[108]	O-H-deformation			
	1020–1040	[108]	Si-OH deformation			
	3200–3700	[108]	O-H-stretching	3200–3700	[108]	O-H-st.

Table 5.8 contd. ...

...Table 5.8 contd.

Compound	Infrared			Raman		
	[cm^{-1}]	Lit.	Comment	[cm^{-1}]	Lit.	Comment
Sulphur Compounds						
S-S	450–550	[139]	S-S-st.			
C-S	600–700	[139]	C-S-st.			
SO_4^{2-}	610–680	[139]				
	1105 ± 25, 1080–1130	[108, 113], [114, 139]		1080–1130	[108]	
Titanium Compounds						
TiO_2 Anatase	500–600	[66]	Ti-O	143, 145	[67], [93, 62]	
	1620–1630	[66]	O-H-def.	400	[92, 93, 62]	
	3100–3600	[66]	O-H-st.	515, 520	[93, 62], [92]	
				640, 650	[93, 62], [92]	
TiO_2 Rutile				236	[62]	
				446, 450, 452	[67], [25, 92], [62]	
				610, 616, 620	[67], [62], [92]	

Sym.: symmetric, asym.: asymmetric, def.: deformation, st.: stretching, ben.: bending, wag.: wagging, o.d.: own data, *Polyphenylene sulphide (PPS)

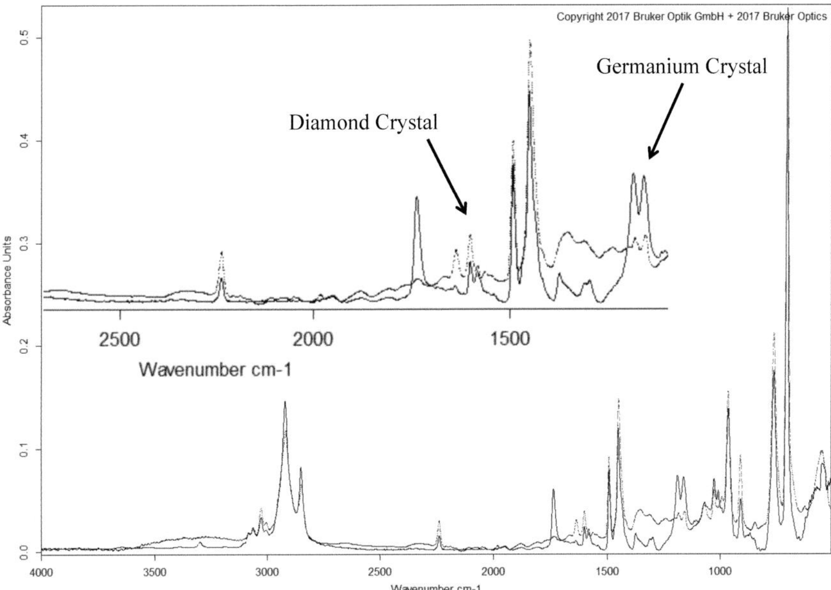

Fig. 5.1 ATR spectra of ABS (Acrylonitrile-Butadiene-Styrene-Copolymer) measured with a diamond crystal and a germanium crystal in the ATR device from KIMW Bruker ATR-IR Polymer, Kunststoff und Additiv Bibliothek, Bruker Optik GmbH. With permission.

5.3.3 *UV-VIS Data*

Table 5.9 UV-VIS absorption, reflection and band gap data of corrosion products, conversion layers and common coating compounds. (Literature with similar data collected in one bracket and the order of the values is similar to the order of the literature.)

Compound	Absorption λ_{max} [nm]	Literature
Aqueous Solutions		
Water (adsorbed)	167	[44, 114]
CTTS (NaCl) in water	185	[61]
CTTS (Cl^-) in water	185–186	o.d.
CTTS (OH^-) in water	187	o.d.
CTTS (CO_3^{2-}) in water	183	o.d.

Table 5.9 contd. ...

...Table 5.9 contd.

Compound	Absorption λ_{max} [nm]	Literature
Carbon Compounds		
$CaCO_3$ (Nano particles)	205/265	[150]
CO_3^{2-}/HCO_3^-	203/255	o.d.
CH (Pentane)	190	[119]
R-OH (Methanol)	183	[114]
R-O-R (Diethyl Ether)	189	[114]
R-COOH (Acetic Acid)	204	[114, 119, 113]
Toluol	261, 206.5	[119, 113]
p-Xylol	216	[119]
Ph-OH	210.5, 211, 270	[113], [114, 119], [113]
Ph-O^-	235, 236, 235/287	[114], [119], [113]
Styrene	248, 248/282/291	[119], [113]
Ph-COOH	230/270	[113]
Ph-$C(CH_3)_2$-Ph	270	[113]
C–C(O)–O–C (Ethyl Acetate)	204, 207	[119], [114, 113]
R-$CONH_2$ (Acetamide)	205	[114, 119, 113]
Polyether Polyurethane	253.7	[117]
R-CHO (Formaldehyde)	310	[119]
R-CO-R (Acetone)	275, 279	[119], [113]
4-Methyl-2-pentanone	283	[119]
Acrylic Acid	200	[119]
Methacrylic Acid	206	[119]
Ethyl Methacrylate	210	[119]
Oxirane	171.3	[147]
1,3,5-Triazine	264, 384	[147]
Imidazole	206	[146]
HMMM (Melamine Methoxylated)	216.5 (in acetonitrile)	[159]

Table 5.9 contd. ...

...Table 5.9 contd.

Compound	Absorption/Reflection λ_{max} [Absorption Edge] [nm]	Lit.	Eg [eV]	Literature
Cerium Compounds				
CeO_2	387, [350]	[120], [23]	3.2, 3.53–3.6, 5.5	[120], [23], [133, 148]
Chromium Compounds				
Cr^{3+}, $[Cr(H_2O)_6]^{3+}$	576–586, 583, 589, 590	[151], [152], o.d., [44, 113]		
	412–426, 416, 423, 425	[151], [44, 113], o.d., [152]		
$Cr(OH)_3$			2.43	[133, 148]
CrO_4^{2-}	277	[152]		
	315	[152]		
CrO_3	263, 395, 551, 789	[46]	2.0	[133, 148]
Cr_2O_3			3.30–3.55	[133, 148]
Copper Compounds				
Cu^{2+}	833	[44, 113]		
CuOH			1.97	[133, 148]
$Cu(OH)_2$			1.80	[133, 148]
Cu_2O	237, [488]	[60], [62]	1.86, 2.54–2.6	[133, 148], [64]
	314	[60]		
	380	[60]		
	462	[60]		
	550	[60]		
CuO			1.40	[133]
Iron Compounds				
Fe^{2+}	1000	[44]		
Fe^{3+}	714	[44, 113]		
$Fe(OH)_2$			2.10	[133, 148]
$Fe(OH)_3$			1.95	[133, 148]
FeO			2.40	[133]

Table 5.9 contd. ...

...Table 5.9 contd.

Compound	Absorption/Reflection λ_{max} [Absorption Edge] [nm]	Lit.	Eg [eV]	Literature
Iron Compounds				
α-FeOOH Goethite	285, 286	[74], [73]		
	305–340, 364, 365	[63], [74], [73]		
	434	[73, 74]		
	480, 481	[74], [73]		
	649	[74]		
	917	[74]		
γ-FeOOH Lepidocrocite	239, 240	[74], [73]		
	304, 305	[74], [73]		
	359, 360	[74], [73]		
	434	[74]		
	485	[74]		
	649	[74]		
β-FeOOH Akaganeite	230	[74]		
	290	[74]		
	502	[74]		
	908	[74]		
δ-FeOOH	217	[74]		
	330	[74]		
	880	[74]		
Fe_2O_3	[590]	[20]	1.90–1.95, 2.1	[133, 148], [20]
α-Fe_2O_3 Hematite	319	[73, 74]		
	380	[73, 74]		
	404, 405	[74], [73]		
	444	[73, 74]		
	529	[73, 74]		

Table 5.9 contd. ...

...Table 5.9 contd.

Compound	Absorption/Reflection λ_{max} [Absorption Edge] [nm]	Lit.	Eg [eV]	Literature
Iron Compounds				
	580	[63]		
	649	[74]		
	884	[74]		
γ-Fe_2O_3 Maghemite	250	[73]		
	315	[73]		
	370	[73]		
	480, [550]	[63], [59]		
Fe_3O_4 Magnetite	300	[74]		
	1400	[74]		
Molybdenum Compounds				
MoO_3*nH_2O	390	o.d.	2.95–3.10	[133]
Nickel Compounds				
$Ni(OH)_2$			2.31	[133, 148]
NiO			3.43, 3.80	[133], [148]

Compound	Absorption λ_{max} [nm]	Lit.
Nitrogen Compounds		
R-NH_2 (Ethylamine)	210	[114]
Ph-NH_2 (Aniline)	230, 230/280	[114, 119], [113]
Ph-NH_3^+	203, 203/254	[114, 119], [113]
R-NH-R (Diethyl Amine)	193	[114]
R_3-N (Triethyl Amine)	213	[114]

Table 5.9 contd. ...

...Table 5.9 contd.

Compound	Absorption/Reflection λ_{max} [Absorption Edge] [nm]	Lit.	Eg [eV]	Literature
Silicon Compounds				
Si	[1130]	[20]	1.1	[20]
Sulphur Compounds				
SO_4^{2-}	200	[113]		
Tin Compounds				
SnO_2	[350]	[20]	3.5, 3.6	[20], [41]
Titanium Compounds				
Ti^{3+}	492	[44]		
TiO_2 Anatase	[387]	[28, 67, 120]	3.18, 3.2	[66], [47, 67, 120, 133, 148]
TiO_2 Rutile	409, [410], [413]	[120], [20], [28, 67]	3.0, 3.05	[20, 67, 120], [133, 148]
Zinc Compounds				
ZnO Zincite	[385], 387, [390]	[45], [120], [20]	3.2, 3.34	[20, 45, 120], [133, 148]
$Zn_5(OH)_8Cl_2$, Simonkolleite	416–437	[65]	2.84–2.98	[65]
Zinc Oxo Hydroxides	260	[60]		
$Zn(OH)_2$	205	o.d.	1.82, 2.39	o.d., [133, 148]
Zirconium Compounds				
ZrO_2	[220], [230]	[137], [48]	4.7–4.8, 5.0–7.0, 5.1–5.2, 5.68	[133, 148], [138], [137], [52]
$Zr(OH)_4$			2.75	[118]
$ZrO(OH)_2$	182–184	o.d.*1	5.88–5.90	o.d.

o.d.: own data

5.4 Mathematical Relations of Complex Numbers

Complex number

$$z=x+iy=|z|\cdot e^{i\varphi}=|z|\cos\varphi+i|z|\sin\varphi=|z|(\cos\varphi+i\sin\varphi) \wedge i=\sqrt{-1} \wedge i^2=-1 \quad \text{Eq. 5.1}$$

Real and imaginary part

$$\text{Re}(z)=x=|z|\cos\varphi \quad \text{Eq. 5.2}$$

$$\text{Im}(z)=y=|z|\sin\varphi \quad \text{Eq. 5.3}$$

Phase angle

$$\tan\varphi=\frac{y}{x}=\frac{\text{Im}(z)}{\text{Re}(z)} \quad \text{Eq. 5.4}$$

$$\varphi=\arctan\frac{y}{x}=\arctan\frac{\text{Im}(z)}{\text{Re}(z)} \quad \wedge x>0 \quad \text{Eq. 5.5}$$

Conjugated complex number

$$\bar{z}=x-iy=|z|\cdot e^{-i\varphi} \quad \text{Eq. 5.6}$$

$$z+\bar{z}=x+iy+x-iy=2x=2\text{Re}(z) \quad \text{Eq. 5.7}$$

$$\text{Re}(z)=\frac{z+\bar{z}}{2} \quad \text{Eq. 5.8}$$

$$\text{Im}(z)=\frac{z-\bar{z}}{2i} \quad \text{Eq. 5.9}$$

$$\overline{(z_1+z_2)}=\bar{z}_1+\bar{z}_2 \quad \text{Eq. 5.10}$$

Absolute value of a complex number

$$|z|=\sqrt{x^2+y^2}=r \quad \text{Eq. 5.11}$$

$$|z|^2=x^2+y^2=z\cdot\bar{z} \quad \text{Eq. 5.12}$$

Addition/Subtraction of complex numbers

$$z_1 + z_2 = (x_1 + x_2) + i(y_1 + y_2) \quad \text{Eq. 5.13}$$

$$z_1 - z_2 = (x_1 - x_2) + i(y_1 - y_2) \quad \text{Eq. 5.14}$$

Multiplication of complex numbers

$$a \cdot z = a \cdot (x + i y) = a x + i a y \quad \text{Eq. 5.15}$$

$$z_1 z_2 = (x_1 + i y_1)(x_2 + i y_2) = x_1 x_2 + i x_1 y_2 + i x_2 y_1 + i^2 y_1 y_2$$
$$z_1 z_2 = (x_1 x_2 - y_1 y_2) + i(x_1 y_2 + x_2 y_1) \quad \text{Eq. 5.16}$$

$$z_1 z_2 = (|z_1| \cdot e^{i\phi_1})(|z_2| \cdot e^{i\phi_2}) = |z_1||z_2| \cdot e^{i(\phi_1 + \phi_2)} \quad \text{Eq. 5.17}$$

$$z \cdot \bar{z} = (x + i y)(x - i y) = x^2 - i x y + i x y + y^2 = x^2 + y^2 \quad \text{Eq. 5.18}$$

Division of complex numbers

$$\frac{z_1}{z_2} = \frac{z_1 \overline{z_2}}{z_2 \overline{z_2}} = \frac{(x_1 + i y_1)(x_2 - i y_2)}{|z_2|^2} = \frac{1}{|z_2|^2}(x_1 x_2 - i x_1 y_2 + i y_1 x_2 - i^2 y_1 y_2) \quad \text{Eq. 5.19}$$

$$\frac{z_1}{z_2} = \frac{1}{|z_2|^2}(x_1 x_2 + y_1 y_2) + \frac{1}{|z_2|^2} i(-x_1 y_2 + y_1 x_2) \quad \text{Eq. 5.20}$$

$$\frac{z_1}{z_2} = \frac{|z_1|}{|z_2|} \cdot e^{i(\varphi_1 - \varphi_2)} \quad \text{Eq. 5.21}$$

To raise to the power of complex numbers

$$z^n = (|z| \cdot e^{i\varphi})^n = |z|^n \cdot e^{i n \varphi} \quad \text{Eq. 5.22}$$

$$z^n = [|z|(\cos\varphi + i \sin\varphi)]^n = |z|^n [\cos(n\varphi) + i \sin(n\varphi)] \quad \text{Eq. 5.23}$$

Moivre´s theorem

$$\left(e^{i\varphi}\right)^{n}=e^{in\varphi}\Rightarrow\left(\cos\varphi+i\sin\varphi\right)^{n}=\cos\left(n\varphi\right)+i\sin\left(n\varphi\right)$$ Eq. 5.24

Roots of complex numbers

$$\sqrt{i}=\frac{1}{\sqrt{2}}+\frac{1}{\sqrt{2}}i$$ Eq. 5.25

$$\sqrt[n]{z}=\sqrt[n]{|z|}\cdot e^{\frac{i\varphi}{n}+\frac{2\pi k}{n}}\ \mathit{with}\,k\in\left[0,1.......,n-1\right]$$ Eq. 5.26

Main value

$$\sqrt{z}=\sqrt{|z|}\cdot e^{\frac{1}{2}i\varphi}$$ Eq. 5.27

$$\sqrt[n]{|z|\left(\cos\varphi+i\sin\varphi\right)}=|z|^{\frac{1}{n}}\left[\cos\frac{\varphi}{n}+i\sin\frac{\varphi}{n}\right]$$ Eq. 5.28

Logarithmic calculus of complex numbers

$$\log z=\ln|z|+i\varphi$$ Eq. 5.29

5.5 Literature

1. "Solubility Product Constants" in D. R. Lide, ed., CRC Handbook of Chemistry and Physics, 89th Edition, CRC Press/Taylor and Francis, Boca Raton, FL. If
2. "Standard Thermodynamic Properties of Chemical Substances" in D. R. Lide, ed., CRC Handbook of Chemistry and Physics, 89th Edition, CRC Press/Taylor and Francis, Boca Raton, FL
3. H. Kaesche, Die Korrosion der Metalle, 3. neubearb. u. erw. Aufl. 1990, Springer Verlag, Heidelberg, 2011, English Edition: H. Kaesche, Corrosion of Metals, Springer Verlag, Berlin-Heidelberg, 2003, S. 35ff
4. P. W. Atkins, Physical Chemistry, Oxford University Press, 1978
5. Encyclopaedia of Electrochemistry, Edited by A.J. Bard and M. Stratmann, 2007 Wiley-VCH Verlag GmbH & Co. KGaA, Weinheim, Vol. 4, Corrosion and Oxide Films, Chap. 1.2 G. S. Frankel, D. Landolt, Thermodynamics of Electrolytic Corrosion, 9–24
6. W. Feitknecht, Werkst. Korros. 1 (1965) 15–26
7. D. A. Probst, G. Henderson, J. Chem. Educ. 73, 10 (1996) 962–964
8. W. Feitknecht, P. Schindler, Pure and Appl. Chem. 6, 2 (1963) 125–206
9. R. D. Armstrong, M. F. Bell, Electrochem. 4 (1974) 1–17
10. Thesis (PhD), T. Ludwig, Kupfer- und Zinkentfernung aus Niederschlagsabfluss von Dächern in einem Eisen-Korrosionssystem, Technical University Berlin, 2007
11. C. Veder, Rutschungen und ihre Sanierung, Springer Verlag, Wien, New York, 1979

12. "Thermodynamic Properties of Aqueous Systems" in D. R. Lide, ed., CRC Handbook of Chemistry and Physics, 89th Edition, CRC Press/Taylor and Francis, Boca Raton, FL.
13. Aqueous Systems at Elevated Temperatures and Pressures: Physical Chemistry in Water, Steam and Hydrothermal Solutions, D.A. Palmer, R. Fernandez-Prini, A.H. Harvey Eds. Elsevier, 2004, D.J. Wesolowski, S. E. Ziemniak, L. M. Anovitz, M. L. Machesky, P. Benezeth, D. A. Palmer, Solubility and surface adsorption characteristics of metal oxide, 493–595
14. Thesis (PhD), W. Beyer, Zum anodischen Verhalten von Zirkonium in sauren Lösungen in potentialbereichen mit Passivität und Lochfraß-Korrosion, Nürnberg University, 1976
15. M. Pourbaix, Atlas of Electrochemical Equilibria in Aqueous Solutions, Pergamon, New York, 1966
16. M. Pourbaix, Atlas of Electrochemical Equilibria in aqueous solutions, National Association of Corrosion engineers, Houston Texas, 1974
17. "Electrochemical Series" in D. R. Lide, ed., CRC Handbook of Chemistry and Physics, 89th Edition, CRC Press/Taylor and Francis, Boca Raton, FL. If
18. E. Barsoukov, J. Ross Macdonald, Impedance Spectroscopy Theory, Experiment, and Applications Second Edition, Wiley-Interscience, Hoboken, 2005
19. Thesis (PhD), P. Keller, Kupfer- und Zinkentfernung aus Niederschlagsabfluss von Dächern in einem Eisen-Korrosionssystem, Heinrich-Heine-University Düsseldorf, 2006
20. C. M. A. Brett, A. M. O. Brett, Electrochemistry, Principles, Methods and Applications, Oxford University Press, New York, 1993
21. P. H. Rieger, Electrochemistry, Second Edition, Chapman & Hall, New York, London, 1994
22. R. W. Revie, H. H. Uhlig, Corrosion and Corrosion Control, John Wiley & Sons, Hoboken New Jersey, 2008
23. B. Elidrissia, U. M. Addoua, M. Regraguia, C. Montyb, A. Bougrinea, A. Kachouanea, Thin Solid Films 379 (2000) 23–27
24. B. Rossenbeck, P. Ebbinghaus, M. Stratmann, G. Grundmeier, Corros. Sci. 48 (2006) 3703–3715
25. M. Etienne, M. Dossot, J. Grausem, G. Herzog, Anal. Chem. 86, 22 (2014) 11203–11210
26. H. N. Shubha, T. V. Venkatesha, K. Vathsala, M. K. Pavitra, M. K. Punith Kumar, Appl. Mater. Interfaces 5 (2013) 10738–10744
27. L. Zedler, M. D. Hager, U. S. Schubert, M. J. Harrington, M. Schmitt, J. Popp, B. Dietzek, Materials Today, 17, 2 (2014) 57–69
28. X. Wei-Xing, Z. Shu, F. Xian-Cai, J. Phys. Chem. Solids 59, 9 (1998) 1647–1658
29. C. A. Gervasi, P. A. Palacios, Ind. Eng. Chem. Res. 52 (2013) 9115–9120
30. E. Pere, H. Cardy, O. Cairon, M. Simon, S. Lacombe, Vib. Spectrosc. 25 (2001) 163–175
31. G. Kurbatov, E. Darque-Ceretti, M. Aucouturier, Surface Modification Technologies VIII, Edited by T.S. Sudarshan, M. Jeandin, The Institute of Materials, 1995, 235–239
32. H. B. Xue, Y. F. Cheng, Mater. Corr. 61, 9 (2010) 756–761
33. S. Nagarajan, M. Mohana, P. Sudhagar, V. Raman, T. Nishimura, S. Kim, Y. S. Kang, N. Rajendran, ACS Appl. Mater. Interfaces 4 (2012) 5134–5141
34. G. Yea, F. Courtecuissea, X. Allonasa, C. Leya, C. Croutxe-Barghorna, P. Rajaa, P. Taylorb, G. Bescondb, Prog. Org. Coat. 73 (2012) 366–373
35. L. G. J. van der Ven, R. T. M. Leijzer, K. J. van den Berg, P. Ganguli, R. Lagendijk, Prog. Org. Coat. 58 (2007) 117–121
36. Thesis (PhD), F. Lesch, Einfluss von Substrateigenschaften auf die Abscheidung übergangsmetallsalzhaltiger Korrosionsschutzmaterialien, Friedrich-Alexander-University Erlangen-Nürnberg, 2014
37. M. Dabala Áa, L. Armelaob, A. Buchbergera, I. Calliari, Appl. Surf. Sci. 172 (2001) 312–322

38. W. Guixiang, Z. Milin, W. Ruizhi, Appl. Surf. Sci. 258 (2012) 2648–2654
39. M. Stamm, Polymer Surfaces and Interfaces, Characterization, Modification and Applications, Springer, Berlin, 2008
40. S. Feliu Jr., C. Maffiotte, J. C. Galván, A. Pardo, M. Concepción Merino, R. Arrabal, The Open Surface Science Journal 3 (2011) 1–14
41. D. Boulainine, A. Kabir, I. Bouanane, B. Boudjema, G. Schmerber, J. Electron. Mater. 45, 8 (2016) 4357–4363
42. Z. Luo, Z. Zhang, W. Wang, W. Liu, Surf. Coat. Technol. 203 (2009) 1516–1522
43. Thesis (PhD), B. Schinkinger, Schichtanalyse und elektrochemische Untersuchungen zur Abscheidung dünner SiO_2– und Organosilanschichten auf verzinktem Stahl, Ruhr-University Bochum, 2004
44. F. W. Küster, A. Thiel, Rechentafeln für die Chemische Analytik, 106. Auflage, Walter de Gruyter, Berlin New York, 2008
45. J. G. Quinones-Galvan, H. Totozintle-Huitle, L. A. Hernandez-Hernanddez, J. S. Arias-Ceron, F. de Moure-Flores, A. Hernandez-Hernandez, E. Campos-Gonzalez, A. Guillen-Cervantes, O. Zelaya-Angel, J. J. Araiza, Materials Research Express, 1 (2014) 1–9
46. Z. M. Hanafi, F. M. Ismail, A. K. Mohamed, Z. Phys. Chem. 205 (1998) 193–198
47. R. Lopez, R. Gomez, J. Sol-Gel Sci. Technol. 61 (2012) 1–7
48. R. R. Pareja, R. L. Ibanez, F. Martin, J. R. Ramos-Barrado, D. Leinen, Surf. Coat. Technol. 200 (2006) 6606–6610
49. M. G. S. Ferreira, C. A. Melendres, Electrochemical and Optical Techniques for the Study and Monitoring of Metallic Corrosion, Kluwer Academic Publisher, 1991, Z. Szklarska-Smialowska, Methods for studying corrosion inhibition, 545–570
50. E. Diler, B. Lescop, S. Rioual, G. Nguyen Vien, D. Thierry, B. Rouvellou, Corros. Sci. 79 (2014) 83–88
51. Analytical Methods in Corrosion Science and Engineering, H.-H. Strehblow, P. Marcus, X-Ray Photoelectron Spectroscopy in Corrosion Research, Taylor & Francis, CRC Press, 2006, 1–37
52. A. Wold, K. Dwight, Solid State Chemistry, Chapman and Hall, New York, 1993
53. T. L. Barr, J. Phys. Chem. 82, 16 (1978) 1801–1810
54. V. I. Parvulescu, H. Bonnemann, V. Parvulescu, U. Endruschat, A. Rufinska, C. W. Lehmann, B. Tesche, G. Poncelet, Appl. Cat. A 214 (2001) 273–287
55. B. M. Reddy, B. Chowdhury, P. G. Smirniotis, Appl. Cat. A. 211 (2001) 19–30
56. W. Fürbeth, M. Stratmann, Corros. Sci. 43 (2001) 207–227
57. Thesis (PhD), A. Leng, University Düsseldorf, 1995
58. Handbook of X-ray Photoelectron Spectroscopy C. D. Wanger, W. M. Riggs, L. E. Davis, J. F. Moulder and G. E. Muilenberg Perkin-Elmer Corp., Physical Electronics Division, Eden Prairie, Minnesota, USA, 1979
59. G. Gunasekaran, R. Natarajan, N. Palaniswamy, Corros. Sci. 43 (2001) 1615–1626
60. M. B. Valcarce, S. R. De Sanchez, M. Vazquez, J. Mater. Sci. 41 (2006) 1999–2007
61. M. Dornbusch, S. Kirsch, C. Henzel, C. Deschamps, S. Overmeyer, K. Cox, M. Wiedow, U. Tromsdorf, M. Dargatz, U. Meisenburg, Prog. Org. Coat. 89 (2015) 332–343
62. M. Odziemkowski, Spectroscop. Prop. Inorg. Organomet. Compd. 40 (2009) 385–449
63. B. Beden, Materials Science Forum, Vols. 192–194, pp 277–290, Trans Tech Publications, Switzerland, 1995
64. B. Balamurugan, B. R. Mehta, Thin Solid Films 396 (2001) 90–96
65. Y. Zhu, X. Zhang, Z. Lan, H. Li, X. Zhang, Q. Li, Materials and Design 93 (2016) 503–508
66. R. Beranek, H. Kisch, Supplementary Information, Royal Society of Chemistry, S1–S10
67. D. A. H. Hanaor, C. C. Sorrell, J. Mater. Sci. 46 (2011) 855–874
68. F. Rueda, J. Mendialdua, A. Rodriguez, R. Casanova, Y. Barbaux, et al., J. Electron Spectrosc. Relat. Phenom. 82 (1996) 135

69. A. I. Minyaev, I. A. Denisov, V. E. Soroko, V. A. Konovalov, Zhurnal Prikladnoi Khimii 59 (1986) 339
70. D. Brion, Appl. Surf. Sci. 5 (1980) 133
71. G. C. Allen, M.T. Curtis, A. J. Hooper, P. M. Tucker, J. Chem. Soc. Dalton Trans. (1974) 1525
72. D. D. Hawn, B. M. De Koven, Surf. Interface Anal. 10 (1987) 63
73. D. M. Sherman, T. D. Waite, American Mineralogist, 70 (1985) 1262–1269
74. R. M. Cornell, U. Schwertmann, The Iron Oxides, Structure, Properties, Reactions Occurrence and Uses, Wiley-VCH GmbH & Co. KGaA, Weinheim, 2003
75. J. F. Moulder, W. F. Stickle, P. E. Sobol, K. D. Bomben, Handbook of X-ray Photoelectron Spectroscopy, Perkin Elmer, Minnesota, 1992
76. T. K. Rout, N. Bandyopadhyay, Anti-Corrosion Methods and Materials 54/1 (2007) 16–20
77. N. Hammouda, H. Chadli, G. Guillemot, K. Belmokre, Advances in Chemical Engineering and Science 1 (2011) 51–60
78. T. Prosek, D. Thierry, C. Taxe, J. Maixner, Corros. Sci. 49 (2007) 2676–2693
79. Thesis (PhD), Y.-Y. Lee, Untersuchung der Porosität dünner Schichten auf Eisen, Philipps-University Marburg, 2003
80. T. F. da Conceicao, N. Scharnagl, W. Deitzel, D. Hoeche, K.U. Kainer, Corros. Sci. 53 (2011) 712–719
81. T. Hirano, A. Koseki, M. Mizoguchi, M. Waki, H. Nakamura, Proceedings of the Electrical Electronics Insulation Conference, 17th 1985, New York, 189–191
82. R. B. King, Encyclopaedia of Inorganic Chemistry, 10 Volume Set, 2. Ed., Wiley & Sons, 2005
83. F. Khelifa, Woodhead Publishing series in metals and surface engineering, 64 (2014) 423–458
84. Handbook of thin film materials, edited by H.S. Nalwa, Volume 2: Characterisation and Spectroscopy of Thin Films, Academic Press, 2002, F. Urs, ATR Spectroscopy of thin films, 191–229
85. N. Ellhalawany, M. M. Saleeb, M. K. Zahran, Prog. Org. Coat. 77 (2014) 548–556
86. J. A. Jan, J. Iglesias, O. Adames, Hyperfine Interact 224 (2014) 279–288
87. G. Lefevre, Adv. Colloid Interface Sci. 107 (2004) 109–123
88. W. Zhang, R. Smith, C. Lowe, J. Coat. Technol. Res. 6, 3 (2009) 315–328
89. N. J. Everall, JCT Coatings Tech 2005, 46–52
90. K. P. J. Williams, I. P. Hayward, Proceedings of the International Conference on Advances Surfaces Treatment: Research & Applications (ASTRA), November, 2003, Hyderabad, India, 3-6, 435–441
91. T. Devine, Proceedings of Corrosion/97, Research Topical Symposia, New Orleans, March 1997, 131–162
92. Kirk-Othmer encyclopaedia of chemical technology, E. W. Smith, Raman scattering, Vol. 21. Wiley, Hoboken, 2006, 321–330
93. NATO ASI Series, Electrochemical and Optical Techniques for the Study and Monitoring of Metallic Corrosion. Edited by M.G.S. Ferreira, C.A. Melendres, C.A. Melendres, Laser Raman Spectroscopy, Principles and Applications to corrosion Studies, Kluwer Academic Publishers, Netherlands, 1991, 355–388
94. Y. Yu, X. Wang, Y. Cao, X. Hu, Appl. Surf. Sci. 172 (2001) 260–264
95. Y. Y. Kharitonov, M. M. Bekhit, S. F. Belevskii, Russ. J. Inorg. Chem. 28, 12 (1993) 1803–1806
96. A. Syamal, M. M. Singh, Ind. J. Chem. 37A (1998) 350–354
97. A. Syamal, M. M. Singh, Ind. J. Chem. 32A (1993) 42–48
98. Gmelins Handbuch der Anorganischen Chemie, 42, 1958
99. M. Keddam, A. Hugot-Le-Goff, H. Takenouti, D. Thierry, M. C. Arevalo, Corros. Sci. 33, 8 (1992) 1243–1252

100. M. C. Bernard, A. Hugot-Le-Goff, D. Massinon, N. Phillips, Corros. Sci. 35, 5-8 (1993) 1339–1349
101. W. Yang, C. Liu, Z. Zhang, Y. Liu, S. Nie, Electronic Supplementary Material (ESI) for RSC Advances, The Royal Society of Chemistry 2014, 1–5
102. W. R. Fischer, U. Schwertmann, Clays and Clay Minerals, 23 (1975) 33–37
103. P. Kavitha, R. Ramesh, M. R. Rajan, C. Stella, INDIAN JOURNAL OF RESEARCH, 4, 12 (2015) 91–93
104. H. C. Liese, American Mineralogist, 52 (1967) 1198–1205
105. T. N. Brusentsova, R. E. Peale, D. Maukonen, G. E. Harlow, J. S. Boesenberg, D. Ebel, American Mineralogist 95 (2010) 1515–1522
106. K. A. RODGERS, Clay Minerals 28 (1993) 85–99
107. O. K. Sricastava, E.A. Secco, Can. J. Chem. 46 (1967)
108. G. Socrates, Infrared and Raman Characteristic Group Frequencies, John Wiley & Sons, Chichester, New York, 2013
109. X. Wang, L. Andrews, J. Phys. Chem. A 110, 33 (2006) 10035
110. J. W. Kauffman, R. H. Hauge, J. L. Margrave, J. Phys. Chem. 89, 16 (1985) 3541
111. X. Wang, L. Andrews, J. Phys. Chem. A 109, 47 (2005) 10689
112. H. Elderfield, J. D. Hem, MINERALOGICAL MAGAZINE 39 (1973) 89–96
113. M. Wächter, Tabellenbuch der Chemie, Wiley-VCH, Weinheim, 2012
114. M. Hesse, H. Meier, B. Zeeh, Spektroskopische Methoden in der organischen Chemie, Thieme Verlag, Stuttgart, 2005
115. T. Hirayama et al. Prog. Org. Coat. 20 (1991) 81–96
116. M. Dornbusch, R. Rasing, U. Christ, Epoxy Resins, Vincentz Network, Hanover, 2016
117. Y. Wang, Z. Sun, J. Tian, H. Wang, H. Wang, Y. Ji, MATERIALS SCIENCE (MEDŽIAGOTYRA). 22, 2 (2016) 290–294
118. F. Di Quarto, M.C. Romano, M. Santamaria, S. Piazza, C. Sunseri, Russ. J. Electrochem. 36, 11 (2000) 1203–1208
119. J. B. Lambert, S. Gronert, H. F. Shurvell, D. A. Lightner, Spektroskopie, Strukturaufklärung in der Organischen Chemie, 2. Ed., Pearson, München, 2012
120. Thesis (PhD), H. Althues, Lumineszierende, transparente Nanokomposite—Synthese und Charakterisierung Technical University Dresden, 2007
121. H. Zhang, X. Li, C. Du, H. Qi, Y. Huang, J. Raman Spectrosc. 40 (2009) 656–660
122. W. Fürbeth, M. Stratmann, Corros. Sci. 43 (2001) 207–227
123. W. Fürbeth, M. Stratmann, Corros. Sci. 43 (2001) 243–254
124. J. Kasperek, M. Lenglet, Revue metallurgie 94, 5 (1997) 713–719
125. D. Thierry, D. Massinon, J. Electrochem. Soc. 138, 3 (1991) 879–880
126. B. X. Huang, P. Tornatore, Y.-S. Li, Electrochim. Acta 46 (2000) 671–679
127. D.-L. Lee, T. Kang, H.-J. Sohn, H.-J. Kim, Mater. Trans. 43, 1 (2002) 49–54
128. J. J. Marcinko, C. J. Moriarty, Molding system with self-releasing moveable member, WO 002003026878A1
129. G. Wranglen, An Introduction to Corrosion and Protection of Metals, Institut för Metallskydd, Stockholm, 1972
130. Encyclopedia of Electrochemistry, Edited by A.J. Bard and M. Stratmann, 2007 Wiley-VCH Verlag GmbH & Co. KGaA, Weinheim, Vol. 4, Corrosion and Oxide Films, Chapt. 7 G. S. Frankel, M. Rohwerder, Electrochemical Techniques for Corrosion, 687–723
131. E. Bardal, Corrosion and Protection, Springer Verlag, London, 2004
132. G. Beranger, The book of Steel, J.-C. Charbonnier, Chap. 13, Corrosion, 335–362, Tec&Doc, Londres, 1996
133. P. Marcus, F. Mansfeld, Analytical Methods in Corrosion Science and Engineering, CRC Press, 2006
134. A. Roustila, J. Chene, C. Severac, J. Alloys Compd. 356-357 (2003) 330–335
135. A. A. M. Ali, M. I. Zaki, Thermochim. Acta 336 (1999) 17–25
136. C. Huang, Z. Tang, Z. Zhang, J. Am. Ceram. Soc., 84 [7] 1637–38 (2001)
137. L. P. Borilo, L. N. Spivakova, Am. J. Mat. Sci. 2012, 2(4), 119–124

138. R. Espinoza-Gonzalez, E. Mosquera, I. Moglia, R. Villarroel, V. M. Fuenzalida, Ceramics International (2014), http://dx.doi.org/10.1016/j.ceramint.2014.07.034
139. B. Stuart, Infrared Spectroscopy: Fundamentals and Applications, Wiley & Sons, Chichester, 2004
140. M. Dornbusch, T. Biehler, M. Conrad, A. Greiwe, D. Momper, L. Schmidt, M. Wiedow, JUnQ, 6, 2, 1–7, 2016
141. H. F. Gao, H. Q. Tan, J. Li, Y. Q. Wang, J. Q. Xun, Surf. Coat. Technol. 212 (2012) 32–36
142. K. A. Saburov, Kh. M. Kamilov, Chem. Nat. Compd. 25 (1989) 695–698
143. P. A. Estep, C. Karr, American Mineralogist 53 (1968) 305–309
144. RRUFF, http://rruff.info/Dawsonite/R050645
145. E. G. Rochow, Einführung in die Chemie der Silikone, Verlag Chemie, Weinheim, 1952
146. J. A. Joule, G. F. Smith, Heterocyclic Chemistry, VNR, New York, 1972
147. A. R. Katritzky, Handbook of Heterocyclic Chemistry, Pergamon Press, Oxford, 1985
148. F. Di Quarto, F. La Mantia, M. Santamaria, Eds. S.-I. Pyun, J.-W. Lee, Modern Aspects of Electrochemistry No. 46, Progress in Corrosion Science and Engineering I, Springer, Heidelberg, 2009
149. H.H. Uhlig, The Corrosion Handbook, John Wiley & Sons, New York, 1953
150. A. Ghadami, J. Ghadam, M. Idrees, Iran. J. Chem. Chem. Eng. 32, 3 (2013) 27–35
151. O.b. Suarez, J. b. Olaya, M. b. Suarez, S. Rodil, J. Chil. Chem. Soc. 57, 1 (2012) 977–982.
152. L. Leita, A. Margon, A. Pastrello, I. Arcon, M. Contin, D. Mosetti, Soil humic acids may favour the persistence of hexavalent chromium in soil, Environmental pollution (Barking, Essex 1987) 157 (6) (2009) 1862–1866.
153. Thesis (PhD), S. M. Noëlle Cambier, Atmospheric corrosion of coated steel; relationship between laboratory and field testing, Ohio State University, 2014
154. H. Vaškova, V. Křesalek, Math. Models Methods Appl. Sci. 7, 5 (2011) 1197–1204
155. M. W. Urban, Infrared and Raman Spectroscopy and Imaging in Coating Analysis, Encyclopedia of Analytical Chemistry, R. A. Meyets (Ed.), John Wiley & Sons, Chichester, 2000, 1756-1773
156. J. C. Cruz, T. A. Osswald, J. Plastics Technol. 4, 3 (2008) 2–21
157. H. Janik, B. Pałys, Z. S. Petrovic, Macromol. Rapid Commun. 24 (2003) 265–268
158. I. Yilgör, E. Yilgör, G. L. Wilkes, Polymer 58 (2015) A1–A36
159. J. Wysoglad, J.-E. Ehlers, T. Lewe, M. Dornbusch, J. Gutmann, Conformational study of melamine crosslinkers and spectroscopical comparison of HMMM molecules by practical measurements and quantum chemical calculations, J. Mol. Struct. 1166 (2018) 456–469

Index

F

G

H

I

K

L

M

N

O

P

R